Silicene-Based Layered Materials

Essential properties

Silicene-Based Layered Materials

Essential properties

Shih-Yang Lin
Department of Physics, National Chung Cheng University, Chiayi, Taiwan

Hsin-Yi Liu
Center for Micro/Nano Science and Technology, National Cheng Kung University, Tainan, Taiwan

Duy Khanh Nguyen
Laboratory of Applied Physics, Advanced Institute of Materials Science, Ton Duc Thang University, Ho Chi Minh City, Vietnam and Faculty of Applied Sciences, Ton Duc Thang University, Ho Chi Minh City, Vietnam

Ngoc Thanh Thuy Tran
Hierarchical Green-Energy Materials (Hi-GEM) Research Center, National Cheng Kung University, Tainan, Taiwan

Hai Duong Pham
Department of Physics, National Cheng Kung University, Tainan, Taiwan

Shen-Lin Chang
Department of Physics, National Chiao Tung University, Hsinchu, Taiwan

Chiun-Yan Lin
Department of Physics, National Cheng Kung University, Tainan, Taiwan

Ming-Fa Lin
Department of Physics/QTC/Hi-GEM, National Cheng Kung University, Tainan, Taiwan

IOP Publishing, Bristol, UK

ISBN 978-0-7503-3299-6 (ebook)
ISBN 978-0-7503-3297-2 (print)
ISBN 978-0-7503-3300-9 (myPrint)
ISBN 978-0-7503-3298-9 (mobi)

DOI 10.1088/978-0-7503-3299-6

Version: 20200701

IOP ebooks

British Library Cataloguing-in-Publication Data: A catalogue record for this book is available from the British Library.

Published by IOP Publishing, wholly owned by The Institute of Physics, London

IOP Publishing, Temple Circus, Temple Way, Bristol, BS1 6HG, UK

US Office: IOP Publishing, Inc., 190 North Independence Mall West, Suite 601, Philadelphia, PA 19106, USA

Contents

Foreword

As unaccustomed as I am to writing forwards, I have nonetheless been called upon to speak no ill, yet write candidly on behalf of efforts by my dear and friend colleague, Prof. Lin present text-book compendium in a critical area of condensed matter physics. Thus, first, might I venture to say that herein lies, or it's preferably hoped, stands an exposition of collaborative works, culled from what is a rapidly growing subfield within a state of The Solid State. 'Twas also an honorably fatiguing calculus of toil, and monographic concerns in which it is hoped will occupy but a sliver in the spectrum of what is arguably one of the most intriguing areas of condensed matter physics: The remarkably confounding, yet slowly unraveling enigma of graphene and her delicately layered, at times tortious incarnations, that give rise to remarkable spectra of opto-electronic, atomistic, molecular and electro-magnetic phenomena. Their selection, treatment and submission of topics, are such that one ought to be pleased. The methods utilized by Professors Lin *et al*, are to put it mildly, astonishingly novel. I have had the privilege of seeing these techniques applied by a number of these practitioners, in several ways that have yet to appear in published form. What is contained in this work, are glorious harbingers of a trough, which I am certain will usher in a wave of low-dimensionally monumentous discoveries for which there are no Lorentz functional analogues.

The preceding lengthy preamble, tries to lay out in brief; the authors' motivations, purpose and aspirations for having collected, and in some places written this labor of love. They introduce *ab initio* approaches to the determination of various electronic and electro-optical properties of several types of graphitic systems: semiconductors, their hetero-structures, and geometric arrangements, topological rearrangements and doping in relation to the same. One of their methods even yielded an intriguing alternative to the Tight-binding approximation approach in determining the Eigen-energies of molecular ensembles in uniform magnetic fields. For choice servings too numerous to heap here, I refer the reader to the table of contents for a menu of tantalizing ideas. And if herewith, they are deemed worthy by peers and students alike, then the writers' wish that this harvest be received as some small contribution in the nourishment of scientific inquiry; would have been humbly fulfilled.

Prof. Godfrey Gumbs

Distinguished Professor of Physics

Hunter College of the City University of New York (CUNY)

Author biographies

Shih-Yang Lin

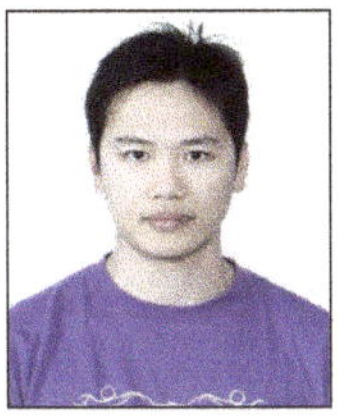

Shih-Yang Lin received his PhD in physics in 2015 from the National Cheng Kung University (NCKU), Taiwan. Since 2015, he has been a postdoctoral researcher at NCKU. Recently, he works as an exchange scholar in University of Houston. His scientific interests are focused on low-dimensional group IV materials and its derivatives using first-principle calculations.

Hsin-Yi Liu

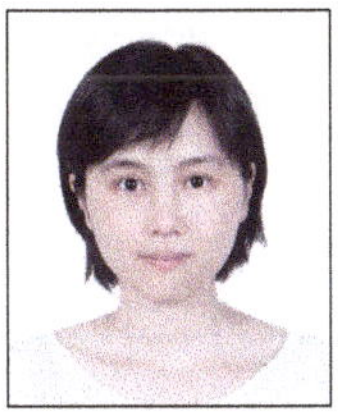

Hsin-Yi Liu received her PhD in chemistry in 2014 Technical University Berlin (TU-Berlin), Germany. She had been a postdoctoral researcher in 2014 at National Taiwan University (NTU), Taiwan. Since 2015, she has worked as a postdoctoral researcher at National Cheng Kung University (NCKU). Recently, her scientific research is focused on low-dimensional group IV materials such as mono- & bilayer materials, nanotubes and nanoribbons. They are investigated by use of first-principle calculations.

Duy Khanh Nguyen

Duy Khanh Nguyen received his PhD degree in Physics in 2019 from the National Cheng Kung University (NCKU), Taiwan. Since 2019, he works as a full-time researcher at Advanced Insitute of Materials Science, Ton Duc Thang University (TDTU), Vietnam. Recently, he also works as a specially appointed researcher at Graduate School of Engineering, Osaka University, Japan. His main research interests are to design computational materials using the first-principles calculations.

Ngoc Thanh Thuy Tran

Ngoc Thanh Thuy Tran obtained her PhD in physics in 2017 from the National Cheng Kung University (NCKU), Taiwan. Afterward, she began to work as a postdoctoral researcher and then an assistant researcher at Hierarchical Green-Energy Materials (Hi-GEM) Research Center, NCKU. Her scientific interest is focused on the fundamental properties of 2D materials and rechargeable battery materials by means of the first-principle calculations.

Hai Duong Pham

Hai Duong Pham earned his PhD degree in July 2020 in the Department of Physics from National Cheng Kung University (NCKU), Taiwan. After finishing the PhD program, he continued to work as a researcher at NCKU. His current scientific interests focus on graphene-like systems using first-principle calculations and tight binding method.

Shen-Lin Chang

Shen-Lin Chang obtained his PhD in 2014 in physics from the National Cheng Kung University (NCKU), Taiwan. Since 2014, he has been a postdoctoral researcher at NCKU and NCTU. His main scientific interests are in the field of condensed matter physics. Most of his research is focused on the electronic and magnetic properties of one-dimensional nanomaterials.

Chiun-Yan Lin

Chiun-Yan Lin obtained his PhD in 2014 in physics from the National Cheng Kung University (NCKU), Taiwan. Since 2014, he has been a postdoctoral researcher in the department of physics at NCKU. His main scientific interests are in the field of condensed matter physics, modeling and simulation of nanomaterials. Most of his research is focused on the electronic, optical and many-particle properties of low-dimensional nanomaterials.

Ming-Fa Lin

Ming-Fa Lin is a distinguished professor in the Department of Physics, National Cheng Kung University, Taiwan. He received his PhD in physics in 1993 from the National Tsing-Hua University, Taiwan. His main scientific interests focus on essential properties of carbon related materials and low-dimensional systems. He is a member of American Physical Society, American Chemical Society, and the Physical Society of Republic of China (Taiwan).

Shih-Yang Lin, Hsin-Yi Liu, Duy Khanh Nguyen, Ngoc Thanh Thuy Tran, Hai Duong Pham, Shen-Lin Chang, Chiun-Yan Lin and Ming-Fa Lin

Chapter 1

Introduction

Since the first discovery of monolayer graphene through mechanical exfoliation of a graphite surface in 2004, other group-IV and group-V systems have been synthesized successfully on specific substrates, such as silicene on Ag(111), Ir(111) and ZrB_2 surfaces [1–3], and germanene on Pt(111), Au(111) and Al(111) surfaces [4–6]. The C, Si, Ge, Sn and Pb atoms are expected to present four half-occupied orbitals in condensed-matter systems, respectively, (2s, $2p_x$, $2p_y$, $2p_z$), (3s, $3p_x$, $3p_y$, $3p_z$), (4s, $4p_x$, $4p_y$, $4p_z$), (5s, $5p_x$, $5p_y$, $5p_z$) and (6s, $6p_x$, $6p_y$, $6p_z$). In general, they can generate π, σ and sp^3 chemical bonding. Such atomic configurations are quite different from those in an isolated state, illustrating clearly that material surfaces can provide very active chemical environments. Varied fundamental properties are achieved easily by modulating various critical factors. Lattice symmetries, planar, buckled and curved geometries [7], normal and unusual stacking configurations (simple and greatly enlarged unit cells) [8], chemical adatom absorption [9], guest-atom substitution [10], defects and vacancies [11], quantum confinement and edge structures [12], hetero-junctions (substrates) [13] and interlayer coupling of composite subsystems play important roles in diverse physical, chemical and material phenomena, e.g. the AA- (see figure 2.2) [14], AB- [8] and ABC-enriched [8] few-layer graphene systems. The basic quasiparticle properties, the single- and multi-orbital hybridizations in the different chemical bonds, and the adatom- and/or host-atom-induced spin configurations appear to be the main focuses of study. To date, systematic theoretical investigations have been conducted on the optimal geometries, electronic structures, magnetic properties, optical spectra, quantum transports and Coulomb excitations for three-dimensional (3D) graphites [15], two-dimensional (2D) graphene-related systems [16] and one-dimensional (1D) graphene nanoribbons [17]. On the experimental side, a lot of model predictions have been examined fully using various high-resolution spectroscopy methods (discussed in chapter 3). In the current book, a theoretical framework built from first-principles calculations is developed

doi:10.1088/978-0-7503-3299-6ch1

thoroughly to comprehend the greatly varied fundamental properties in silicene-based materials, in particular for the diverse charge- and spin-created phenomena. Such emergent materials include single-layer group-IV systems, sliding bilayer silicenes, hydrogen-, halogen-, alkali- and metal-atom-adsorbed silicenes, B-, C- and N-substituted silicene systems, Si-atom adsorption and substitution on graphenes, Si-adlayer and bilayer AB-bt silicene on Ag(111), and Si–, Ge–, Sn– and Pb–graphene bilayer composites.

Both numerical methods and phenomenological models are available in exploring the basic sciences of physics, chemistry and materials. The numerical methods, including first-principles calculations, molecular dynamics [18] and quantum Monte Carlo approaches [19], are utilized frequently in detailed investigations of fundamental properties with complicated single- and many-particle interactions arising from charge and spin density distributions. It is well known that the Vienna *ab initio* simulation package (VASP, discussed in section 2.1) is accurate and reliable for most condensed-matter materials, since suitable crystal potentials, electron–electron Coulomb interactions and spin-dependent interactions are covered fully in the evaluation processes. Through detailed and careful analyses, the various calculated results can be unified to achieve concise physical and chemical pictures in order to understand thoroughly the variety of physical phenomena. The theoretical frameworks developed in our previous [20] and current book, can provide the full information for comprehending the rich and unique properties of emergent 2D materials. Specifically, the proposed single- and multi-orbital hybridizations and the atom-created spin–orbit coupling are responsible for the significant differences between monolayer group-IV and group-V materials. It should be noted that the VASP calculations cannot deal with certain open issues, such as the magneto-electronic properties, magneto-optical selection rules, quantum Hall conductivities and magnetoplasmon modes. That is to say, the VASP calculations are too cumbersome to solve the rich magnetic quantization phenomena exactly. Instead, the generalized tight-binding model, the effective-mass approximation, the Fermi golden rule, the static Kubo formula and the modified random-phase approximation (RPA) are frequently used to determine electronic [21], optical [22], transport [23], Coulomb-excitation [24] and decay properties [25], respectively. However, such methods are unable to explore the basic issues surrounding the total ground state energy and the optimal geometric structure. For example, significant parameters for the hopping integrals in these models could be obtained from good numerical fitting for the low-lying energy bands along the high-symmetry points between the tight-binding model and the first-principles method, e.g. the first pair of valence and conduction bands in the AA- and AB-bt bilayer silicene systems [26]. In addition, the generalized tight-binding model and the modified RPA have been merged together successfully to understand the many-body magneto-electronic Coulomb excitations in emergent layered systems [24]. However, as the low-energy band structures are very complicated (the extremely non-uniform environments in unit cells survive), it is almost impossible to simulate them using the tight-binding model.

Various synthesis methods and high-resolution experimental measurements are being developed further to verify the theoretical predictions and identify diverse phenomena. Generally speaking, the growth methods are very sensitive to the types of atoms and/or molecules and the geometric structures in the emergent layered materials. For example, 2D few-layer graphene systems have been produced successfully using mechanical exfoliation, chemical vapor deposition (CVD) and the electrostatic manipulation of scanning tunneling microscopy (STM) [27]. Furthermore, 1D graphene nanoribbons can be cut directly from graphene layers (a top-down method) through lithographic patterning and etching [28, 29], sono-chemical breaking [30], metal-catalyzed cutting [31] and oxidization reactions [32]. In examining the fundamental properties, low-energy electron diffraction (LEED) and reflected high-energy electron diffraction (RHEED) (details in chapter 3) [33], tunneling electron microscopy (TEM) [34] and STM [35], angle-resolved photo-emission spectroscopy (ARPES) [35], scanning tunneling spectroscopy (STS) [36], the superconducting quantum interference device (SQUID) [37], optical absorption, reflection and transmission spectroscopy [38], electron energy loss spectroscopy and light inelastic scattering [39], quantum transport apparatus [40], and differential thermal calorimeter and laser flash analysis [41] are useful in understanding, respectively, the surface morphologies, the side and top views of the optimal geometric structures, the occupied quasiparticle energy spectra and lifetimes, the special van Hove singularities, the induced magnetic fields (the net magnetic moments), the optical absorption spectra (the vertical excitation structures), the single-particle excitations and plasmon modes, the unique Hall conductivities, and the temperature- and magnetic-field-dependent heat capacities. Furthermore, spin-polarized STM, ARPES and STS can identify the spatial spin configurations [42], the spin-split energy bands [43] and the spin-decomposed density of states [44].

Few-layer group-IV materials, which consist of monoelemental carbon [45, 46], silicon [47, 48], germanium [49, 50], tin [51, 52] and lead [9, 10] atoms, have been produced successfully in experimental laboratories. Layered graphene systems can be synthesized using various methods, while Si-, Ge-, Sn- and Pb-related 2D materials are grown frequently on commensurate substrates using molecular beam epitaxy (MBE). For the latter materials, the significant chemical bonding between the group-IV adlayer and substrate might play critical roles in the fundamental properties, such as the reconstruction of surface and substrate morphologies [53], and a weakly or strongly modified Dirac-cone-like band structure [54] or even its complete absence [55]. Moreover, rich and unique phenomena have been identified in graphene, such as different stacking configurations [56, 57] and layer numbers [58, 59], the strongest mechanical response ability [60, 61], diverse magneto-electronic states [70], various inter-Landau-level optical transitions [62], unusual quantum Hall conductivities [63–65], and multiple electron–hole excitations and plasmon modes [66, 67]. These results are expected to vary greatly in other group-IV systems. On the theoretical side, first-principles calculations, the effective-mass approximation and the tight-binding model are utilized frequently to explore the electronic properties (band structures, wave functions and densities of states). The first method is highly suitable for studying the total ground state energy, multi- and single-orbital hybridizations, and spin-related configurations under optimal geometric structures; that is, it can provide concise physical and chemical pictures to explain diverse phenomena (details in [20] and this book). Direct combination of the first-principles method with the effective-mass approximation and the tight-binding method might be very useful in understanding low-energy physics. For example, the reliable intralayer and interlayer hopping integrals in few-layer graphene systems could provide exact and full information on the electronic [45], optical [68], magnetic [69], transport [68] and Coulomb-excitation properties [15, 71]. Specifically, the diverse magnetic quantization phenomena, which are created by a uniform perpendicular magnetic field ($B_z\hat{z}$), could be solved exactly by the generalized tight-binding model, but not the low-energy perturbation method under the irregular or complicated band structures [72–76]. First-principles calculations cannot deal with any B_z-related external fields, since such fields lead to very large unit cells associated with the Peierls phases of the vector potentials. The geometric, electronic and magnetic properties of monolayer group-IV materials and the differences between the theoretical models are worthy of systematic study. In addition, further investigations need to be conducted on the various physical, chemical and material properties of Si-, Ge-, Sn- and Pb-related materials.

Layered structures, stacking configurations and layer numbers play critical roles in greatly varying the fundamental properties and enhancing potential applications. To date, there have been a lot of experimental [47, 77–79] and theoretical [45, 50, 51, 80–82] studies on graphene-related materials. Such studies illustrate clearly that the rich and unique phenomena are dominated by the π bonding of carbon-2p$_z$ orbitals. For example, few-layer graphenes have been identified that present AAA [83], ABA [56], ABC [56, 83], AAB [57], twisted [84], sliding [69], modulated [85] and turbostratic stacking [86]. Most unusual electronic structures are examined using high-resolution STS and ARPES measurements (discussed in sections 3.2 and 3.3). Such systems may belong to the band-overlap semimetals with low-density free electrons and holes, or the narrow-gap semiconductors. Furthermore, the stacking-enriched low-lying band structures include the Dirac-cone linear [85], parabolic [84], sombrero-shaped [87] and oscillatory energy dispersions [26], in which uniform and non-uniform dispersions create diverse magnetic quantization phenomena. Interestingly, dramatic transformations between the first and the second energy bands are achieved by the relative shift of the two carbon-honeycomb lattices [88], thus leading to well-behaved, perturbed and undefined Landau levels (the latter has never been observed in any layered material [13]). In general, the sliding, twisted and modulated bilayer graphene systems exhibit, respectively, semi-metallic [84], zero-gap semiconducting (monolayer-like) and metallic behaviors [88] under specific models. On the other hand, the sliding bilayer silicenes in this work are expected to display more complicated phenomena, as a result of the strong coupling among the buckled structures, stacking configurations, multi-orbital hybridizations (not only the Si-3p$_z$ orbitals), intralayer and interlayer hopping integrals, and spin–orbit interactions [89]. The sublattice-, layer-, orbital- and spin-decomposed contributions to the fundamental properties and the significant cooperative and/or competitive relationships with displacements will be the focus of study, e.g. the semimetal–semiconductor transition [88] and the transformation of non-magnetic and magnetic configurations [26]. To date, only bilayer AB-bottom–top silicene has been verified to survive the experimental synthesis process using high-resolution TEM [47]. Moreover, the theoretical predictions [77], conducted using the generalized tight-binding model, show unusual magneto-electronic properties, e.g. valley- and spin-dominated Landau levels, and rich magneto-optical absorption spectra [85].

Chemical modifications are a very efficient method to create drastic changes and dramatic transitions in the physical, chemical and material phenomena, in particular for adatom chemisorption [90] and guest-atom substitution [91]. A silicene-based honeycomb lattice is capable of providing a very active chemical environment, since it possesses an observable buckled structure and the dangling $3p_z$ orbitals. Chemical bonds should be generated easily at its surface, e.g. silicene oxides [92] and hydrogenated silicenes [93]. It is well known that hydrogen atoms, with one 1s orbitals, have rather strong chemical interactions with other atoms. Such active adatoms are utilized frequently to stabilize free boundaries with a lot of unbound orbitals, e.g. the H-saturated edge structures of finite-length carbon nanotubes [94] and graphene nanoribbons [95], that are expected to exist in silicene-related 1D materials. They could survive within the open spaces of emergent materials and would be very useful in energy storage [96]. H-atoms are very suitable for research into the basic science owing to the concise orbital hybridization under the single-orbital case, such as the sp^3 chemical bonding in graphene and graphane [97], and hydrogenated graphene nanoribbons [98]. Moreover, hydrogen-modulated energy gaps could be achieved through variation of its concentration, e.g. the wide range of E_g ~ 0.0–3.5 eV was achieved under various hydrogenations of graphene systems [97]. Similar theoretical studies have been generalized to hydrogen-adatom chemisorption on silicene surfaces [99]. Compared to pristine silicenes and hydrogenated graphenes, the enhanced buckling and specific chemical bonding are predicted to play critical roles in producing the geometric and electronic properties. For example, both Si–H-bond-dominated valence bands at the middle energy and H-distribution-induced localized energy bands across the Fermi level are characteristic of hydrogenated silicene systems, since such phenomena are revealed in the general cases.

The group-VII halogen atoms are outstanding candidates for the chemisorption of 2D emergent materials, since they possess quite high electron affinities. Such guest atoms have five electrons in the outermost p_x, p_y and p_z orbitals; furthermore, adatom-induced spin configurations might appear after chemical adsorption. The significant chemical bonding which is formed with the host atoms will induce large electron transfers and thus create diverse fundamental properties. Importantly, fluorinated [100–102], chlorinated [103, 104], brominated [105] and iodinated [106] graphene systems have been produced successfully using specific chemical methods. For high adatom concentrations, the flourinated systems exhibit semiconducting behaviors and the three other systems exhibit metallic behaviors, based on high-resolution measurements of the transport properties [107], optical spectra [108] and occupied electronic energy bands [109]. To date, very few theoretical and experimental studies have been conducted on the fundamental properties of halogenated silicene systems, illustrating clearly the great challenges in the sample growth of pristine materials. Previous predictions show unusual geometric, electronic and magnetic properties of halogenated silicene [110, 111]. The whole range of halogenation effects, which arise from the multi-orbital Si–X, Si–Si and X–X chemical bonds and adatom-created magnetism, will be investigated thoroughly in this book, in particular for the adsorption conditions and mechanisms determining feature-rich energy bands, band gaps, free carrier densities, ferromagnetic or non-magnetic configurations, and van Hove singularities. Furthermore, the important differences between fluorination and other halogenations will be discussed in detail.

Metal-adatom chemisorption on the silicene surface is expected to present opposite doping behaviors (n-type free conduction electrons) [112–114], which are completely different from the π-type phenomena under various halogenations [110, 114]. Furthermore, sodium, magnesium and aluminum possess, respectively, one, two and three outermost electron orbitals (specific atomic configurations) in providing sufficient metallization, i.e. the Na-, Mg- and Al-adsorbed silicenes can represent typical metallic systems. It is well known that alkali graphite intercalation compounds have been investigated both experimentally and theoretically for more than forty years. Such layered materials are very good conductors, with electrical conductivities and free carrier densities [115–117] such as those in copper. These results illustrate high charge transfer from the alkali to the carbon atoms. According to up-to-date applications in commercialized products, Li– and Li^+–graphite compounds serve as the best anode materials in Li^+-based batteries [118–120]. Recently, first-principles predictions were performed for 1D Li-, Na-, K-, Rb- and Cs-related graphene nanoribbons with various adsorption cases using careful analyses, indicating the full transfer (the dramatic transformation) of their half-occupied s-orbitals into conduction carriers [121]. Similar behavior has been deduced to appear in 2D graphene systems with p- and s-bonding of carbon [20, 122]. To date, Al-dependent graphene systems have been one of the most widely studied materials, since the high sensitivity of Al atoms is useful in environmental and energy engineering, such as in toxic gas sensors [123], potential hydrogen storage materials [124, 125] and aluminum-ion batteries [126]. Based on first-principles studies [127], there exist significant hybridizations between the (3s, $3p_x$, $3p_y$) and $2p_z$ orbitals in the Al–C bonds, where such phenomena are responsible for the quasi-rigid shifts of energy bands, the free electrons in the conduction Dirac cone, and the Al-dominated valence and conduction bands. Specifically, the Al-induced high free carrier density is comparable to that of alkali doping. Thus Na-, Mg- and Al-doped silicene materials will be able to create various orbital hybridizations, vary the fundamental properties and provide more information for potential applications.

The chemisorption and substitution of silicon atoms on graphene are very interesting topics, since first-principles predictions are available for understanding the important differences between the two kinds of chemical modifications [128, 129] and for providing more information about other approaches for the formation of silicon–carbide compounds [130]. In the case of a carbon-created honeycomb lattice, silicon atoms are regarded as adatoms and guest atoms under chemisorption and substitution, respectively. The adatoms present optimal positions at the bridge sites above the graphene surface [131], while the guest atoms present Si- and C-co-dominated A and B sublattices with deformed hexagons [132]. Both silicon and carbon atoms have quite active dangling bonds, leading to significant multi-orbital hybridizations in the Si–C bonds. Most importantly, the planar bonding between Si and C atoms is expected to be stronger and more complicated compared to the perpendicular configuration. That is to say, the π and σ bonding of graphene will be changed drastically by Si substitution or Si chemisorption. This is reflected directly in the varied properties, spatial charge distributions, atom-dominated energy bands, and atom- and orbital-projected densities of states. Furthermore, their gradual transformation could be achieved through a Si- or C-substituted graphene or silicene honeycomb lattice (detailed discussions in chapters 9 and 10).

The significant atomic interactions between a group-IV adlayer and substrate can create other types of diverse phenomena. Layered 2D materials are grown frequently on various metals or insulators using MBE, which is sensitive to the lattice constant of the semi-infinite substrate. For example, C-, Si-, Ge-, Sn- and Pb-few-layer materials have been synthesized on, respectively, Ag(111), Ir(111)/ZrB_2R(0001) [1–3], Pt(111), Al(111)/Au(111) [4–6], and Bi_2Te_3 (111) [133]. Their surface morphologies can be examined accurately with simultaneous experimental observations using both STM and LEED, RHEED or TEM, and direct evidence of the top and side views has been obtained from such measurements (discussed in chapter 3). Most cases show clear surface reconstructions in the adlayer and substrate [134], while only a few cases retain the original hexagon as the unit cell or reveal an incommensurate moiré structure [135]. Moreover, up-to-date ARPES measurements, performed for the occupied electronic energy spectra near the Fermi level, have identified the presence, modification and absence of Dirac-cone-like band structures [35]. Interestingly, the group-IV adlayer and substrate atoms are capable of inducing non-negligible chemical bonding, thus leading to diverse lattice symmetries and electronic properties. To explore the substrate effects fully, the commensurate moiré superlattice and a sufficiently thick substrate need to be included thoroughly in the first-principles calculations. Monolayer silicene–Ag (111) is chosen as a model material for study, focusing on the critical roles of the Si–Ag, Si–Si and Ag–Ag chemical bonds. How to vary the fundamental properties using enlarged unit cells, multi-orbital hybridizations and Ag layers is analysed in detail.

The composite condensed-matter systems, which consist of two different monolayer subsystems, are formed through intrinsic interlayer atomic interactions (e.g. significant van der Waals interactions). The various coupled materials are expected to exhibit diverse phenomena. Furthermore, they are highly suitable for the experimental verification of well-known 2D theories. For example, an insulating boron-nitride monolayer, with a large band gap $E_g > 3.0$ eV, has been synthesized successfully above a graphene surface using the CVD growth method [136]. The interlayer hopping integrals from the dangling $2p_z$ orbitals are weak, and significant differences appear between the insulating and zero-gap behaviors. As a result, the low-lying gapless Dirac cone of monolayer graphene is predicted to barely be affected by the interlayer hopping integrals. On the other hand, these integrals play a critical role in the moiré superlattice structure. The low-energy physical properties appear to be dominated by the electronic states in single-layer graphene, e.g. measurements of magneto-transport properties verify a periodic Aharonov–Bohm effect due to the magnetic flux quantum, owing mainly to the greatly enlarged unit cell. Interestingly, when group-IV monolayer materials are combined, more complicated phenomena might emerge, which is related closely to the combination of the planar and buckled or buckled and buckled structures. Such materials may have certain meta-stable superlattice structures under careful first-principles calculations, in which the stacking symmetries determine the spatial charge distributions (the π and σ bonding) and vary the fundamental properties. A C-, Ge-, Sn- or Pb-monolayer on silicene could serve as a model system in exploring the critical mechanisms from single- and multi-orbital hybridizations and spin–orbit coupling.

Chapters 1–14 of this book explore fully the diverse fundamental properties of silicene-related 2D systems through first-principles calculations, in particular for the critical factors, buckled and planar geometries, moiré superlattices [135], single- and multi-orbital hybridizations, spin–orbit coupling and spin-dependent many-particle interactions. Concise physical and chemical pictures are determined carefully from the developed theoretical framework, as described in chapter 2. Other phenomenological models, e.g. the tight-binding model and the effective-mass approximation, are compared to the current models in detail. Chapter 3 discusses the various high-resolution experimental measurement methods, including LEED and RHEED [33], TEM [34], STM and SP-STM [35, 42], STS and SP-STS [36, 44], ARPES and SP-ARPES [35, 43], and SQUID [37]. Monolayer group-IV systems, with the four half-occupied orbitals of s, p_x, p_y and p_z, are investigated in terms of their diverse phenomena under different atomic numbers in chapter 4. Whether the low-lying energy bands depend on the π and/or the π and σ orbitals is the subject of study. Furthermore, other diverse physical phenomena are discussed in detail. The main focus of chapter 5 is the dramatic transformations in fundamental properties through a relative shift in bilayer silicene, e.g. the zero- and finite-gap transition, the defined or undefined π- and σ-band widths, and the existence or absence of intralayer and interlayer spin–orbit interactions. The quite strong Si–H bonds in hydrogenated silicene systems, illustrated clearly in chapter 6, are analysed thoroughly to obtain unusual results in terms of buckling-enhanced honeycomb lattices, atom-dominated band properties, charge redistributions after hydrogenation, and van Hove singularities in the atom- and orbital-projected densities of states. The complicated relationships between the adatom chemisorption cases (distribution configurations and concentrations) and band properties (semiconductors or semimetals) will be clarified from the multi-orbital hybridizations in Si–H bonds. Chapter 7 explores fully the strong chemisorption effects in halogenated silicene materials, in particular for the significant differences between fluorination and other halogenations. The p-type metals and finite-gap semiconductors, the spin-split and spin-degenerate energy bands, and the ferromagnetic, antiferromagnetic and non-magnetic configurations are expected to depend strongly on the halogen adatoms, concentrations and distributions. Many kinds of electronic and magnetic behaviors will be investigated and classified according to the critical mechanisms (the orbital hybridizations in X–Si bonds and the halogen-induced spin configurations). Another type of charge transfer, n-type doping, which is calculated and analysed carefully in chapter 8, arises through the metal-adatom adsorption of Na, Mg and Al. They have quite different atomic configurations and single- and multi-orbital chemical bonding, and diverse fundamental properties. The relationship between 2D free carrier density and Fermi-momentum states is analysed carefully. Whether the ferromagnetic configurations are revealed under specific adatom and adsorption cases needs to be investigated thoroughly through detailed calculations. In chapter 9, we cover the drastic changes in the planar σ and perpendicular π bonding in silicene created by chemical substitution using guest atoms of B, C and N. These binary compounds could provide more information about diverse phenomena. Chapter 10 illustrates clearly the significant differences between chemisorption and substitution effects as the Si adatoms and Si guest atoms on graphene increase from the dilute to full cases. Si-adlayer and bilayer AB-bt silicene on a Ag(111) substrate, which are explored thoroughly in chapter 11, are highly suitable for obtaining a full understanding of the unusual types of chemical bonding (the intralayer and interlayer Si–Si, Si–Ag and Ag–Ag bonds) and their critical roles in surface reconstructions and fundamental properties (e.g. the semiconductor–semimetal transitions). There are important differences between the monolayer and bilayer cases in terms of magnetic properties. The composite bilayer systems in chapter 12, which consist of Si–, Ge–, Sn– and Pb–graphene, are good candidates for studying the various moiré superlattices associated with very complicated interlayer atomic interactions (significant van der Waals interactions). The modified multi-Dirac-cone structures, which might appear in different stable valleys, are a very interesting topic of study. Specific examinations of these structures are conducted on the low-lying electronic properties and their relationships to extremely non-uniform environments. Finally, chapters 13, 14 and 15 present our concluding remarks, open issues/potential applications, and problems, respectively.

References

[1] Kiraly B, Iski E V, Mannix A J, Fisher B L, Hersam M C and Guisinger N P 2013 Solid-source growth and atomic-scale characterization of graphene on Ag(111) *Nat. Commun.* **4** 2804

[2] Coraux J, N 'Diaye A T, Busse C and Michely T 2008 Structural coherency of graphene on Ir(111) *Nano Lett.* **8** 565–70

[3] Walker L S, Marotto V R, Rafiee M A, Koratkar N and Corral E L 2011 Toughening in graphene ceramic composites *ACS Nano* **5** 3182–90

[4] Li L *et al* 2014 Buckled germanene formation on Pt(111) *Adv. Mater.* **26** 4820–4

[5] Zhuang J *et al* 2017 Cooperative electron–phonon coupling and buckled structure in germanene on Au(111) *ACS Nano* **11** 3553–9

[6] Derivaz M, Dentel D, Stephan R, Hanf M C, Mehdaoui A, Sonnet P and Pirri C 2015 Continuous germanene layer on Al(111) *Nano Lett.* **15** 2510–6

[7] Peng Q, Wen X and De S 2013 Mechanical stabilities of silicene *RSC Adv.* **3** 13772–81

[8] Kamal C, Chakrabarti A, Banerjee A and Deb S K 2013 Silicene beyond mono-layers—different stacking configurations and their properties *J. Phys. Condens. Matter* **25** 085508

[9] Lin X and Ni J 2012 Much stronger binding of metal adatoms to silicene than to graphene: a first-principles study *Phys. Rev.* B **86** 075440

[10] Hussain T, Kaewmaraya T, Chakraborty S and Ahuja R 2016 Defect and substitution-induced silicene sensor to probe toxic gases *J. Phys. Chem.* C **120** 25256–62
[11] Özçelik V O, Gurel H H and Ciraci S 2013 Self-healing of vacancy defects in single-layer graphene and silicene *Phys. Rev.* B **88** 045440
[12] Song Y L, Zhang Y, Zhang J M and Lu D B 2010 Effects of the edge shape and the width on the structural and electronic properties of silicene nanoribbons *Appl. Surf. Sci.* **256** 6313–7
[13] Gao N, Li J C and Jiang Q 2014 Bandgap opening in silicene: effect of substrates *Chem. Phys. Lett.* **592** 222–6
[14] Pflugradt P, Matthes L and Bechstedt F 2014 Unexpected symmetry and AA stacking of bilayer silicene on Ag(111) *Phys. Rev.* B **89** 205428
[15] Ho J H, Lu C L, Hwang C C, Chang C P and Lin M F 2006 Coulomb excitations in AA- and AB-stacked bilayer graphites *Phys. Rev.* B **74** 085406
[16] Wang X F and Chakraborty T 2007 Coulomb screening and collective excitations in a graphene bilayer *Phys. Rev.* B **75** 041404
[17] Wu J Y, Chen S C, Roslyak O, Gumbs G and Lin M F 2011 Plasma excitations in graphene: their spectral intensity and temperature dependence in magnetic field *ACS Nano* **5** 1026–32
[18] Pei Q X, Zhang Y W, Sha Z D and Shenoy V B 2013 Tuning the thermal conductivity of silicene with tensile strain and isotopic doping: a molecular dynamics study *J. Appl. Phys.* **114** 033526
[19] Borowik P, Thobel J L and Adamowicz L 2016 Monte Carlo study of electron transport in monolayer silicene *Semicond. Sci. Technol.* **31** 115004
[20] Tran N T T, Lin S Y, Lin C Y and Lin M F 2017 *Geometric and Electronic Properties of Graphene-related Systems: Chemical Bonding Schemes* (Boca Raton, FL: CRC Press)
[21] Berdiyorov G R, Neek-Amal M, Peeters F M and van Duin A C 2014 Stabilized silicene within bilayer graphene: a proposal based on molecular dynamics and density-functional tight-binding calculations *Phys. Rev.* B **89** 024107
[22] Shyu F L 2017 Field-modulated low-energy electronic and optical properties of armchair silicene nanoribbons *Physica* E **85** 117–24
[23] Zhou B, Zhou B, Zeng Y, Zhou G and Duan M 2016 Tunable electronic and transport properties for ultranarrow armchair-edge silicene nanoribbons under spin–orbit coupling and perpendicular electric field *Phys. Lett.* A **380** 282–7
[24] Wu J Y, Chen S C and Lin M F 2014 Temperature-dependent Coulomb excitations in silicene *New J. Phys.* **16** 125002
[25] Zaminpayma E and Nayebi P 2016 Band gap engineering in silicene: a theoretical study of density functional tight-binding theory *Physica* E **84** 555–63
[26] Do T N, Shih P H, Gumbs G, Huang D, Chiu C W and Lin M F 2018 Diverse magnetic quantization in bilayer silicene *Phys. Rev.* B **97** 125416
[27] Allen M J, Tung V C and Kaner R B 2009 Honeycomb carbon: a review of graphene *Chem. Rev.* **110** 132–45
[28] Wang X and Dai H 2010 Etching and narrowing of graphene from the edges *Nat. Chem.* **2** 661
[29] Tapasztó L, Dobrik G, Lambin P and Biró L P 2008 Tailoring the atomic structure of graphene nanoribbons by scanning tunnelling microscope lithography *Nat. Nanotechnol.* **3** 397

[30] Wu Z S, Ren W, Gao L, Liu B, Zhao J and Cheng H M 2010 Efficient synthesis of graphene nanoribbons sonochemically cut from graphene sheets *Nano Res.* **3** 16–22

[31] Wang J, Ma L, Yuan Q, Zhu L and Ding F 2011 Transition-metal-catalyzed unzipping of single-walled carbon nanotubes into narrow graphene nanoribbons at low temperature *Angew. Chem. Int. Ed.* **50** 8041–5

[32] Ma L, Wang J and Ding F 2012 Strain-induced orientation-selective cutting of graphene into graphene nanoribbons on oxidation *Angew. Chem. Int. Ed.* **51** 1161–4

[33] Hattori A N *et al* 2011 Formation of wide and atomically flat graphene layers on ultraprecision-figured 4H-SiC (0001) surfaces *Surf. Sci.* **605** 597–605

[34] Borysiuk J, Bożek R, Strupiński W, Wysmołek A, Grodecki K, Stępniewski R and Baranowski J M 2009 Transmission electron microscopy and scanning tunneling microscopy investigations of graphene on 4H-SiC (0001) *J. Appl. Phys.* **105** 023503

[35] Vogt P *et al* 2012 Silicene: compelling experimental evidence for graphene-like two-dimensional silicon *Phys. Rev. Lett.* **108** 155501

[36] Li G, Luican A and Andrei E Y 2009 Scanning tunneling spectroscopy of graphene on graphite *Phys. Rev. Lett.* **102** 176804

[37] Girit C, Bouchiat V, Naaman O, Zhang Y, Crommie M F, Zettl A and Siddiqi I 2008 Tunable graphene dc superconducting quantum interference device *Nano Lett.* **9** 198–9

[38] Mak K F, Ju L, Wang F and Heinz T F 2012 Optical spectroscopy of graphene: from the far infrared to the ultraviolet *Solid State Commun.* **152** 1341–9

[39] Lu J, Loh K P, Huang H, Chen W and Wee A T 2009 Plasmon dispersion on epitaxial graphene studied using high-resolution electron energy-loss spectroscopy *Phys. Rev.* B **80** 113410

[40] Wang L *et al* 2017 Industrial production of ultra-stable sulfonated graphene quantum dots for Golgi apparatus imaging *J. Mater. Chem.* B **5** 5355–61

[41] Townsend D I and Tou J C 1980 Thermal hazard evaluation by an accelerating rate calorimeter *Thermochim. Acta* **37** 1–30

[42] Wortmann D, Heinze S, Kurz P, Bihlmayer G and Blügel S 2001 Resolving complex atomic-scale spin structures by spin-polarized scanning tunneling microscopy *Phys. Rev. Lett.* **86** 4132

[43] Riley J M *et al* 2014 Direct observation of spin-polarized bulk bands in an inversion-symmetric semiconductor *Nat. Phys.* **10** 835

[44] Pietzsch O, Kubetzka A, Bode M and Wiesendanger R 2004 Spin-polarized scanning tunneling spectroscopy of nanoscale cobalt islands on Cu(111) *Phys. Rev. Lett.* **92** 057202

[45] Shao Y *et al* 2018 Epitaxial growth of flat antimonene monolayer: a new honeycomb analogue of graphene *Nano Lett.* **18** 2133

[46] Pham K D and Nguyen C V 2018 First principles calculations of the geometric structures and electronic properties of van der Waals heterostructure based on graphene, hexagonal boron nitride and molybdenum diselenide *Diam. Relat. Mater.* **88** 151

[47] Yaokawa R, Ohsuna T, Morishita T, Hayasaka Y, Spencer M J S and Nakano H 2016 Monolayer-to-bilayer transformation of silicenes and their structural analysis *Nat. Commun.* **7** 10657

[48] Shu H, Cao D, Liang P, Wang X, Chen X and Lu W 2014 Two-dimensional silicene nucleation on a Ag(111) surface: structural evolution and the role of surface diffusion *Phys. Chem. Chem. Phys.* **16** 304

[49] Shih P H, Chiu Y H, Wu J Y, Shyu F L and Lin M F 2017 Coulomb excitations of monolayer germanene *Sci. Rep.* **7** 40600

[50] Dávila M E and Lay G L 2016 Few layer epitaxial germanene: a novel two-dimensional Dirac material *Sci. Rep.* **6** 20714

[51] Zhu F F *et al* 2015 Epitaxial growth of two-dimensional stanene *Nat. Mater.* **14** 1020

[52] Liao M, Zang Y, Guan Z, Li H, Gong Y and Xue Q K 2018 Superconductivity in few-layer stanene *Nat. Photonics* **14** 344

[53] Andzane J *et al* 2019 Effect of graphene substrate type on formation of Bi_2Se_3 nanoplates *Sci. Rep.* **9** 4791

[54] Sadeddine S *et al* 2017 Compelling experimental evidence of a Dirac cone in the electronic structure of a 2D silicene layer *Sci. Rep.* **7** 44400

[55] Chen R B, Chen S C, Chiu C W and Lin M F 2017 Optical properties of monolayer tinene in electric fields *Sci. Rep.* **7** 1849

[56] Ping J and Fuhrer M S 2012 Layer number and stacking sequence imaging of by transmission electron microscopy *Nano Lett.* **12** 4635

[57] Lin C Y, Huang B L, Ho C H, Gumbs G and Lin M F 2018 Geometry-diversified Coulomb excitations in trilayer AAB stacking graphene *Phys. Rev.* B **98** 195442

[58] Ohta T, Bostwick A, McChesney J L, Seyller T, Horn K and Rotenberg E 2007 Interlayer interaction and electronic screening in multilayer graphene investigated with angle-resolved photoemission spectroscopy *Phys. Rev. Lett.* **98** 206802

[59] Coletti C *et al* 2013 Revealing the electronic band structure of trilayer graphene on SiC: an angle-resolved photoemission study *Phys. Rev.* B **88** 155439

[60] Guinea F, Katsnelson M and Geim A 2010 Energy gaps and a zero-field quantum Hall effect in graphene by strain engineering *Nat. Phys.* **6** 30

[61] Wong J H, Wu B R and Lin M F 2012 Strain effect on the electronic properties of single layer and bilayer graphene *J. Phys. Chem.* C **116** 8271

[62] Yuan S, Roldn R and Katsnelson M I 201 Landau level spectrum of ABA-and ABC-stacked trilayer graphene *Phys. Rev.* B **84** 125455

[63] Bao W *et al* 2011 Stacking-dependent band gap and quantum transport in trilayer graphene *Nat. Phys.* **7** 948

[64] Zhang Y, Tan Y W, Stormer H L and Kim P 2005 Experimental observation of the quantum Hall effect and Berry's phase in graphene *Nature* **438** 201

[65] Zhang L, Zhang Y, Camacho J, Khodas M and Zaliznyak I 2011 The experimental observation of quantum Hall effect of $l = 3$ chiral quasiparticles in trilayer graphene *Nat. Phys.* **7** 953

[66] Politano A and Chiarello G 2014 Plasmon modes in graphene: status and prospect *Nanoscale* **6** 10927

[67] Jiao N, Kang S, Han K, Shen X and Wang W 2019 Shape effects on graphene magneto-plasmons *Phys. Rev.* B **99** 195447

[68] Koshino M 2013 Stacking-dependent optical absorption in multilayer graphene *New J. Phys.* **15** 015010

[69] Huang Y K, Chen S C, Ho Y H, Lin C Y and Lin M F 2014 Feature-rich magnetic quantization in sliding bilayer graphenes *Sci. Rep.* **4** 7509

[70] Lin C Y, Wu J Y, Chiu Y H and Lin M F 2015 Magneto-electronic properties of multilayer graphenes *Phys. Chem. Chem. Phys.* **17** 26008

[71] Shyu F L and Lin M F 2000 Loss spectra of graphite-related systems: a multiwall carbon nanotube, a single-wall carbon nanotube bundle, and graphite layers *Phys. Rev.* B **62** 8508
[72] Lin C Y, Chen R B, Ho Y H and Lin M F 2018 *Electronic and Optical Properties of Graphite-Related Systems* (Boca Raton, FL: CRC Press)
[73] Lin C Y, Do T N, Huang Y K and Lin M F 2017 *Optical Properties of Graphene in Magnetic and Electric Fields* (Bristol: IOP Publishing)
[74] Chen S C, Wu J Y, Lin C Y and Lin M F 2017 *Theory of Magnetoelectric Properties of 2D Systems* (Bristol: IOP Publishing)
[75] Lin C Y, Wu J Y, Chiu C W and Lin M F 2019 *Coulomb Excitations and Decays in Graphene-Related Systems* (Boca Raton, FL: CRC Press)
[76] Lin C Y, Do T N, Wu J Y, Shih P H, Lin S Y, Ho C H and Lin M F 2019 *Diverse Quantization Phenomena in Emergent Layered Materials* (Boca Raton, FL: CRC Press)
[77] Li H, Fu H X and Meng S 2015 Silicene: from monolayer to multilayer—a concise review *Chin. Phys.* B **24** 086102
[78] Lee J K, Lee S C, Ahn J P, Kim S C, Wilson J I and John P 2008 The growth of AA graphite on (111) diamond *J. Chem. Phys.* **129** 234709
[79] Novoselov K S *et al* 2004 Electric field effect in atomically thin carbon film *Science* **306** 666
[80] Liu H *et al* 2014 Phosphorene: an unexplored 2D semiconductor with a high hole mobility *ACS Nano* **8** 4033
[81] Yang Q Q *et al* 2018 Phosphorene: 2D bismuthene fabricated via acid-intercalated exfoliation showing strong nonlinear near-infrared responses for mode-locking lasers *Nanoscale* **10** 21106
[82] Pumera M and Sofer Z 2017 2D monoelemental arsenene, antimonene, and bismuthene: beyond black phosphorus *Adv. Mater.* **29** 1605299
[83] Zhou L *et al* 2018 Surface structure of few layer graphene *Carbon* **136** 255
[84] Sharpe A L, Fox E J, Barnard A W, Finney J, Watanabe K and Kastner M A 2019 Emergent ferromagnetism near three-quarters filling in twisted bilayer graphene *Science* **10** 1126
[85] Chiu Y H, Lai Y H, Ho J H, Chuu S and Lin M F 2008 Electronic structure of a two-dimensional graphene monolayer in a spatially modulated magnetic field: Peierl's tight-binding model *Phys. Rev.* B **77** 045407
[86] Uemura K, Ikuta T and Maehashi K 2018 Turbostratic stacked CVD graphene for high-performance devices *Jpn J. Appl. Phys.* **57** 030311
[87] Pham K D, Phuc H V, Hieu N N, Hoi B D and Nguyen C V 2018 Electronic properties of GaSe/MoS_2 and GaS/$MoSe_2$ heterojunctions from first principles calculations *AIP Adv.* **8** 075207
[88] Thuy Tran N T, Lin S Y, Glukhova O E and Lin M F 2015 Configuration-induced rich electronic properties of bilayer graphene *J. Phys. Chem.* C **119** 10623
[89] Lin S Y, Chang S L, Thuy Tran N T, Yang P H and Lin M F 2015 H–Si bonding-induced unusual electronic properties of silicene: a method to identify hydrogen concentration *Phys. Chem. Chem. Phys.* **17** 26443
[90] Chen L, Hu H, Ouyang Y, Pan H Z, Sun Y Y and Liu F 2011 Atomic chemisorption on graphene with Stone–Thrower–Wales defects *Carbon* **49** 3356–61
[91] Wang H *et al* 2011 Doping monolayer graphene with single atom substitutions *Nano Lett.* **12** 141–4
[92] Wang R, Pi X, Ni Z, Liu Y, Lin S, Xu M and Yang D 2013 Silicene oxides: formation, structures and electronic properties *Sci. Rep.* **3** 3507

[93] Osborn T H, Farajian A A, Pupysheva O V, Aga R S and Voon L L Y 2011 *Ab initio* simulations of silicene hydrogenation *Chem. Phys. Lett.* **511** 101–5

[94] Novaes F D, Rurali R and Ordejón P 2010 Electronic transport between graphene layers covalently connected by carbon nanotubes *ACS Nano* **4** 7596–602

[95] Linden S *et al* 2012 Electronic structure of spatially aligned graphene nanoribbons on Au (788) *Phys. Rev. Lett.* **108** 216801

[96] Jose D and Datta A 2011 Structures and electronic properties of silicene clusters: a promising material for FET and hydrogen storage *Phys. Chem. Chem. Phys.* **13** 7304–11

[97] Rosas J H, Gutiérrez R R, Escobedo-Morales A and Anota E C 2011 First principles calculations of the electronic and chemical properties of graphene, graphane, and graphene oxide *J. Mol. Model.* **17** 1133–9

[98] Soriano D, Munoz-Rojas F, Fernández-Rossier J and Palacios J J 2010 Hydrogenated graphene nanoribbons for spintronics *Phys. Rev.* B **81** 165409

[99] Houssa M, Scalise E, Sankaran K, Pourtois G, Afanas' Ev V V and Stesmans A 2011 Electronic properties of hydrogenated silicene and germanene *Appl. Phys. Lett.* **98** 223107

[100] Robinson J T *et al* 2010 Properties of fluorinated graphene films *Nano Lett.* **10** 3001–5

[101] Feng W, Long P, Feng Y and Li Y 2016 Two-dimensional fluorinated graphene: synthesis, structures, properties and applications *Adv. Sci.* **3** 1500413

[102] Leenaerts O, Peelaers H, Hernández-Nieves A D, Partoens B and Peeters F M 2010 First-principles investigation of graphene fluoride and graphane *Phys. Rev.* B **82** 195436

[103] Sahin H and Ciraci S 2012 Chlorine adsorption on graphene: chlorographene *J. Phys. Chem.* C **116** 24075–83

[104] Fan L, Zhang H, Zhang P and Sun X 2015 One-step synthesis of chlorinated graphene by plasma enhanced chemical vapor deposition *Appl. Surf. Sci.* **347** 632–5

[105] Li Y *et al* 2012 Synthesis of partially hydrogenated graphene and brominated graphene *J. Mater. Chem.* **22** 15021–4

[106] Šimek P *et al* 2015 Towards graphene iodide: iodination of graphite oxide *Nanoscale* **7** 261–70

[107] Tahara K, Iwasaki T, Furuyama S, Matsutani A and Hatano M 2013 Asymmetric transport property of fluorinated graphene *Appl. Phys. Lett.* **103** 143106

[108] Karlicky F and Otyepka M 2013 Band gaps and optical spectra of chlorographene, fluorographene and graphane from G_0W_0, GW_0 and GW calculations on top of PBE and HSE06 orbitals *J. Chem. Theory Comput.* **9** 4155–64

[109] Ho K I *et al* 2014 Fluorinated graphene as high performance dielectric materials and the applications for graphene nanoelectronics *Sci. Rep.* **4** 5893

[110] Gao N, Zheng W T and Jiang Q 2012 Density functional theory calculations for two-dimensional silicene with halogen functionalization *Phys. Chem. Chem. Phys.* **14** 257–61

[111] Zhang W B, Song Z B and Dou L M 2015 The tunable electronic structure and mechanical properties of halogenated silicene: a first-principles study *J. Mater. Chem.* C **3** 3087–94

[112] Sahin H and Peeters F M 2013 Adsorption of alkali, alkaline-earth, and 3D transition metal atoms on silicene *Phys. Rev.* B **78** 085423

[113] Xu Q, Yang G M, Fan X and Zheng W T 2019 Adsorption of metal atoms on silicene: stability and quantum capacitance of silicene-based electrode materials *Phys. Chem. Chem. Phys.* **21** 4276–85

[114] Quhe R *et al* 2012 Tunable and sizable band gap in silicene by surface adsorption *Sci. Rep.* **2** 853

[115] Dresselhaus M S and Dresselhaus G 2002 Intercalation compounds of graphite *Adv. Phys.* **51** 1–186
[116] Xu J *et al* 2017 Recent progress in graphite intercalation compounds for rechargeable metal (Li, Na, K, Al)-ion batteries *Adv. Sci.* **4** 1700146
[117] Smart M C *et al* 1999 Irreversible capacities of graphite in low-temperature electrolytes for lithium-ion batteries *J. Electrochem. Soc.* **146** 3963–9
[118] Wu Y P, Rahm E and Holze R 2003 Carbon anode materials for lithium ion batteries *J. Power Sources* **114** 228–36
[119] Yoo E, Kim J, Hosono E, Zhou H S, Kudo T and Honma I 2008 Large reversible Li storage of graphene nanosheet families for use in rechargeable lithium ion batteries *Nano Lett.* **8** 2277–82
[120] Lee E and Persson K A 2012 Li absorption and intercalation in single layer graphene and few layer graphene by first principles *Nano Lett.* **12** 4624–8
[121] Lin Y T, Lin S Y, Chiu Y H and Lin M F 2017 Alkali-created rich properties in graphene nanoribbons: chemical bondings *Sci. Rep.* **7** 1722
[122] Sun M, Tang W, Ren Q, Wang S, Du Y and Zhang Y 2015 First-principles study of the alkali earth metal atoms adsorption on graphene *Appl. Surf. Sci.* **356** 668–73
[123] Ao Z M, Yang J, Li S and Jiang Q 2008 Enhancement of CO detection in Al doped graphene *Chem. Phys. Lett.* **461** 276–9
[124] Ao Z M and Peeters F M 2010 High-capacity hydrogen storage in Al-adsorbed graphene *Phys. Rev.* B **81** 205406
[125] Ao Z M, Jiang Q, Zhang R Q, Tan T T and Li S 2009 Al doped graphene: a promising material for hydrogen storage at room temperature *J. Appl. Phys.* **105** 0 74307
[126] Lin M C, Gong M, Lu B, Wu Y, Wang D Y, Guan M and Dai H 2015 An ultrafast rechargeable aluminium-ion battery *Nature* **520** 324
[127] Lin S Y, Lin Y T, Tran N T T, Su W P and Lin M F 2017 Feature-rich electronic properties of aluminum-absorbed graphenes *Carbon* **120** 209–18
[128] Aktürk E, Ataca C and Ciraci S 2010 Effects of silicon and germanium adsorbed on graphene *Appl. Phys. Lett.* **96** 123112
[129] Houmad M, Zaari H, Benyoussef A, El Kenz A and Ez-Zahraouy H 2015 Optical conductivity enhancement and band gap opening with silicon doped graphene *Carbon* **94** 1021–7
[130] Ceballos-Mendivil L G *et al* 2015 Synthesis of silicon carbide using concentrated solar energy *Sol. Energy* **116** 238–46
[131] Hu C H, Zheng Y, Zhang Y, Wu S Q, Wen Y H and Zhu Z Z 2011 Electronic and magnetic properties of silicon adsorption on graphene *Solid State Commun.* **151** 1128–30
[132] Chen Y *et al* 2013 Silicon-doped graphene: an effective and metal-free catalyst for NO reduction to N_2O? *ACS Appl. Mater. Interfaces* **5** 5994–6000
[133] Xu Y, Tang P and Zhang S C 2015 Large-gap quantum spin Hall states in decorated stanene grown on a substrate *Phys. Rev.* B **92** 081112
[134] Lalmi B, Oughaddou H, Enriquez H, Kara A, Vizzini S, Ealet B and Aufray B 2010 Epitaxial growth of a silicene sheet *Appl. Phys. Lett.* **97** 223109
[135] Feng B *et al* 2012 Evidence of silicene in honeycomb structures of silicon on Ag(111) *Nano Lett.* **12** 3507–11
[136] Son M, Lim H, Hong M and Choi H C 2011 Direct growth of graphene pad on exfoliated hexagonal boron nitride surface *Nanoscale* **3** 3089–93

**Shih-Yang Lin, Hsin-Yi Liu, Duy Khanh Nguyen, Ngoc Thanh Thuy Tran,
Hai Duong Pham, Shen-Lin Chang, Chiun-Yan Lin and Ming-Fa Lin**

Chapter 2

The theoretical frameworks

Many emergent 2D materials have been synthesized successfully in experimental laboratories, such as the group-IV [1, 2] and group-V [3, 4] materials and their composite [5] layered condensed-matter systems. Furthermore, rich and unique physical, chemical and material phenomena have been clearly identified in high-resolution experimental measurements. Current theoretical models might have difficulty in dealing with these diverse and complicated properties. For example, first-principles calculations are unable to explore fully the rich magnetic quantization phenomena including the magneto-electronic states [6], magneto-optical absorption spectra and selection rules [7], unusual quantum Hall transport [8], and the inter-Landau-level excitations and magnetoplasmon modes [9]. How to parameterize generalized tight-binding models suitably will become one of the mainstream topics of the basic physical sciences, in particular for various intrinsic interactions and non-uniform environments. The current theoretical frameworks for fundamental properties are worthy of systematic development.

2.1 The first-principles method

This book is devoted to developing a theoretical framework to unify various physical quantities from first-principles evaluations. Concise physical and chemical pictures are proposed for a thorough understanding of the geometric, electronic and magnetic properties in the presence of specific critical factors, e.g. being very sensitive to the layer numbers, stacking configurations, adatom or molecule chemical absorption, guest-atom substitution, heterojunctions and composite structures. Silicene-based emergent materials are chosen as the model material to demonstrate diverse phenomena.

Computer modeling and simulation serve as indispensable methods in basic science research. Numerical calculations need to solve the energy spectra and wave functions of quasiparticles in condensed-matter systems exactly and simultaneously.

doi:10.1088/978-0-7503-3299-6ch2

All quasiparticles frequently experience specific scattering events through wave pocket forms, e.g. the electron-crystal-potential and electron–electron Coulomb interactions. It is very difficult to deal with a Schrödinger equation with single- and many-particle interactions. Some approximate models have been developed to achieve reliable solutions. Among them, first-principles calculations are a dominant method in the quantum-mechanical simulation of periodic systems.

Until now, first-principles calculations have been suitable for and efficient in solving very complicated intrinsic interactions under a series of approximations and simplifications to obtain the eigenvalues and eigenfunctions, and thus reliable geometric, electronic and magnetic properties. In recent decades, the Vienna *ab initio* simulation package (VASP) [10] has been used to compute an approximate solution to the many-body Schrödinger equation, either within the density functional theory (DFT), solving the Kohn–Sham equations [11], or within the Hartree–Fock approximation, evaluating the Roothaan equations [12]. The calculated results include the optimal geometries of regular lattices and moiré superlattices, energy bands, wave-vector-dependent wave functions, spatial charge densities and their differences in independent and isolated subsystems, van Hove singularities in the density of states, net magnetic moments, and ferromagnetic, anti-ferromagnetic and non-magnetic spin configurations. Also, other numerical methods are very powerful for specific topics, e.g. the molecular dynamics [13] approach in simulating dynamic growth and structural transformation, and quantum Monte Carlo approaches [14] in determining the charge-density-induced exchange and correlation interactions of emergent materials.

All quasiparticle properties, such as energy bands, decay rates, electronic excitations, screening charges and effective Coulomb potentials, are determined mainly by their charges and spins. Most of these could be solved using the static and time-dependent first-principles methods. In particular, the static methods (e.g. VASP) can be used for geometric structures, electronic properties and spin configurations. In this book, the rich fundamental properties of silicene-related materials at zero temperature are explored thoroughly using first-principles DFT and VASP simulations [10]. Furthermore, the local density approximation (LDA) is chosen for the model studies. The projector augmented wave approach (PAW) (pseudopotentials in [15, 16]) is available for characterizing the electron–ion crystal potential. Importantly, the exchange-correlation energy, which is associated with the electron–electron Coulomb interactions, is evaluated using the Perdew–Burke–Ernzerhof functional through the generalized gradient approximation (using quantum Monte Carlo to obtain the charge-density-dependent many-particle energy in [14]). Specifically, the spin-dependent single- and many-particle interactions need to be taken into account for the significant materials and adatoms, such as the spin–orbit coupling in Si- [17, 18], Ge- [18], Sn- [18, 19], Pb- [20], GaAs- [21], Sb- [22] and Bi-induced [20] monolayer systems. The Hubbard-like on-site Coulomb interactions might play important roles in determining the total ground state energies, optimal geometries, band structures, Van Hove singularities, atom- and boundary-induced magnetic configurations, and magnetic moments. For example, zigzag graphene nanoribbons, with edge-created spin arrangements, are predicted to have a ferromagnetic configuration along the same boundary, but an anti-ferromagnetic configuration across the ribbon center [23]. For monolayer silicene and graphene, a vacuum space of 15 Å is inserted between periodic images to avoid the interactions of neighboring cells. The cutoff energy of the Block wave functions, which are built from complete plane waves, is set to be ~500 eV. To reach the optimal geometry with the lowest ground state energy, all the atoms of a condensed-matter system will experience relaxation processes by adjusting their positions. Then, the most stable configuration is utilized as an initial input for further calculation of the fundamental physical properties.

For the specific calculations of electronic properties and optimal geometric structures on monolayer silicene, the first Brillouin zones are sampled by, respectively, $100 \times 100 \times 1$ and $12 \times 12 \times 1$ k-points via the Monkhorst–Pack scheme. Also, these points are sufficient to obtain reliable atom-, orbital- and spin-projected densities of states. The convergence of the Hellmann–Feynman force on each atom is about 0.01 eV Å^{-1}, accompanied by a total energy difference of $\Delta E < 10^{-5}$ eV. The above-mentioned calculation process is shown in the flow chart of figure 2.1.

For structure-enriched and adatom- or guest-atom-doped silicene-related systems, the current first-principles calculation results include the ground state energies, binding energies, interlayer distances, bond lengths, bond angles, buckled and planar honeycomb lattices, optimal positions and heights of adatoms, host- and guest-created structures, moiré superlattices (due to chemisorption or substitution, heterojunctions and interlayer composites), atom-dominated and spin-split or spin-degenerate band structures, adatom-induced energy gaps and n–p-type doping, spatial charge distributions and their variation after chemical modification, spin density arrangements, magnetic moments, and atom-, orbital- and spin-decomposed van Hove singularities in the densities of states. Self-developed analyses are very useful in dealing with the initial data from first-principles evaluations, e.g. the weak energy bands in 3D energy–wave-vector space [24]. Interestingly, how many kinds of electronic properties (finite-gap semiconductors, zero-gap semiconductors, semimetals and metals) and magnetic configurations (non-magnetism, ferromagnetism and anti-ferromagnetism) are present is worth detailed analysis. It should be noted that there might exist diverse relationships between the above-mentioned electronic and magnetic properties. All the numerical calculations become more complicated and difficult compared to those carried out for few-layer graphene materials. The critical factors arise from the non-negligibly buckled structures and significant spin-dependent interactions, e.g. the distinct bilayer AA-bt, AA-bb, AB-bt and AB-bb silicene systems [25], with negligible or significant spin–orbit coupling. The theoretical predictions for silicene-based systems are compared thoroughly to high-resolution measurements from TEM, LEED, STM, ARPES, SP-ARPES, STS, SP-STS, SQUID and optical spectroscopy (details in chapter 3). Such studies are able to provide suitable and reliable information about potential applications.

Significant single- and multi-orbital hybridization in chemical bonds (e.g. Si–X bonds) is proposed to account for the diverse geometric and electronic properties through the careful use of data and analyses. The outer atomic orbitals of the host and guest atoms are included simultaneously in the numerical calculations. The

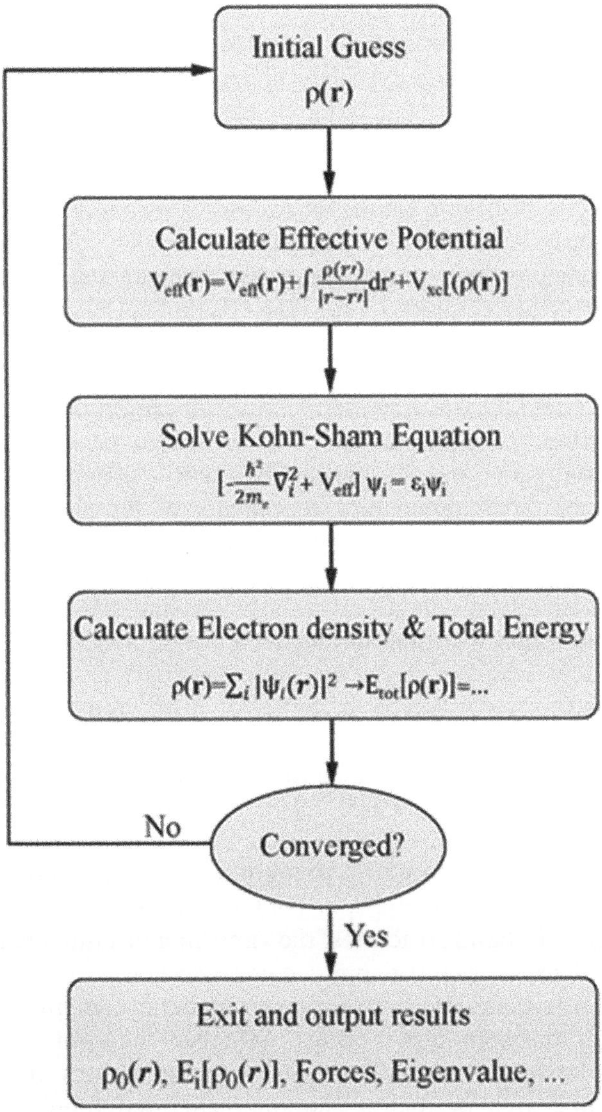

Figure 2.1. A flow chart of the first-principles VASP calculations.

dramatic changes in the chemical bonding in few-layer buckled and planar honeycomb lattices and the critical orbital hybridization of the same or different atoms will be investigated in detail, e.g. the well-behaved or undefined π and σ bonding, the important differences between monolayer and bilayer silicene systems in spatial charge distributions, and the diverse orbital interactions (the intralayer and interlayer interactions) due to chemical modification. It should be noted that single- and multi-orbital chemical bonding can be found clearly in the atom-dominated energy bands, the spatial charge distributions before and after chemical reactions, and the atom- and orbital-projected van Hove singularities (discussed in chapters 4–11). Interestingly, the orbital hybridization in condensed-matter systems is expected to be quite different from that in isolated chemical bonds. They seem to be responsible for the rich and unique geometric structures and electronic properties of silicene-related systems. This point of view could be further generalized to any condensed-matter system, such as the unusual and complicated orbital hybridization in the different moiré superlattices of lithium-ion-based cathode and anode materials, which are currently the subject of significant investigation.

To explore single-particle and many-particle phenomena fully, first-principles calculations are further combined with other theoretical models. These composite methods, which include relativistic spin–orbit coupling [17], spin-dependent on-site Coulomb interactions [26], optical vertical excitations [27], electronic excitations and Coulomb decay rates [28], are very useful in the full exploration of, respectively, the electronic properties, the spin-split states and ferromagnetic, anti-ferromagnetic and non-magnetic configurations, the various frequency-dependent absorption structures, the frequency and momentum dependence of the electron–hole pairs and plasmon modes, and the wave-vector-enriched quasiparticle lifetimes. In general, up-to-date calculations show that the direct combination of different models presents a lot of advantages and disadvantages simultaneously, therefore modifications are required to allow further development.

The spin configurations of electron or hole quasiparticles in the various group-IV-related materials vary greatly in terms of layer numbers, stacking configurations, adatoms in chemisorption, guest atoms under substitution and heterofunctions. According to VASP calculations, relativistic spin–orbit coupling is characterized by the single-particle form, and spin-created electron–electron Coulomb interactions are evaluated within the Hubbard-like model using the mean-field approach (the standard strength U) [29]. The non-magnetic, anti-ferromagnetic and ferromagnetic spin configurations are examined thoroughly through the spin-split band structures, the vanishing or finite magnetic moments, the spatial spin densities and the spin-decomposed density of states. It should be noted that the non-magnetic and antiferromagnetic configurations are distinguished through the spin distributions. Whether the spin configurations are driven by the host atoms, guest atoms, edge structures or heterojunctions (substrates) is one area of focus. The spin-generated fundamental properties can be verified using spin-polarized ARPES [30], STM [31], STS [32], optical spectroscopy [33] and quantum transport devices [34]. In addition, spin configurations can also be investigated using the phenomenological Hubbard model (the tight-binding model plus the spin-dependent on-site Coulomb interactions) [26]. For example, anti-ferromagnetic configurations, induced by the boundary-carbon spins of zigzag graphene nanoribbons, are found in the spin-degenerate band structure and density of states, and the opposite spin distributions across their centers.

The direct combination of first-principles calculations with the dynamic Kubo formula within linear or nonlinear responses is very useful in understanding the fundamental optical properties. When a condensed-matter system is perturbed by an electromagnetic wave, the occupied electronic states are excited vertically to unoccupied states because of the almost vanishing photon momenta. The optical absorption, reflection and transmission spectra are dominated by single-particle transitions or many-body excitonic effects (electron–hole pairs with attractive Coulomb potentials), which are very sensitive to the intrinsic band properties (e.g. band curvatures, wave functions and charge densities) and the external environment (temperature, gate voltage, magnetic field and carrier doping). The numerical calculations related to the many-particle scattering processes appear to be very difficult and complicated. After first-principles calculations of wave-vector-dependent energy spectra and wave functions, they could be utilized as an input for the linear Kubo formula (Fermi's golden rule due to dipole perturbations) [35]. Then, the real and imaginary parts of the transverse dielectric functions and the energy loss functions under the long wavelength limit depend strongly on the excitation frequency, where the prominent structures in the real and imaginary parts represent, respectively, the prominent electron–hole excitations and plasmon modes [36]. In addition, the threshold absorption frequency, the optical gap, might be different to or larger than the energy gap.

While the first-principles method is linked to the random-phase approximation, the time-dependent LDA is available to explore dynamic excitation processes via the electron–electron Coulomb interactions. The longitudinal dielectric functions, which can characterize the charge screening ability under any Coulomb field, are evaluated directly from the momentum- and frequency-dependent polarization functions [26]. Obviously, the band-structure effects have been included in the numerical calculations, where the electronic energy spectra and the Coulomb matrix elements dominate the main features of the single- and many-particle excitations. The $(\mathbf{q}, \omega)$-phase diagrams are the focus of study, since they are responsible for the different regions (boundaries) of the electron–hole excitations and the $\mathbf{q}$-dispersion relations of plasmon modes (the existence and termination of collective excitations). The above-mentioned framework is suitable for monolayer 2D materials. However, how to achieve varied excitation phase diagrams for few-layer condensed-matter systems becomes an open issue, since the orbital-dependent hopping integrals and Coulomb interactions, with the intralayer and interlayer forms, need to be taken into consideration simultaneously. To overcome this special difficulty completely, the Bloch wave functions, which are evaluated from the first-step first-principles calculations, can be transformed or projected into the sublattice-dependent wavefunctions. That is, the electronic states are built from the different sublattices. Then, the layer-dependent polarization function tensor (the longitudinal dielectric function tensor) is well defined under the modified random-phase approximation (details in [37]). Moreover, the dimensionless screened response function, characterized in previous studies [37], is available for exploring complete dynamic excitation processes fully, such as the multi-plasmon modes and distinct electron–hole pair regions in layered graphene systems in the presence or absence of carrier doping.

2.2 The generalized tight-binding model and others

How to solve various Hamiltonians is a mainstream topic in the basic sciences. It is very interesting in obtaining a full understanding of the diverse magnetic quantizations in condensed-matter systems, in particular for emergent layered materials [38]. First-principles methods are unable to explore such phenomena because of the giant unit cell under the periodical Peierls phase of the vector potential (discussed later) [38]. The significant Hamiltonians might possess complex effects arising from multi- and single-orbital hybridization, atom- and layer-dependent spin–orbit interactions, uniform and non-uniform magnetic fields, electric fields, composite fields, gate voltages, intralayer and interlayer hopping integrals, the number of layers, stacking configurations, planar, buckled, curved and folded lattices, hybridized structures, specific dimensionalities, adatom-chemisorption and guest-atom doping. The generalized tight-binding model, which is built from the subenvelope functions of different sublattices, is developed to include all the critical interactions simultaneously. The magnetically quantized energy spectra and sublattice-dependent subenvelope functions can be computed very efficiently using the exact diagonalization method even under a rather large Hamiltonian matrix with complex elements. This model is suitable for and reliable in conducting systematic studies on 3D graphites [39], 2D few-layer graphenes [40], silicene-related systems [41], 1D graphene nanoribbons [42] and carbon nanotubes [43]. It has been generalized successfully to other layered materials, e.g. germanene and tinene [44, 45], MoS_2 [46], phosphorene [47], GaAs [48] and bismuthene [49]. Moreover, the generalized tight-binding model is combined directly with the single- and many-particle modified theories to investigate the other fundamental physical properties, such as the magneto-optical spectra and selection rules, quantum Hall conductivities and magneto-Coulomb excitations. The details of the generalized tight-binding model are discussed in [38].

The critical parameters in the most stable geometric structures and intrinsic interactions, the bond lengths, the buckled angles, the interlayer distances, the heights of different sublattices, the atom- and stacking-induced site energies, the intralayer and interlayer hopping integrals, and the atom- and environment-dependent spin–orbit coupling, which are necessary for the generalized tight-binding model, will be obtained from careful analyses based on the first-principles calculation results. This is achieved by fitting the low-lying energy bands close to the high-symmetry wave vectors, such as the good agreement in the AAA- (figure 2.2(a)) [50], ABA- (figure 2.2(b)) [51], ABC- (figure 2.2(c)) [52] and AAB-stacked [53] graphene systems, AB- [17] and AA-bt [54] bilayer silicene materials, single-layer germanene [55], tinene [55] and bismuthene [49], and few-layer phosphorene [56]. Ten important parameters, which include one intralayer hopping integral and nine atomic interactions due to the carbon-$2p_z$ orbitals, are utilized to agree with the unusual low-lying band structure in the trilayer AAB-stacked graphene with a rather narrow band gap (E_g < 10 meV). This unique stacking configuration is responsible for the oscillatory, sombrero-shaped and parabolic energy dispersions (not shown). In order to fit the first pair of abnormal valence and conduction bands well for the buckled

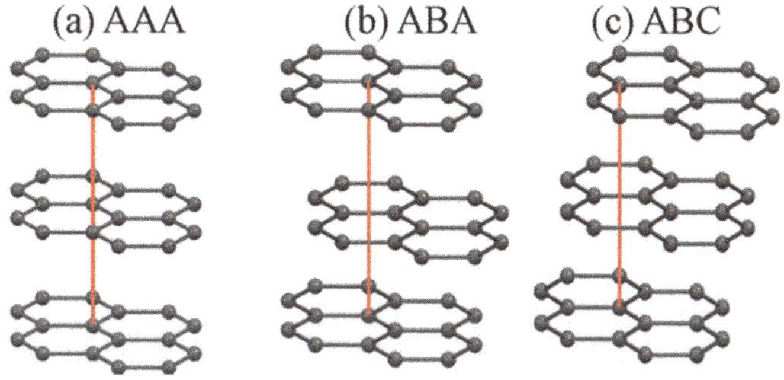

Figure 2.2. The geometric structures of (a) AAA-, (b) ABA- and (c) ABC-stacked graphene systems.

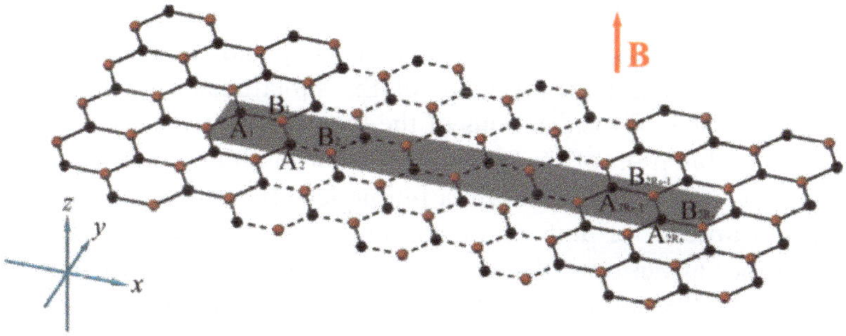

Figure 2.3. The enlarged rectangular unit cell of a monolayer honeycomb lattice due to a uniform perpendicular magnetic field.

bilayer AB- and AA-bt silicene systems (not shown), the nearest vertical hopping integral is deduced to be much higher than the intralayer one, since the relative position vector of two $3p_z$ orbitals is, respectively, perpendicular and parallel to the center of mass hopping integral. Spin–orbit coupling is absent under AA stacking, while layer-dependent coupling arises in AB buckled structures. Until now, it has been very difficult or impossible to obtain a set of good parameters for two pairs of irregular energy bands in any bilayer silicene system. The key reason might reflect directly the fact that the multi-orbital sp^3 bonding, the complicated interlayer hopping integrals and the layer-dependent spin–orbit coupling under significant buckling effects need to be taken into account simultaneously.

The generalized tight-binding model is able to predict the varied magnetic quantization phenomena in emergent 2D materials, such as the rich and unique Landau levels created by few-layer graphene [7, 9], silicene [27, 42], germanene [28], tinene [44], antimonene [22] and bismuthene [49]. The main features of Landau levels, the field-strength-dependent electronic energy spectra and the spatial probability distributions, have been investigated thoroughly. The magnetic Hamiltonian (with a huge dimension) of monolayer graphene is chosen as a typical example, and is shown in figure 2.3 for a monolayer honeycomb lattice (graphene, silicene or germanene). A uniform perpendicular magnetic field $B_z\hat{z}$ creates an extra Peierls phase in the generalized tight-binding function through the vector potential $\mathbf{A} = [0, B_z, 0]$. The B_z-induced spatial period is assumed to be the integral time of the honeycomb lattice (the gray rectangle in figure 2.3). The well-known phase is denoted by $G_R = \frac{2\pi}{\phi_0}\int A \cdot dl$, where $\phi_0 = hc/e$ is the magnetic flux quantum and $\phi = B_z\,(3\sqrt{3}b^2)/2$ corresponds to the magnetic flux through a hexagon. There are $4R_B$ ($R_B = \phi_0/\phi$) carbon atoms in total in an enlarged unit cell. Consequently, the magnetic Hamiltonian matrix, which is based on the tight-binding functions of the periodical $2p_z$ orbitals, is a $16R_B \times 16R_B$ Hermitian matrix. In general, its dimension is giant under the observable magnetic-field strength in a laboratory. For example, the Hamiltonian is a 50 000-dimensional matrix at $B_z = 10$ T. It should be noted that the magnetic Hamiltonian matrix elements might be real or imaginary numbers, and they are very sensitive to intrinsic geometric structures, orbital hybridization and spin–orbit interactions. The magneto-electronic energy spectra and wave functions are solved numerically using the efficient exact diagonalization method (using the band-like method under rearrangement of the sublattice-dependent tight-binding functions and the spatial localizations of the magnetic probability distributions).

The direct combination of the generalized tight-binding model and the dynamic Kubo formula is very useful to obtain a full understanding of optical excitations in the presence or absence of uniform or non-uniform magnetic and electric fields, in particular for the various optical selection rules. The critical electric dipole moment, which determines the absorption intensity of the available vertical transition channels, could be evaluated accurately from the layer-dependent subenvelope functions under any external fields through the gradient approximation. Furthermore, the huge calculation barriers, which are related closely to the numerical calculation of magneto-optical spectra, are reduced greatly by the localization features of Landau levels [7, 9]. To date, stacking-, layer-number-, dimension- and field-enriched optical absorption spectra have been studied systematically for 2D few-layer graphenes with different layer numbers and stacking configurations [57], three kinds of 3D graphites [58], radius- and chiral-angle-dependent 1D carbon nanotubes [59], semiconducting graphene nanoribbons [60], Si-substituted graphenes [61], layered silicene [27, 42] and phosphorene systems [47]. The predicted results seem to show diverse magneto-optical and optical selection rules, which are examined thoroughly through the spatial distribution symmetries of the initial and final Landau levels. Interestingly, complete studies on trilayer AAA-, ABA-, ABC- and AAB-stacked graphenes and sliding bilayer systems present rich and unique magneto-optical and optical spectra, owing mainly to the complex relations among the interlayer hopping integrals, magnetic quantization and Coulomb potential energies. The single, twin and double magneto-absorption peaks originate from, respectively, the symmetric, asymmetric and splitting Landau-level energy spectra. The different magneto-optical selection rules are created by the three kinds of Landau levels in sliding bilayer graphenes. The specific AAB stacking without mirror symmetry about the ($z = 0$)-plane, displays unique optical absorption spectrum compared to the other stacking types. Moreover, the frequent Landau-level anti-crossing in the magnetic- and electric-field dependence leads to extra absorption peaks and lower intensities. The unusual magneto-electronic states in monolayer and AB-bt bilayer silicene systems [62] and few-layer phosphorene materials [63] are predicted to exhibit absorption phenomena.

The linear static Kubo formula, combined with the generalized tight-binding model, is available for full exploration of the unusual quantum Hall effects (QHEs) for few-layer graphenes with higher- and lower-symmetry stacking configurations, as well as the emergent 2D materials. The developed theoretical model will be very useful in the exact identification of the magneto-electronic selection rules under the static case; that is, the effective transition channels in the magneto-transport properties can be checked thoroughly by the spatial symmetries of the initial and final Landau-level states. The dependence of quantum conductivity on the Fermi energy, magnetic field and electric field are investigated in detail. The feature-rich magneto-electronic states in the above-mentioned layered systems seem to create their extraordinary magneto-transport properties. For example, the sliding bilayer systems, with well-behaved, perturbed and undefined Landau levels, present unusual QHEs, including integer and non-integer conductivities, zero and non-vanishing conductivities at the neutral point, well-like, staircase and composite quantum structures, and different step heights. Furthermore, reduced conductivity, complex plateaus and abnormal structures are revealed in AAB-stacked trilayer graphene, with splitting and perturbed Landau levels. These results can be recognized to clearly arise from the crossing and anti-crossing Landau-level energy spectra and the spatial oscillation modes. The theoretical predictions of the varied quantum properties could be generalized to mainstream layered systems, in which careful calculations with huge computation times and physical analyses associated with exact scattering pictures are required in thorough investigations, e.g. the current studies on the unusual quantum transport properties of bilayer AB-bt silicene and phosphorene systems [62, 63].

Our previous studies and book present a thoroughly modified theory of Coulomb excitations and decay rates in graphene-related systems, in which the theoretical framework combines the layer-dependent random-phase approximation and the generalized tight-binding model [64]. This framework can deal with many critical factors associated with different lattice symmetries, layer numbers, dimensions, stacking configurations, orbital hybridization, intralayer and interlayer hopping integrals, spin–orbit coupling, temperatures, electron or hole doping, gate voltages, magnetic fields and mechanical strains. The layer-dependent tensors of bare polarization functions and dielectric functions are obtained for layered graphene systems with specific numbers of layers and stacking configurations. Most importantly, the dimensionless screened response function, which characterizes the intrinsic ability of charge screening, is redefined through the inelastic scattering theory. Obviously, there exist rich and unique electronic excitation phenomena due to the different energy bands and wave functions in the various condensed-matter systems, as indicated by the diverse momentum and frequency phase diagrams. The calculated results, accompanied by concise physical pictures, illustrate clearly the very important roles of the electron–electron Coulomb interactions, which are capable of accounting for up-to-date experimental measurements. Current studies focusing on layered graphenes include static charge screening (Friedel oscillations), the dynamic propagation of plasmon waves in 2D planes and Coulomb decay rates. The above-mentioned model could be generalized to other emergent 2D materials with detailed calculations and investigations, such as layered silicene, germanene, tinene, phosphorene, antimonene, bismuthene and MoS_2.

In general, the tight-binding model can be reduced to the **k·p** approximation (the effective-mass approximation [65]) in the absence of certain hopping integrals. The former is able to provide a whole band structure, while the latter is suitable only for low-lying valence and conduction bands about a stable valley. Specifically, analytical energy spectra and wave functions, relative to the high-symmetric point, might be achieved under the effective-mass method with concise atomic interactions. Their rich relationships have been illustrated successfully in graphene-related materials. That is, they are both utilized to explore the fundamental properties of few-layer graphenes, carbon nanotubes and graphene nanoribbons [66]. Theoretical [67] and experimental [68] studies have verified that monolayer graphene exhibits a pair of linearly interesting valence and conduction bands at the Fermi level, the bilayer AA and AB stacking cases exhibit two pairs of vertically separated Dirac-cone structures and parabolic energy dispersions, respectively, and the AAA-, ABA-, ABC- and AAB-stacked trilayer graphenes exhibit three split Dirac cones, one modified Dirac-cone structure and two pairs of parabolic bands, one pair of partially flat, sombrero-shaped and linear bands, and one pair of oscillatory, sombrero-like and parabolic energy dispersions, respectively. Interestingly, the electronic properties are varied greatly by the number of layers and stacking configurations. Only the uniform electronic structures are well characterized by the **k·p** approximation; therefore, this method is qualitatively reliable in describing the fundamental properties of monolayer and bilayer graphenes. The rich magnetic quantization behaves in this manner [69–73].

The **k·p** approximation becomes more difficult in layered silicene and germanene systems, since the buckling honeycomb lattices, the non-orthogonal π and σ bonding and the interlayer multi-orbital hybridization are largely enhanced by an increasing layer number (details in chapters 4 and 5). This method, which considers the nearest-neighbor π bonding and the weak, but significant, spin–orbit coupling in monolayer silicene and germanene, presents a slightly separated Dirac-cone structure near the K and K′ valley, as observed in the tight-binding model [74] and first-principles calculations [75]. However, for bilayer silicenes with specific stacking configurations and buckling structures, the first-principles results in chapter 5 clearly show strong distortions of Dirac-cone π bands at the K and K′ point, shallower σ bands initiated from the Γ point, their co-dominance in the low-lying electronic energy spectra, π- and σ-band anti-crossing, non-well-behaved π- and σ-band widths, and *e* spin-split bands in specific stacking symmetries. The unusual band structures make the effective-mass approximation impossible to use in exploring the rich and unique phenomena in few-layer group-IV systems. Obviously, many reliable parameters need to be included in the tight-binding model, such as the intralayer and interlayer multi-orbital hybridization and the layer-dependent spin–orbit coupling. This might be too cumbersome to be addressed fully through current theoretical models. The current tight-binding model for AB-bt bilayer silicene [27] has $2p_z$-orbital-related hopping integrals and spin–orbit interactions, and is only suitable for the fundamental properties related to the first pair of energy bands nearest to the Fermi level. Under the complex chemical and physical environments in layered silicene, germanene, tinene and plumbene [66], phenomenological models must overcome technical barriers to obtain a good fit with the first-principles band structures, and would thus become very useful in proposing concise mechanisms for diverse properties.

References

[1] Vogt P *et al* 2012 Silicene: compelling experimental evidence for graphene-like two-dimensional silicon *Phys. Rev. Lett.* **108** 155501

[2] Dávila M E, Xian L, Cahangirov S, Rubio A and Le Lay G 2014 Germanene: a novel two-dimensional germanium allotrope akin to graphene and silicene *New J. Phys.* **16** 095002

[3] Ji J *et al* 2016 Two-dimensional antimonene single crystals grown by van der Waals epitaxy *Nat. Commun.* **7** 13352

[4] Zhang J L *et al* 2016 Epitaxial growth of single layer blue phosphorus: a new phase of two-dimensional phosphorus *Nano Lett.* **16** 4903–8

[5] Brent J R *et al* 2015 Tin (II) sulfide (SnS) nanosheets by liquid-phase exfoliation of herzenbergite: IV–VI main group two-dimensional atomic crystals *J. Am. Chem. Soc.* **137** 12689–96

[6] Lai Y H, Ho J H, Chang C P and Lin M F 2008 Magnetoelectronic properties of bilayer Bernal graphene *Phys. Rev.* B **77** 085426

[7] Ho Y H, Chiu Y H, Lin D H, Chang C P and Lin M F 2010 Magneto-optical selection rules in bilayer Bernal graphene *ACS Nano* **4** 1465–72

[8] Zhang Y, Tan Y W, Stormer H L and Kim P 2005 Experimental observation of the quantum Hall effect and Berry's phase in graphene *Nature* **438** 201

[9] Wu J Y, Chen S C, Roslyak O, Gumbs G and Lin M F 2011 Plasma excitations in graphene: their spectral intensity and temperature dependence in magnetic field *ACS Nano* **5** 1026–32

[10] Kresse G and Furthmüller J 1996 Efficient iterative schemes for *ab initio* total-energy calculations using a plane-wave basis set *Phys. Rev.* B **54** 11169

[11] Castro A, Marques M A and Rubio A 2004 Propagators for the time-dependent Kohn–Sham equations *J. Chem. Phys.* **121** 3425–33

[12] Lee C M 1974 Spectroscopy and collision theory. III. Atomic eigenchannel calculation by a Hartree–Fock–Roothaan method *Phys. Rev.* A **10** 584

[13] Kresse G and Hafner J 1993 *Ab initio* molecular dynamics for liquid metals *Phys. Rev.* B **47** 558

[14] Foulkes W M C, Mitas L, Needs R J and Rajagopal G 2001 Quantum Monte Carlo simulations of solids *Rev. Mod. Phys.* **73** 33

[15] Blöchl P E 1994 Projector augmented-wave method *Phys. Rev.* B **50** 17953

[16] Kresse G and Joubert D 1999 From ultrasoft pseudopotentials to the projector augmented-wave method *Phys. Rev.* B **59** 1758

[17] Liu F, Liu C C, Wu K, Yang F and Yao Y 2013 $d + id'$ chiral superconductivity in bilayer silicene *Phys. Rev. Lett.* **111** 066804

[18] Matthes L, Pulci O and Bechstedt F 2013 Massive Dirac quasiparticles in the optical absorbance of graphene, silicene, germanene, and tinene *J. Phys. Condens. Matter* **25** 395305

[19] Matthes L, Pulci O and Bechstedt F 2014 Optical properties of two-dimensional honeycomb crystals graphene, silicene, germanene, and tinene from first principles *New J. Phys.* **16** 105007
[20] Bihlmayer G, Blügel S and Chulkov E V 2007 Enhanced Rashba spin–orbit splitting in Bi/Ag (111) and Pb/Ag (111) surface alloys from first principles *Phys. Rev.* B **75** 195414
[21] Dresselhaus P D, Papavassiliou C M A, Wheeler R G and Sacks R N 1992 Observation of spin precession in GaAs inversion layers using antilocalization *Phys. Rev. Lett.* **68** 106
[22] Zhao M, Zhang X and Li L 2015 Strain-driven band inversion and topological aspects in antimonene *Sci. Rep.* **5** 16108
[23] Son Y W, Cohen M L and Louie S G 2006 Half-metallic graphene nanoribbons *Nature* **444** 347
[24] Lin S Y, Chang S L, Shyu F L, Lu J M and Lin M F 2015 Feature-rich electronic properties in graphene ripples *Carbon* **86** 207–16
[25] Meng L *et al* 2013 Buckled silicene formation on Ir (111) *Nano Lett.* **13** 685–90
[26] Wu J Y, Chen S C and Lin M F 2014 Temperature-dependent Coulomb excitations in silicene *New J. Phys.* **16** 125002
[27] Do T N, Gumbs G, Shih P H, Huang D, Chiu C W, Chen C Y and Lin M F 2019 Peculiar optical properties of bilayer silicene under the influence of external electric and magnetic fields *Sci. Rep.* **9** 624
[28] Shih P H, Chiu C W, Wu J Y, Do T N and Lin M F 2018 Coulomb scattering rates of excited states in monolayer electron-doped germanene *Phys. Rev.* B **97** 195302
[29] Anisimov V I, Aryasetiawan F and Lichtenstein A I 1997 First-principles calculations of the electronic structure and spectra of strongly correlated systems: the LDA+ U method *J. Phys. Condens. Matter* **9** 767
[30] Lin C L *et al* 2013 Substrate-induced symmetry breaking in silicene *Phys. Rev. Lett.* **110** 076801
[31] Lin C L *et al* 2012 Structure of silicene grown on Ag (111) *Appl. Phys. Express* **5** 045802
[32] Li G, Luican A and Andrei E Y 2009 Scanning tunneling spectroscopy of graphene on graphite *Phys. Rev. Lett.* **102** 176804
[33] Mak K F, Ju L, Wang F and Heinz T F 2012 Optical spectroscopy of graphene: from the far infrared to the ultraviolet *Solid State Commun.* **152** 1341–9
[34] Wang L *et al* 2017 Industrial production of ultra-stable sulfonated graphene quantum dots for Golgi apparatus imaging *J. Mater. Chem.* B **5** 5355–61
[35] Nomura K and MacDonald A H 2007 Quantum transport of massless Dirac fermions *Phys. Rev. Lett.* **98** 076602
[36] Gao W, Shu J, Qiu C and Xu Q 2012 Excitation of plasmonic waves in graphene by guided-mode resonances *ACS Nano* **6** 7806–13
[37] Lin C Y, Wu J Y, Chiu C W and Lin M F 2019 *Coulomb Excitations and Decays in Graphene-related Systems* (Boca Raton, FL: CRC Press)
[38] Lin C Y, Do T N, Huang Y K and Lin M F 2017 *Optical Properties of Graphene in Magnetic and Electric Fields* (Bristol: IOP Publishing)
[39] Safran S A and DiSalvo F J 1979 Theory of magnetic susceptibility of graphite intercalation compounds *Phys. Rev.* B **20** 4889
[40] Lin S Y, Ho Y H, Huang Y C and Lin M F 2012 Magneto-electronic specific heat of graphene *J. Phys. Soc. Jpn* **81** 084602

[41] Ezawa M 2012 Topological phase transition and electrically tunable diamagnetism in silicene *Eur. Phys. J.* B **85** 363
[42] Ezawa M 2007 Metallic graphene nanodisks: electronic and magnetic properties *Phys. Rev.* B **76** 245415
[43] Ajiki H and Ando T 1993 Magnetic properties of carbon nanotubes *J. Phys. Soc. Jpn* **62** 2470–80
[44] Chen R B, Chen S C, Chiu C W and Lin M F 2017 Optical properties of monolayer tinene in electric fields *Sci. Rep.* **7** 1849
[45] Ezawa M 2015 Photo-induced topological superconductor in silicene, germanene, and stanene *J. Supercond. Novel Magn.* **28** 1249–53
[46] Shanavas K V and Satpathy S 2015 Effective tight-binding model for MX_2 under electric and magnetic fields *Phys. Rev.* B **91** 235145
[47] Ostahie B and Aldea A 2016 Phosphorene confined systems in magnetic field, quantum transport, and superradiance in the quasiflat band *Phys. Rev.* B **93** 075408
[48] Strandberg T O, Canali C M and MacDonald A H 2009 Magnetic properties of substitutional Mn in (110) GaAs surface and subsurface layers *Phys. Rev.* B **80** 024425
[49] Chen S C, Wu J Y and Lin M F 2018 Feature-rich magneto-electronic properties of bismuthene *New J. Phys.* **20** 062001
[50] Zhong X, Amorim R G, Scheicher R H, Pandey R and Karna S P 2012 Electronic structure and quantum transport properties of trilayers formed from graphene and boron nitride *Nanoscale* **4** 5490–8
[51] Koshino M and McCann E 2009 Gate-induced interlayer asymmetry in ABA-stacked trilayer graphene *Phys. Rev.* B **79** 125443
[52] Cong C, Yu T, Sato K, Shang J, Saito R, Dresselhaus G F and Dresselhaus M S 2011 Raman characterization of ABA- and ABC-stacked trilayer graphene *ACS Nano* **5** 8760–8
[53] Do T N, Lin C Y, Lin Y P, Shih P H and Lin M F 2015 Configuration-enriched magneto-electronic spectra of AAB-stacked trilayer graphene *Carbon* **94** 619–32
[54] Shih P H, Do T N, Gumbs G, Huang D, Pham H D and Lin M F 2019 Rich magnetic quantization phenomena in AA bilayer silicene *Sci. Rep.* **9** 1–8
[55] Ezawa M 2015 Monolayer topological insulators: silicene, germanene, and stanene *J. Phys. Soc. Jpn* **84** 121003
[56] Zhou X Y *et al* 2015 Landau levels and magneto-transport property of monolayer phosphorene *Sci. Rep.* **5** 12295
[57] Muñoz W A, Covaci L and Peeters F M 2013 Tight-binding description of intrinsic superconducting correlations in multilayer graphene *Phys. Rev.* B **87** 134509
[58] Lin S Y, Ho Y H, Shyu F L and Lin M F 2013 Electronic thermal property of graphite *J. Phys. Soc. Jpn* **82** 074603
[59] Jiang J, Saito R, Samsonidze G G, Jorio A, Chou S G, Dresselhaus G and Dresselhaus M S 2007 Chirality dependence of exciton effects in single-wall carbon nanotubes: tight-binding model *Phys. Rev.* B **75** 035407
[60] Son Y W, Cohen M L and Louie S G 2006 Energy gaps in graphene nanoribbons *Phys. Rev. Lett.* **97** 216803
[61] Shahrokhi M and Leonard C 2017 Tuning the band gap and optical spectra of silicon-doped graphene: many-body effects and excitonic states *J. Alloys Compd.* **693** 1185–96
[62] Do T N, Shih P H, Gumbs G, Huang D, Chiu C W and Lin M F 2018 Diverse magnetic quantization in bilayer silicene *Phys. Rev.* B **97** 125416

[63] Li L *et al* 2017 Direct observation of the layer-dependent electronic structure in phosphorene *Nat. Nanotechnol.* **12** 21

[64] Chen S C, Wu J Y, Lin C Y and Lin M F 2017 *Theory of Magnetoelectric Properties of 2D Systems* (Bristol: IOP Publishing)

[65] Yu P Y and Cardona M 1996 *Fundamentals of Semiconductors: Physics and Materials Properties* (Berlin: Springer)

[66] Marconcini P and Macucci M 2011 The k·p method and its application to graphene, carbon nanotubes and graphene nanoribbons: the Dirac equation, arXiv: 1105.1351

[67] Lin C Y, Do T N, Huang Y K and Lin M F 2017 *Electronic and Optical Properties of Graphene in Magnetic and Electric Fields* (San Rafael, CA: Morgan and Claypool)

[68] Lauffer P, Emtsev K V, Graupner R, Seyller T, Ley L, Reshanov S A and Weber H B 2008 Atomic and electronic structure of few-layer graphene on SiC (0001) studied with scanning tunneling microscopy and spectroscopy *Phys. Rev.* B **77** 155426

[69] Lin C Y, Chen R B, Ho Y H and Lin M F 2018 *Electronic and Optical Properties of Graphite-Related Systems* (Boca Raton, FL: CRC Press)

[70] Lin C Y, Do T N, Huang Y K and Lin M F 2017 *Optical Properties of Graphene in Magnetic and Electric Fields* (Bristol: IOP Publishing)

[71] Chen S C, Wu J Y, Lin C Y and Lin M F 2017 *Theory of Magnetoelectric Properties of 2D Systems* (Bristol: IOP Publishing)

[72] Lin C Y, Wu J Y, Chiu C W and Lin M F 2019 *Coulomb Excitations and Decays in Graphene-Related Systems* (Boca Raton, FL: CRC Press)

[73] Lin C Y, Do T N, Wu J Y, Shih P H, Lin S Y, Ho C H and Lin M F 2019 *Diverse Quantization Phenomena in Emergent Layered Materials* (Boca Raton, FL: CRC Press)

[74] Ezawa M 2013 Photoinduced topological phase transition and a single Dirac-cone state in silicene *Phys. Rev. Lett.* **110** 026603

[75] Wang Y P and Cheng H P 2013 Absence of a Dirac cone in silicene on Ag (111): first-principles density functional calculations with a modified effective band structure technique *Phys. Rev.* B **87** 245430

Chapter 3

Experimental measurements

Theoretical predictions associated with the geometric structures, spin configurations, density of states (DOS), occupied energy bands and magnetic moments of silicene-based systems can be verified using high-resolution measurements from LEED, TEM, STM, spin-polarized STM, STS, ARPES and SQUID. The basic principles of these experimental devices and recent progress in measured results are discussed in the following.

3.1 LEED and TEM

Low-energy electron diffraction (LEED) is an effective method to determine the surface structure of materials by using a collimated beam of low energy electrons and observation of diffracted electrons as spots on a fluorescent screen, as shown in figure 3.1. Transmission electron microscopy (TEM) is a microscopic technique to resolve the side view of a stable structure. A high-energy electron beam with a uniform current density is able to penetrate through an ultrathin specimen to form an image, owing mainly to the interactions between incident charges and the sample. TEM was first demonstrated by Ernst Ruska and Max Knoll in 1931 [1] and the former received the Nobel Prize in Physics in 1986 for inventing electron microscopy [2]. TEMs can achieve a higher spatial resolution than light microscopes as a result of the smaller de Broglie wavelength of electrons. A standard TEM device, as shown in figure 3.2, is composed of an electron source, the first and second condenser lenses with an aperture, the specimen, an objective lens with an aperture, intermediate and projector lenses, and a fluorescent screen. The set of condenser lenses makes the incident beam focus on a specific sample. Then, the objective lens collects all the scattered electrons after transmission, forms an image of the morphology of the sample and determines the limit of structural resolution. A set of intermediate lenses magnifies this image and projects it on a fluorescent screen, a layer of photographic film or a sensor. Compared to other microscopes, the main advantage of TEM

doi:10.1088/978-0-7503-3299-6ch3

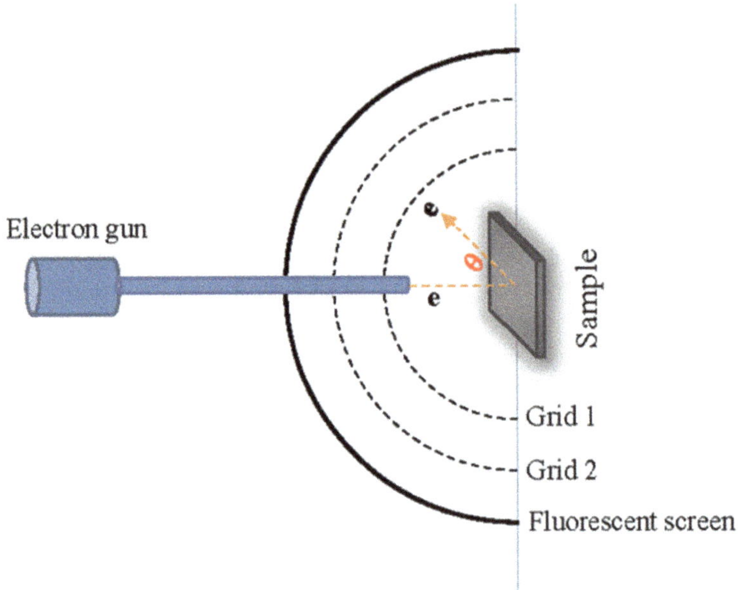

Figure 3.1. Low-energy electron diffraction.

measurements is providing full and accurate information simultaneously in real space (the imaging mode) and reciprocal space (the diffraction mode). The de Broglie wavelength of electrons in TEM is much smaller than the atomic separations in condensed-matter systems. This apparatus is thus suitable for observing crystal structures below atomic sizes. To date, the best spatial resolution for TEM measurements has been about 1–2 Å [3]. In contrast, x-ray diffraction is not suitable for accurate measurements of the nanoscale arrangements of few-layer emergent materials, since the intrinsic response is too weak.

TEM measurements are very powerful in observing the crystal structure directly, examining chemical bonding, locating and identifying the types of defects, and even studying the structural phase transitions. Such measurements possess a very strong electron atomic scattering factor, being ~10 000 times that of x-ray diffraction. This feature provides electron scattering as an advantage in detecting the weakest diffracted spot. However, the resolution of TEM is limited by spherical and chromatic aberrations of the lenses. More sensitive techniques for improving the diffraction resolution have become indispensable. By applying a monochromator and a Cs corrector to a TEM, the structural resolution can reach less than 0.5 Å [4], which is called high-resolution TEM (HRTEM). This method has been utilized successfully and extensively in exploring the crystal structures and lattice imperfections of low-dimensional nanomaterials [5, 6]. The TEM and HRTEM measurements of graphene-related systems have proven highly suitable for identifying diverse nanostructures due to significant carbon-sp^2 bonding, such as the multi-walled cylindrical structures of carbon nanotubes [7, 8], the curved [8–10], folded [11, 12] and scrolled [13–15] profiles of graphene nanoribbons, as well as the stacking configurations and the interlayer

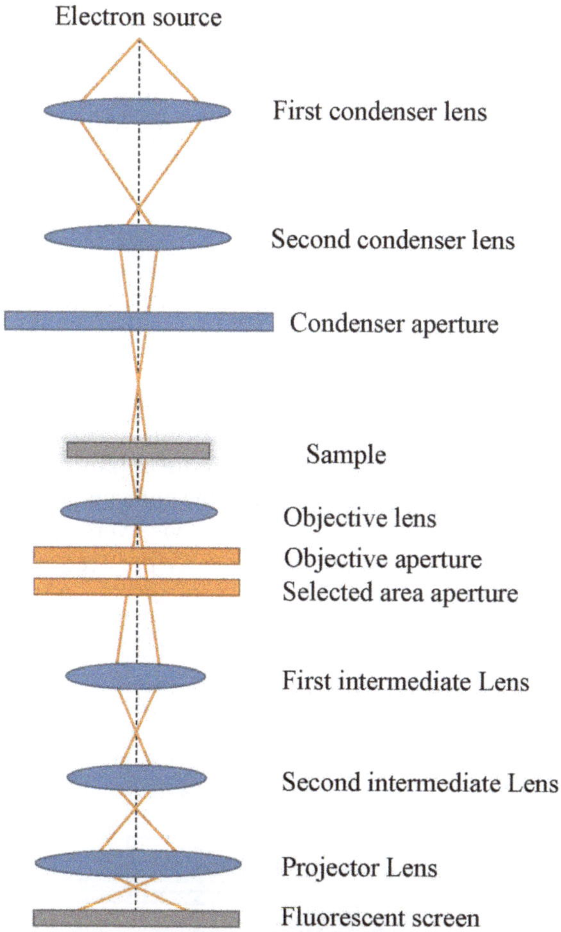

Figure 3.2. Transmission electron microscopy.

distances of layered graphene systems [16–18]. When both TEM and STM measurements (discussed below) are combined, 3D morphology can be achieved under the careful association of top- and side-view geometric structures. For example, the armchair or zigzag structure along the open boundary (the tubular axis) and the stacking configuration between two graphene nanoribbons (two coaxial carbon nanotubes) are examined using, respectively, STM and TEM.

Interestingly, certain 2D emergent materials can also be identified through careful TEM examinations. For example, a previous study shows clearly that there exist three types of bilayer silicenes after treating calcium-intercalated monolayer silicene ($CaSi_2$) with a BF_4^--based ionic liquid. That is to say, bilayer silicene systems might consist of four-, five- and six-membered silicon rings with three nearest neighbors. The six-membered rings correspond to the AB-bottom-top stacking configuration (AB-bt), examined accurately using high-angle annular dark field scanning TEM.

3.2 STM, STS, SP-STM and SP-STS

The scanning tunneling microscope (STM) has become the most important experimental apparatus for resolving the atomic arrangement on surfaces since its first discovery by Binnig and Rohrer in 1982 [19, 20], in particular for emergent 2D lattices [21–26]. This device is capable of characterizing surface topographies with both lateral and vertical atomic resolutions, such as nanoscale bond lengths [27], crystal orientations [28], planar and nonplanar structures [29, 30], step edges [31], local vacancies [32], dislocations [32, 33], adsorbed adatoms and molecules [34, 35], and nanoclusters and nanoislands [36, 37]. The STM device, shown in figure 3.3, consists of a conducting (semiconducting) solid surface and a sharp metal tip, where a spacing of several angströms is manipulated by specific piezoelectric feedback devices. Within the bias voltage between the surface and tip, a very weak current is created through the quantum tunneling effect [19, 20]. It is very sensitive to distance, and is assumed to present an exponential decay form [20]. This current flows from the occupied electronic states of the tip into the unoccupied states of the sample surface under a positive bias voltage ($V > 0$), or vice versa ($V < 0$). Generally, the quantum tunneling current is chosen to serve as the feedback signal. The most

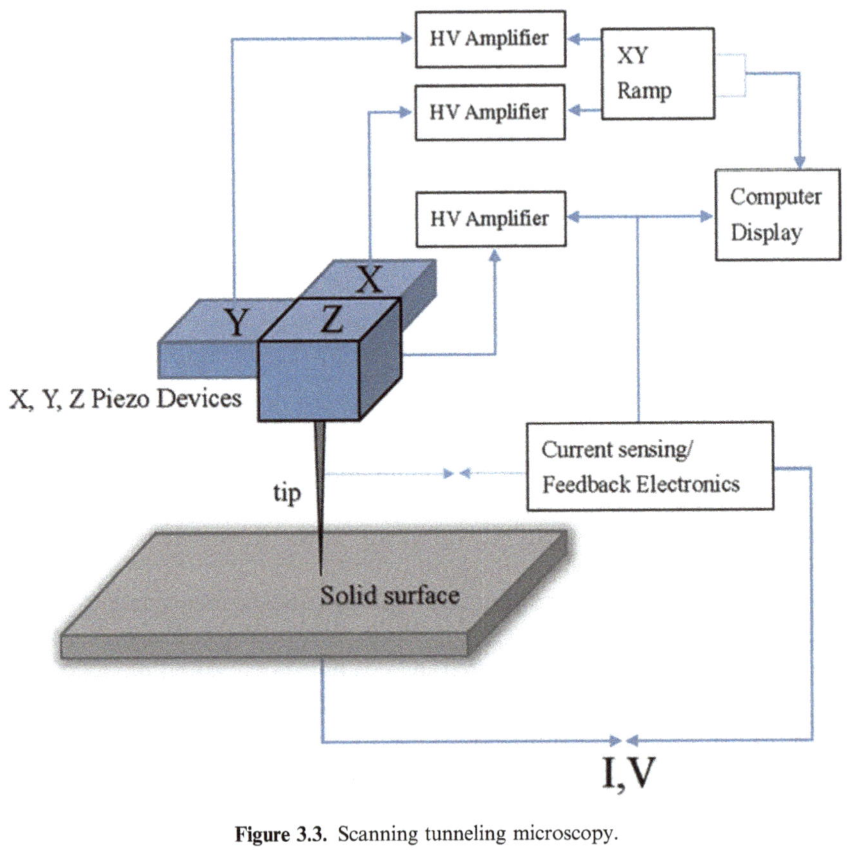

Figure 3.3. Scanning tunneling microscopy.

commonly used mode, the constant tunneling mode (current and voltage), will be operated to resolve the surface structure very accurately by applying a piezoelectric device. By combining a probe (the metal tip) and a precise scanning device, full spectroscopic information on the surface morphology can be obtained at preselected positions under well-defined conditions. The response of the surface charges is affected little by background effects, owing mainly to the ultra-high vacuum environment. The up-to-date spatial resolution of STM measurements can reach ~0.1 Å. In addition to STM, other experimental techniques are available to characterize the geometric properties of low-dimensional materials fully, such as TEM [12, 38], scanning transmission electron microscopy (STEM) [39, 40], atomic force microscopy (AFM) [41] and low-energy electron diffraction (LEED) [42]. Specifically, LEED patterns have confirmed (1 × 1) Ir(111) [43], ($\sqrt{3} \times \sqrt{3}$) adlayer silicone [44] and ($\sqrt{7} \times \sqrt{7}$) Ir(111) [45]. They are very useful in examining the superstructures of emergent layered materials produced through epitaxial growth on various substrates.

High-resolution STM measurements are quite powerful in identifying spatially stable surface morphologies with nanoscale precision. Such sensitive analyses of carbon-related sp^2-bonding systems have verified a variety of crystals, illustrating clearly the complex relationships between the hexagonal lattice, the finite-width quantum confinement, the flexible geometry and the distinct chemical bonding of carbon atoms. From recent observations, graphene nanoribbons exhibit a nanoscale-width planar honeycomb lattice, with non-chiral (armchair and zigzag) [46] or chiral edge structures [47]. Furthermore, honeycomb lattices have been shown to present curved [48], folded [29], scrolled [29] and stacked configurations [49]. It should be noted that folded graphene nanoribbons have also been confirmed by HRTEM measurements (discussed above [12]). Hollow carbon nanotubes possess achiral and chiral arrangements of hexagons on their cylindrical surfaces [7]. Atomic-scale measurements are able to identify directly the AB [50], ABC [51] and AAB [52] stacking configurations of few-layer graphene systems, corrugated substrates and buffer graphene layers [53], rippled graphene islands [54, 55] and adatom distributions on planar and buckled graphene surfaces. Layered bulk graphite has been verified to display 2D networks of local defects on the surface [56], and pyridinic-nitrogen and graphitic-nitrogen structures.

Most importantly, the combination of high-resolution STM and LEED or reflected high-energy electron diffraction (RHEED [67]) experiments has confirmed lattice structures and their reconstructions in the emergent group-IV and group-V 2D materials. Few-layer silicene systems on Ag(111), Ir(111) and $ZrB_2(0001)$ appear to present, respectively, (4 × 4) [57, 58], ($\sqrt{3} \times \sqrt{3}$) [45, 59] and (2 × 2) honeycomb lattices [60]. Furthermore, distinct superstructures have been revealed in the substrate surfaces. The formation of chemical bonds between the silicon adlayer and substrate (the passivation of Si-$3p_z$-induced dangling bonds) is responsible for the varied supercells and thus the weakly or strongly modified Dirac-cone structures. Similar studies have been conducted on monolayer germanene, including an adlayer on Pt(111), Al(111), Ag(111) and Au(111), respectively, with 3 × 3 [61], (2 × 2) [62],

(2 × 2) [63] and (3 × 3) [64] unit cells. Interestingly, a single Pb monolayer deposited successfully on aluminum oxide super-thin film grown on Ag(111) has been identified to present a moiré superlattice ($\sqrt{37} \times \sqrt{37}$ (R = 25.3°)) [65] using STM and LEED. There also exists incommensurate moiré modulation in its growth with Au(111) [66]. Moreover, a biatomic-layer honeycomb lattice of 2D tinene on Bi_2Te_3(111), without reconstruction structures, has been observed using both STM and RHEED measurements. Two kinds of experimental techniques associated with first-principles calculations are suitable for determining the thickness and buckling height of each adlayer, where the measured results are, respectively, (3.35 Å, 0), (2.51 Å, 0.46 Å), (2.66 Å, 0.66 Å), (2.93 Å, 0.85 Å) and (3.10 Å, 0.92 Å) for C-, Si-Ge, Sn- and Pb-layered materials.

Scanning tunneling spectroscopy (STS), shown in figure 3.4, is a modified form of STM with more sensitive operation with the absence of feedback devices. This apparatus can characterize the energy-dependent density of electronic states in condensed-matter systems, according to the specific relationship between the tunneling current and bias voltage. *I–V* curves are measured from a probing tip at the same height during the scanning of a whole surface. The resulting data provide detailed information on the surface electronic states, when the tunneling current is assumed to be proportional to the constant DOS of the probing tip. The normalized

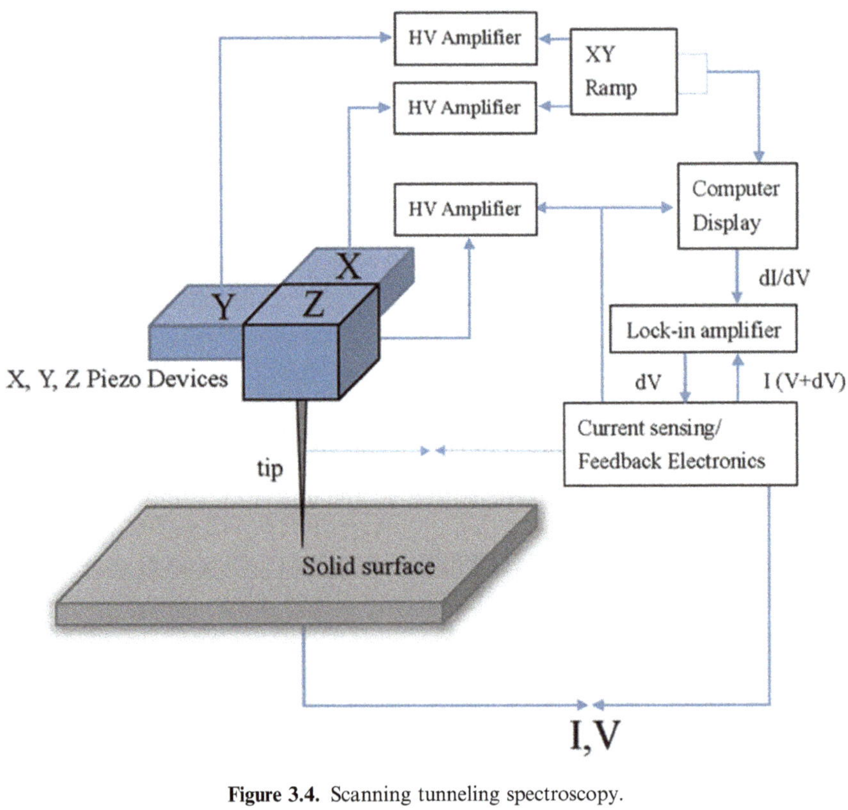

Figure 3.4. Scanning tunneling spectroscopy.

differential conductance of the tip-surface tunneling junction, which is defined as $(dI/dV)/(I/V)$, accounts for the DOS in the surface nanostructure. Its relationship to the bias voltage is interpreted as the energy-dependent DOS. Specifically, the normalized differential conductance, induced by a small ac dV modulation, is examined using a lock-in amplifier. As a result, the noise involved in the measured conductance will be reduced greatly. The highest resolution of current measurements is ~10 pA. STS is available for the identification of the spin-split DOS when magnetic tips are utilized in the experimental observations.

The sensitive STS measurements can provide full information on van Hove singularities and semiconducting, semi-metallic and metallic behaviors, in particular to distinguish the zero- and finite-gap semiconductors. Such studies, which include the energy, number, intensity and form of special structures in the DOS, are very successful in verifying the varied electronic properties in graphene-related materials, such as graphene nanoribbons [46, 47], carbon nanotubes [7], few-layer graphenes [50, 51], adatom-adsorbed graphenes [68, 69] and graphite [52, 56]. The main features of graphene nanoribbons—the width- and edge-dependent energy gaps and the square-root-divergent asymmetric peaks of 1D parabolic energy bands—are identified from precisely defined boundary structures [46, 47]. Similar prominent peaks have been revealed in seamless carbon nanotubes, where these structures exhibit chirality- and radius-dominated band gaps and energy spacings between two neighboring subbands [7]. Even in the presence of curvature effects on the cylindrical surface (the misorientation of $2p_z$ orbitals and the significant hybridization of carbon four orbitals) [70], armchair nanotubes belong to the 1D metals with a finite DOS at the Fermi level (a sufficiently high free carrier density) [71, 72]. A large number of STS measurements conducted on few-layer and adatom-doped graphenes show clear low-lying DOS characteristics, including a V-shaped energy dependence vanishing at the Dirac point in monolayer systems (a zero-gap semiconductor) [73, 74], asymmetry-created symmetric peak structures in twisted bilayer graphenes (discussed in chapter 4) [75], a gate-voltage-created band gap in bilayer AB stacking and tri-layer ABC stacking [76], a prominent peak centered around the Fermi level due to partially flat bands (surface states) in tri-layer and penta-layer ABC stacking [77], a sharp dip structure at E_F accompanied by a pair of asymmetric peaks in tri-layer AAB stacking (a narrow-gap semiconductor with low-lying constant-energy loops) [78] and a red shift of the Dirac point coming from n-type electron doping of Bi adatoms [79]. The measured DOS of AB-stacked (Bernal) graphite is finite near the Fermi level characteristic of the 3D semi-metallic property [80] and presents the splitting of π and π^* strong peaks at deeper or higher state energies [80]. It should be noted that the well-defined π and σ band widths in graphene-related systems are worthy of further STS studies. In addition, there have been no measured STS results for the AAA- and ABC-state graphites to date, owing mainly to the low content of natural graphite [81] and the difficulties in sample growth [50–52]. It should be noted that both STM and STS can be operated on the same sample to characterize geometric structures and electronic properties simultaneously, e.g. hexagon arrangements on cylindrical surfaces, edge orientations, and the radius- and width-dependent energy gaps of carbon nanotubes and graphene nanoribbons [7, 46, 47].

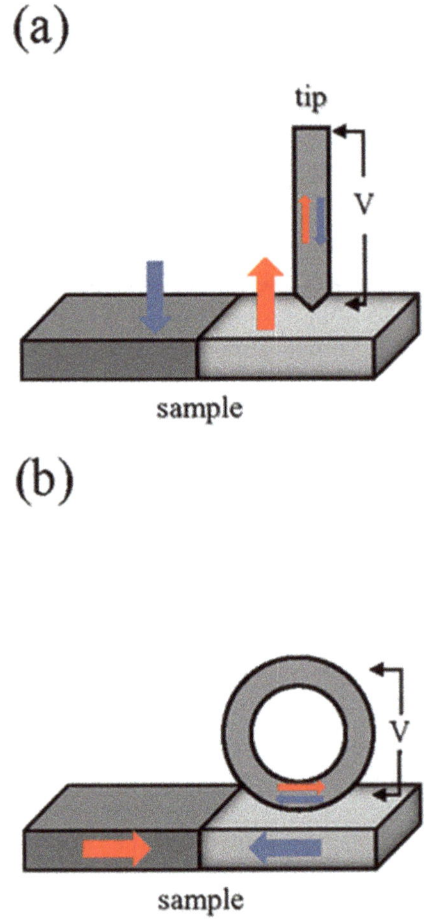

Figure 3.5. Spin-polarized scanning tunneling microscopy.

Spin configurations and their excitations, with atomic resolution, play critical roles in magnetic materials, in particular for ferromagnetic and antiferromagnetic phenomena. Spin-polarized STM (SP-STM) is constructed via a combination of atom-resolution and spin-sensitivity capabilities, where electron spins are measured using a ferromagnetic or antiferromagnetic probe tip, as shown in figure 3.5. SP-STM is suitable for obtaining atomically resolved morphologies of specific spin configurations in a condensed-matter system by recording constant-current images which represent the electron density near the sample surface, and simultaneously the spin polarization of charge carriers arising from the local magnetic moment. It is very sensitive to the spin orientation of tunneling electrons. This magnetic imaging technique was first proposed by Pierce in 1988 [80] and then achieved by Wiesendanger *et al* in 1990 [82]. The experimental set-up of SP-STM is displayed in figure 3.5. The basic concept is that the spin polarization of the tunneling current is a function of the state energy. As a result, the spin polarization of electronic states that contributes to quantum tunneling depends strongly on the sample bias voltage. For example, when a finite negative sample bias (V) is applied between the tip and sample surface, the occupied electronic states in the range of U below the Fermi level will have contributions to the tunneling current. In the tunneling process, such carriers change into unoccupied tip states with a range of V. The specific spin configurations of both the tip and sample states dominate the quantum tunneling. Generally, the spin polarization of the tunneling current varies with the bias voltage.

Currently, SP-STM and SP-STS are providing unprecedented insight into the diverse magnetic properties of mainstream materials, leading to the discovery of new types of magnetic ordering at the nanoscale. SP-STM has been used to determine magnetic domains and domain-wall structures at the high spatial resolution of 10 nm, a length scale inaccessible to other magnetic imaging techniques [81]. High-resolution SP-STM and SP-STS are also available for studying the spin-dependent scattering states around individual impurity adatoms on magnetic substrates. For example, when O atoms are absorbed on Fe substrates, the spin-dependent scattering of Fe electrons with d-like symmetry has been revealed in real space by SP-STS [83]. Current SP-STM can provide detailed information on magnetic phenomena with atomic resolution, such as the spin polarizations of individual adatoms on magnetic surfaces [84]. Moreover, the magnetic skyrmion and ferromagnetic states have been observed using SP-STM measurement for an epitaxial Fe film on an Ir(111) substrate with the application of an electric field [85]. Specifically, SP-STS is very important in understanding magneto-electronic properties, e.g. metallic ferromagnetism and semiconducting antiferromagnetism [86, 87]. Further SP-STM and SP-STS studies are required to verify the rich and unique spin configurations in graphene-related 1D and 2D systems. SP-STS plays a critical role in examining magneto-electronic properties, in particular for host- and guest-atom-induced ferromagnetism and antiferromagnetism [88].

3.3 ARPES and SP-ARPES

Angle-resolved photoemission spectroscopy (ARPES) is the most reliable and efficient method for full exploration of the quasi-particle energies and decay rates of occupied electronic states in various first Brillouin zones [89]. The measured wave-vector-dependent energy dispersions and lifetimes can examine the theoretical predictions from, respectively, first-principles calculations and the tight-binding model [90, 91], and the self-energy method (the screened exchange energy in [91]). It is well known that this form of spectroscopy cannot provide any information about magnetically quantized Landau levels. Generally, the ARPES chamber covers the sample synthesis instrument to measure *in situ* band properties. When a condensed-matter system is illuminated by soft x-rays (figure 3.6), the occupied electrons are excited to unoccupied intermediate states after an electric dipole transition. Photoelectrons are stimulated by incident photons and sufficiently high-energy electrons can escape outside the material surface into the free vacuum. Then such carriers are counted by an angle-resolved energy analyzer. The total momentum of the photoelectrons is estimated from the free-particle model. Furthermore, their

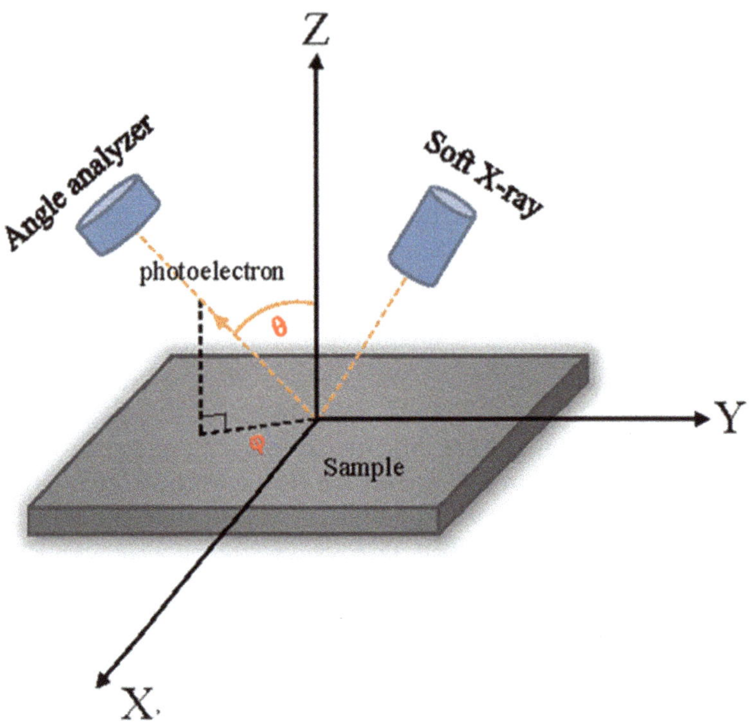

Figure 3.6. Angle-resolved photoemission spectroscopy.

parallel and perpendicular components are evaluated through measured polar and azimuthal angles. Energy is conserved during the photoemission process, while the conservation law is not suitable for the k_z-component momentum under the breaking of translation symmetry along the z-direction. ARPES measurements are available for 2D and quasi-2D systems with negligible energy dispersions normal to the sample surface. However, the non-conservation issue can be overcome by using the important characteristics of the k_z-dependent band structure, as verified for the 3D band structure of bulk graphite at $k_z = 0$ (details in [91]). Interestingly, the ARPES measurements can provide the energy widths of occupied electronic states, illustrating directly the many-body inelastic scattering due to electron–electron and electron–phonon interactions (the latter at finite temperatures [91]). Improvements in energy and momentum have become a critical factor for the investigation of low-dimensional materials. The current highest resolutions for the energy and angular distributions are, respectively, 1 meV and 0.1° in the UV regime [92]. Specifically, measurements of spin-split energy bands can be achieved using magnetic materials as ARPES detectors, e.g. spin-resolved energy dispersions in topological insulators [93], ferromagnetic and antiferromagnetic materials [94], and superconductors [93]. Such experimental studies are very useful for the spin-dominated band structures in silicene-related materials (discussed further in chapters 4–12).

High-resolution ARPES measurements are the only way to examine and identify wave-vector-dependent occupied energy bands and band gaps directly. Measured results have confirmed the feature-rich electronic energy spectra of carbon-related sp^2-bonding systems, as observed under various dimensions [96–101], layer numbers [96, 101–103], stacking configurations [104], substrates [96, 97, 101], and adatom and molecule adsorptions [105, 106]. For example, graphene nanoribbons are verified to possess 1D parabolic energy subbands around the high-symmetry point, in which the energy gaps decline with increasing ribbon width for any chiral or achiral edge structures [107]. For 2D few-layer graphenes, ARPES experiments have identified linear valence Dirac cones in monolayer systems [96, 98, 108], two pairs of parabolic bands in bilayer AB stacking [96, 104], co-existent linear and parabolic dispersions in twisted bilayer materials [102], linear and parabolic bands in tri-layer ABA stacking [96, 101], linear, partially flat and sombrero-shaped bands in tri-layer ABC stacking [101], substrate-created observable energy spacing between the separated valence and conduction bands in bilayer AB stacking [96, 101], substrate-induced oscillatory bands in few-layer ABC stacking [96], and the semimetal–semiconductor transition and tunable low-lying energy bands after adatom or molecule chemisorption on the graphene surfaces [95, 105, 106]. Bernal (AB-stacked) graphite appears to present a 3D band structure, where bilayer- and monolayer-like energy dispersions, respectively, come to exist at $k_z = 0$ and 1 (the K and H corner points in the 3D first Brillouin zone), and there is a strong trigonal warping effect around the K–H axis [91]. Also note that the well-behaved π- and σ-electronic energy spectra along, respectively, the KMΓ and ΓMK directions require further study using ARPES.

Most importantly, ARPES measurements are also conducted on the unusual band structures of the emergent 2D materials, such as the studies on few-layer silicene [108], germanene [63], tinene [67], Pb [66], phosphorene [109], antimonene [110] and bismuthene [111] systems in the presence of substrate bonding. Interestingly, a significant separated valence Dirac bonding initiated from the *K*/*K*′ point, with an energy gap of ~0.6 eV, is observed for the epitaxial silicene sheets on silver(111) and silver(110) substrates [44]. This result shows clearly that unusual intrinsic Si–Ag atomic interactions play critical roles in modifying the free-standing case (a very small band gap from the first-principles calculations). Quasi-multi Dirac-cone band structures have been observed in few-layer germanene systems on Au(111) [112]. However, the graphene-like energy bands might be completely absent under strong chemical bonding at the mixed boundary, such as the electronic energy spectra of silicene–ZrB_2 (0001). Bilayer tinene on Bi_2Te_3 [110] was found to exhibit hole-like metallic behavior near the Γ point, which is quite different from the first-principles prediction in the free-standing case [113]. A metallic property is also identified in Pb monolayer–Au(111), owing mainly to the very strong adlayer–substrate chemical bonding [66].

3.4 SQUID

The superconducting quantum interference device (SQUID), shown in figures 3.7(a)–(c), has attracted a lot of experimental [114] and theoretical [115–120] attention since the first device in the 1960s [115]. SQUIDs can furnish rich and

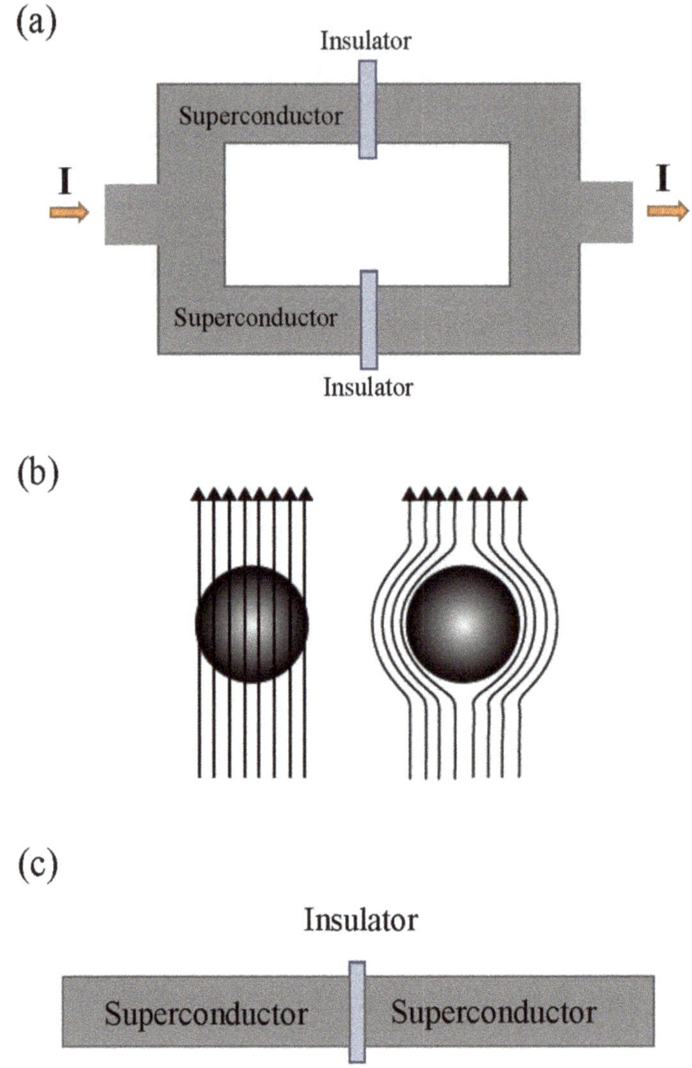

Figure 3.7. (a) The SQUID structure, (b) the Meissen phenomenon and (c) the Josephson junction.

unique measurements and experiments. One famous application is measuring the magnetic flux with super-high precision, based on the Meissner effect [119], Cooper pair tunneling [117, 118] and the Josephson junction [116, 117, 120]. SQUIDs have become instrumental in condensed-matter physics for their precision and versatility. The current resolution for magnetization is about 5×10^{-8} emu [121]. Moreover, the combination of SQUID and SP-STM (section 3.2) is very useful in identifying the diverse spin distribution configurations in emergent ferromagnetic [122] and anti-ferromagnetic [122, 123] materials.

Phenomenological superconducting theories [124] show clearly that the Ginzburg–Landau equations are responsible for the dc and ac supercurrents in Josephson junctions (details in [114]). A single junction composed of a superconductor–insulator–

superconductor, as illustrated in figure 3.7(c), presents a dc supercurrent under a vanishing gate voltage [114]. Furthermore, a static voltage across this junction can create radiofrequency current oscillations. Interestingly, macroscopic long-range quantum interference is generated by a superconducting circuit covering two Josephson junctions through a magnetic flux. The standard SQUID can induce sin- and cos-form oscillatory supercurrents that are very sensitive to the magnetic-field strength.

It is well known that high-precision SQUID measurements can distinguish thoroughly between paramagnetic, diamagnetic, ferromagnetic and antiferromagnetic responses [125]. They are very useful in examining and verifying novel magnetic phenomena by observing temperature- and magnetic-field-dependent magnetization (or magnetic susceptibility). The rich and unique magnetic properties include the weak diamagnetic susceptibility of AB-stacked graphite [126], the paramagnetic and diamagnetic behavior of the metallic and semiconducting carbon nanotubes [127], the paramagnetic response in metals (Al, Sn and Pt) [128], the diamagnetic behaviors under the traditional [129] and high-T_c [130] superconducting states, the antiferromagnets of heavy-fermion compounds [131], the multi antiferromagnetic transitions at different temperatures in ternary intermetallic compounds ($RE_2T_3X_5$) [132], the ferromagnetic transition metals [133], the unusual evolution from bulk antiferromagnetism to 2D ferromagnetism in SiM_2 materials (M = Gd and Eu) [134] and the ferromagnetic or antiferromagnetic configurations in topological insulators [135].

References

[1] Knoll M and Ruska E 1932 Das Elektronenmikroskop *Z. Phys.* **78** 318–39

[2] Ruska E 1987 The development of the electron microscope and of electron microscopy *Rev. Mod. Phys.* **59** 627

[3] Lentzen M 2006 Progress in aberration-corrected high-resolution transmission electron microscopy using hardware aberration correction *Microsc. Microanal.* **12** 191–205

[4] Kisielowski C *et al* 2008 Detection of single atoms and buried defects in three dimensions by aberration-corrected electron microscope with 0.5-Å information limit *Microsc. Microanal.* **14** 469–77

[5] Ray M *et al* 2014 Highly lattice-mismatched semiconductor metal hybrid nanostructures: gold nanoparticle encapsulated luminescent silicon quantum dots *Nanoscale* **6** 2201–10

[6] Ophus C, Rasool H I, Linck M, Zettl A and Ciston J 2017 Automatic software correction of residual aberrations in reconstructed HRTEM exit waves of crystalline samples *Adv. Struct. Chem. Imaging* **2** 15

[7] Iijima S 1991 Helical microtubules of graphitic carbon *Nature* **354** 56–8

[8] Kosynkin D V *et al* 2009 Longitudinal unzipping of carbon nanotubes to form graphene nanoribbons *Nature* **458** 872–6

[9] Cataldo F *et al* 2010 Graphene nanoribbons produced by the oxidative unzipping of single-wall carbon nanotubes *Carbon* **48** 2596–602

[10] Kumar P, Panchakarla L S and Rao C N R 2011 Laser-induced unzipping of carbon nanotubes to yield graphene nanoribbons *Nanoscale* **3** 2127–9

[11] Liu Z, Suenaga K, Harris P J F and Iijima S 2009 Open and closed edges of graphene layers *Phys. Rev. Lett.* **102** 015501

[12] Zhang J, Xiao J, Meng X, Monroe C, Huang Y and Zuo I M 2010 Free folding of suspended graphene sheets by random mechanical stimulation *Phys. Rev. Lett.* **104** 166805
[13] Viculis L M, Mack J J and Kaner R B 2003 A chemical route to carbon nanoscrolls *Science* **299** 1361
[14] Savoskin M V *et al* 2007 Carbon nanoscrolls produced from acceptor-type graphite intercalation compounds *Carbon* **45** 2797–800
[15] Shioyama H and Akita T 2003 A new route to carbon nanotubes *Carbon* **41** 179–81
[16] Campos-Delgado J *et al* 2008 Bulk production of a new form of sp^2 carbon: crystalline graphene nanoribbons *Nano Lett.* **8** 2773–8
[17] Warner J H, Rümmeli M H, Gemming T, Büchner B and Briggs G A D 2009 Direct imaging of rotational stacking faults in few layer graphene *Nano Lett.* **9** 102–6
[18] Lee J K, Lee S C, Py A J, Kim S C, Wilson J I B and John P 2008 The growth of AA graphite on (111) diamond *J. Chem. Phys.* **129** 234709
[19] Binnig G and Rohrer H 1982 Scanning tunnel microscope *Helv. Phys. Acta* **55** 726–35
[20] Binnig G and Rohrer H 1987 Scanning tunneling microscopy—from birth to adolescence *Rev. Mod. Phys.* **59** 615
[21] Marchini S, Günther S and Wintterlin J 2007 Scanning tunneling microscopy of graphene on Ru(0001) *Phys. Rev.* B **76** 075429
[22] Lalmi B *et al* 2010 Epitaxial growth of a silicene sheet *Appl. Phys. Lett.* **97** 223109
[23] Dávila M E, Xian L, Cahangirov S, Rubio A and Lay G L 2014 Germanene: a novel two-dimensional germanium allotrope akin to graphene and silicene *New J. Phys.* **16** 095002
[24] Zhang C D *et al* 2009 Surface structures of black phosphorus investigated with scanning tunneling microscopy *J. Phys. Chem.* C **113** 18823
[25] Ugeda M M *et al* 2014 Giant bandgap renormalization and excitonic effects in a monolayer transition metal dichalcogenide semiconductor *Nat. Mater.* **13** 1091
[26] Wang M X *et al* 2012 The coexistence of superconductivity and topological order in the Bi_2Se_3 thin films *Science* **336** 52
[27] Biegelsen D K, Bringans R D, Northrup J E and Swartz L-E 1990 Reconstructions of GaAs $(\bar{1}\bar{1}\bar{1})$ surfaces observed by scanning tunneling microscopy *Phys. Rev. Lett.* **65** 452
[28] Hirano M, Shinjo K, Kaneko R and Murata Y 1997 Observation of superlubricity by scanning tunneling microscopy *Phys. Rev. Lett.* **78** 1448
[29] Patra N, Wang B and Kral P 2009 Nanodroplet activated and guided folding of graphene nanostructures *Nano Lett.* **9** 3766
[30] Song J J, Zhang H J, Cai Y L, Zhang Y X, Bao S N and He P M 2016 Bottom-up fabrication of graphene nanostructures on Ru(1010) *Nanotechnology* **27** 055602
[31] Vo T H *et al* 2014 Large-scale solution synthesis of narrow graphene nanoribbons *Nat. Commun.* **5** 3189
[32] Vancsó P *et al* 2016 The intrinsic defect structure of exfoliated MoS_2 single layers revealed by scanning tunneling microscopy *Sci. Rep.* **6** 29726
[33] Novello A M *et al* 2015 Scanning tunneling microscopy of the charge density wave in 1T $-TiSe_2$ in the presence of single atom defects *Phys. Rev.* B **92** 081101(R)
[34] Mielke J, Hanke F, Peters M V, Hecht S, Persson M and Grill L 2015 Adatoms underneath single porphyrin molecules on Au(111) *J. Am. Chem. Soc.* **137** 1844
[35] Schouteden K, Debehets J, Muzychenko D, Li Z, Seo J W and Haesendonck C V 2017 Adsorption of Te atoms on Au(111) and the emergence of an adatom-induced bound state *J. Phys.: Condens. Matter* **29** 125001

[36] Eren B *et al* 2016 Activation of Cu(111) surface by decomposition into nanoclusters driven by CO adsorption *Science* **351** 6272

[37] Emmrich M *et al* 2015 Subatomic resolution force microscopy reveals internal structure and adsorption sites of small iron clusters *Science* **348** 6232

[38] Du J, Jones B and Lanagan M 2005 Preparation and characterization of dielectric glass-ceramics in Na_2O–PbO–Nb_2O_5–SiO_2 system *Mater. Lett.* **59** 2821

[39] Mcmullan D 2006 Scanning electron microscopy 1928–1965 *Scanning* **17** 175

[40] Pennycook S J and Boatner L A 1988 Chemically sensitive structure-imaging with a scanning transmission electron microscope 1928–1965 *Nature* **336** 565

[41] Lang K M 2004 Conducting atomic force microscopy for nanoscale tunnel barrier characterization 1928–1965 *Rev. Sci. Instrum.* **78** 2726

[42] Scheibner E J, Germer L H and Hartman C D 1960 Apparatus for direct observation of low-energy electron diffraction patterns *Rev. Sci. Instrum.* **31** 112

[43] Li L *et al* 2013 Two-dimensional transition metal honeycomb realized: Hf on Ir(111) *Nano Lett.* **13** 4671

[44] Vogt P *et al* 2012 Atomic and molecular adsorption on Ir(111) *Phys. Rev. Lett.* **108** 155501

[45] Meng L *et al* 2013 Buckled silicene formation on Ir(111) *Nano Lett.* **13** 685

[46] Enoki T, Fujii S and Takai K 2012 Zigzag and armchair edges in graphene *Carbon* **50** 3141

[47] Tao C *et al* 2011 Spatially resolving edge states of chiral graphene nanoribbons *Nat. Phys.* **7** 616

[48] Kim K, Sussman A and Zettl A 2010 Graphene nanoribbons obtained by electrically unwrapping carbon nanotubes *ACS Nano.* **4** 1362

[49] Klusek Z *et al* 2005 Local electronic edge states of graphene layer deposited on Ir(111) surface studied by STM/CITS *Appl. Surf. Sci.* **252** 1221

[50] Que Y, Xiao W, Chen H, Wang D, Du S and Gao H J 2015 Stacking-dependent electronic property of trilayer graphene epitaxially grown on Ru (0001) *Appl. Phys. Lett.* **107** 263101

[51] Xu R, Yin L J, Qiao J B, Bai K K, Nie J C and He L 2015 Direct probing of the stacking order and electronic spectrum of rhombohedral trilayer graphene with scanning tunneling microscopy *Phys. Rev.* B **91** 035410

[52] Liu Z, Zheng Q S and Liu J Z 2010 Stripe/kink microstructures formed in mechanical peeling of highly orientated pyrolytic graphite *Appl. Phys. Lett.* **96** 201909

[53] Geringer V *et al* 2010 Intrinsic and extrinsic corrugation of monolayer graphene deposited on SiO_2 *Phys. Rev. Lett.* **102** 076102

[54] de Parga A L V *et al* 2008 Periodically rippled graphene: growth and spatially resolved electronic structure *Phys. Rev. Lett.* **102** 056807

[55] Varchon F, Mallet P, Veuillen J-Y and Magaud L 2008 Ripples in epitaxial graphene on the Si-terminated SiC(0001) surface *Phys. Rev.* B **77** 235412

[56] Červenka J, Katsnelson M I and Flipse C F J 2008 Room-temperature ferromagnetism in graphite driven by two-dimensional networks of point defects *Nat. Phys.* **5** 840

[57] Arafune R *et al* 2013 Structural transition of silicene on Ag(111) *Surf. Sci.* **608** 297

[58] Dimoulas A 2015 Silicene and germanene: silicon and germanium in the 'flatland' *Microelectron. Eng.* **131** 68

[59] Wei W, Dai Y, Huang B, Whangbo M H and Jacob T 2015 Loss of linear band dispersion and trigonal structure in silicene on Ir(111) *J. Phys. Chem. Lett.* **6** 1065

[60] Gill T G *et al* 2017 Metallic atomically-thin layered silicon epitaxially grown on silicene/ZrB_2 *2D Mater.* **4** 021015

[61] Li L *et al* 2014 Buckled germanene formation on Pt(111) *Adv. Mater.* **26** 4820

[62] Derivaz M *et al* 2015 Continuous germanene layer on Al(111) *Nano Lett.* **4** 2510

[63] Lin C H *et al* 2018 Single-layer dual germanene phases on Ag(111) *Phys. Rev. Mater.* **2** 024003

[64] Cantero E D *et al* 2017 Growth of germanium on Au(111): formation of germanene or intermixing of Au and Ge atoms? *Phys. Chem. Chem. Phys.* **19** 18580

[65] Dalmas J *et al* 2005 Ordered surface alloy formation of immiscible metals: the case of Pb deposited on Ag(111) *Phys. Rev.* B **72** 155424

[66] Crepaldi A *et al* 2013 Combined ARPES and STM study of Pb/Au(111) moiré structure: one overlayer, two symmetries *Phys. Rev.* B **87** 115138

[67] Zhu F *et al* 2015 Epitaxial growth of two-dimensional stanene *Nat. Mater.* **14** 1020

[68] Hou Z and Terakura K 2015 Effect of nitrogen doping on the migration of the carbon adatom and monovacancy in graphene *J. Phys. Chem.* C **119** 4922

[69] Varns R and Strange P 2008 Stability of gold atoms and dimers adsorbed on graphene *J. Phys. Condens. Matter* **20** 22

[70] Shyu F L, Chang C P, Chen R B, Chiu C W and Lin M F 2003 Stability of gold atoms and dimers adsorbed on graphene *Phys. Rev.* B **67** 045405

[71] Mintmire J W and White C T 1998 Universal density of states for carbon nanotubes *Phys. Rev.* B **81** 2506

[72] Odom T W, Huang J L and Lieber C M 2002 STM studies of single-walled carbon nanotubes *J. Phys. Condens. Matter* **14** 6

[73] Deshpande A, Bao W, Miao F, Lau C N and LeRoy B J 2009 Spatially resolved spectroscopy of monolayer graphene on SiO_2 *Phys. Rev.* B **79** 205411

[74] Hiebel F, Mallet P, Magaud L and Veuillen J-Y 2009 Atomic and electronic structure of monolayer graphene on 6H-SiC($000\bar{1}$)(3×3): a scanning tunneling microscopy study *Phys. Rev.* B **80** 235429

[75] Yan W *et al* 2012 Angle-dependent van Hove singularities in a slightly twisted graphene bilayer *Phys. Rev. Lett.* **109** 126801

[76] Lui C H, Li Z, Mak K F, Cappelluti E and Heinz T F 2011 Observation of an electrically tunable band gap in trilayer graphene *Nat. Phys.* **7** 944

[77] Pierucci D *et al* 2015 Evidence for flat bands near the Fermi level in epitaxial rhombohedral multilayer graphene *ACS Nano* **9** 5432

[78] Lin C Y, Do T N, Huang Y K and Lin M F 2017 *Electronic and Optical Properties of Graphene in Magnetic and Electric Fields* (San Rafael, CA: Morgan and Claypool)

[79] Hu T *et al* 2018 Enhanced n-doping of epitaxial graphene on SiC by bismuth *Appl. Phys. Lett.* **113** 011602

[80] Klusek Z 1999 Investigations of splitting of the π bands in graphite by scanning tunneling spectroscopy *Appl. Surf. Sci.* **151** 251

[81] Bernal J D 1924 The structure of graphite *Proc. R. Soc. Lond.* A **106** 749

[82] Wiesendanger R, Güntherodt H-J, Güntherodt G, Gambino R J and Ruf R 1990 Observation of vacuum tunneling of spin-polarized electrons with the scanning tunneling microscope *Phys. Rev. Lett.* **65** 247

[83] Ravlić R, Bode M, Kubetzka A and Wiesendanger R 2003 Correlation of dislocation and domain structure of Cr(001) investigated by spin-polarized scanning tunneling microscopy *Phys. Rev.* B **67** 174411

[84] Yayon Y, Brar V W, Senapati L, Erwin S C and Crommie M F 2007 Observing spin polarization of individual magnetic adatoms *Phys. Rev.* B **99** 067202
[85] Hsu P J, Kubetzka A, Finco A, Romming N, von Bergmann K and Wiesendanger R 2017 Electric-field-driven switching of individual magnetic skyrmions *Nat. Nanotechnol.* **12** 123
[86] Pietzsch O, Kubetzka A, Bode M and Wiesendanger R 2000 Real-space observation of dipolar antiferromagnetism in magnetic nanowires by spin-polarized scanning tunneling spectroscopy *Phys. Rev. Lett.* **84** 5212
[87] Heinze S, Bode M, Kubetzkal A, Pietzsch O, Nie X, Blüge S and Wiesendanger R 2000 Real-space imaging of two-dimensional antiferromagnetism on the atomic scale *Science* **288** 5472
[88] Brune H and Gambardella P 2009 Magnetism of individual atoms adsorbed on surfaces *Surf. Sci.* **603** 1812
[89] Stepan H 2013 *Photoelectron Spectroscopy: Principles and Applications* (Berlin: Springer)
[90] Lin S Y, Tran N T T, Chang S L, Su W P and Lin M F 2018 *Structure- and Adatom-enriched Essential Properties of Graphene Nanoribbons* (Boca Raton, FL: CRC Press)
[91] Lin C Y, Chen R B, Ho Y H and Lin M F 2018 *Electronic and Optical Properties of Graphite-related Systems* (Boca Raton, FL: CRC Press)
[92] Levy G, Nettke W, Ludbrook B M, Veenstra C N and Damascelli A 2014 Deconstruction of resolution effects in angle-resolved photoemission *Phys. Rev.* B **90** 045150
[93] Qi X L and Zhang S C 2011 Topological insulators and superconductors *Rev. Mod. Phys.* **83** 1057
[94] de Jong S *et al* 2010 Droplet-like Fermi surfaces in the anti-ferromagnetic phase of $EuFe_2As_2$, an Fe-pnictide superconductor parent compound *EPL* **89** 27007
[95] Papagno M *et al* 2011 Large band gap opening between graphene Dirac cones induced by Na adsorption onto an Ir superlattice *ACS Nano* **6** 199–204
[96] Ohta T, Bostwick A, McChesney J L, Seyller T, Horn K and Rotenberg E 2007 Interlayer interaction and electronic screening in multilayer graphene investigated with angle-resolved photoemission spectroscopy *Phys. Rev. Lett.* **98** 206802
[97] Siegel D A, Regan W, Fedorov A V, Zettl A and Lanzara A 2013 Charge-carrier screening in single-layer graphene *Phys. Rev. Lett.* **110** 146802
[98] Bostwick A, Ohta T, Seyller T, Horn K and Rotenberg E 2007 Quasiparticle dynamics in graphene *Nat. Phys.* **3** 36–40
[99] Sugawara K, Sato T, Souma S, Takahashi T and Suematsu H 2006 Fermi surface and edge-localized states in graphite studied by high-resolution angle-resolved photoemission spectroscopy *Phys. Rev.* B **73** 045124
[100] Grüneis A *et al* 2008 Electron–electron correlation in graphite: a combined angle-resolved photoemission and first-principles study *Phys. Rev. Lett.* **100** 037601
[101] Coletti C *et al* 2013 Revealing the electronic band structure of trilayer graphene on SiC: an angle-resolved photoemission study *Phys. Rev.* B **88** 155439
[102] Kim K S *et al* 2013 Coexisting massive and massless Dirac fermions in symmetry-broken bilayer graphene *Nat. Mater.* **12** 887–92
[103] Sutter P, Hybertsen M S, Sadowski J T and Sutter E 2009 Electronic structure of few-layer epitaxial graphene on Ru (0001) *Nano Lett.* **9** 2654–60
[104] Ohta T, Bostwick A, Seyller T, Horn K and Rotenberg E 2006 Controlling the electronic structure of bilayer graphene *Science* **313** 951–4

[105] Zhou S Y, Siegel D A, Fedorov A V and Lanzara A 2008 Metal to insulator transition in epitaxial graphene induced by molecular doping *Phys. Rev. Lett.* **101** 086402
[106] Grüneis A and Denis Vyalikh V 2008 Tunable hybridization between electronic states of graphene and a metal surface *Phys. Rev.* B **77** 193401
[107] Ruffieux P *et al* 2012 Electronic structure of atomically precise graphene nanoribbons *ACS Nano* **6** 6930–5
[108] De Padova P *et al* 2019 New findings on multilayer silicene on Si(111)$\sqrt{3} \times \sqrt{3}$R30°Ag template *Materials* **12** 2258
[109] Ehlen N *et al* 2016 Evolution of electronic structure of few-layer phosphorene from angle-resolved photoemission spectroscopy of black phosphorous *Phys. Rev.* B **94** 245410
[110] Lei T *et al* 2016 Electronic structure of antimonene grown on Sb_2Te_3 (111) and Bi_2Te_3 substrates *J. Appl. Phys.* **119** 015302
[111] Reis F *et al* 2017 Bismuthene on a SiC substrate: a candidate for a high-temperature quantum spin Hall material *Science* **357** 287
[112] Dávila M E and Lay G L 2016 Few layer epitaxial germanene: a novel two-dimensional Dirac material *Sci. Rep.* **6** 20714
[113] Huang C, Zhou J, Wu H, Deng K, Jena P and Kan E 2016 Quantum phase transition in germanene and stanene bilayer: from normal metal to topological insulator *J. Phys. Chem. Lett.* **7** 1919
[114] Hollenhorst J N and Giffard R P 1980 Input noise in the hysteretic rf SQUID: Theory and experiment *J. Appl. Phys.* **51** 1719–25
[115] Weber J 1960 Detection and generation of gravitational waves *Phys. Rev.* **117** 306
[116] Josephson B D 1962 Possible new effects in superconductive tunneling *Phys. Lett.* **1** 251–3
[117] Anderson P W and Rowell J M 1963 Probable observation of the Josephson superconducting tunneling effect *Phys. Rev. Lett.* **10** 230–2
[118] McCumber D E 1968 Effect of ac impedance on dc voltage–current characteristics of superconductor weak-link junctions *J. Appl. Phys.* **39** 3113
[119] Meissne W and Ochsenfeld R 1933 Ein neuer Effekt bei Eintritt der Supraleitfähigkeit *Naturwissenschaften* **21** 787–8
[120] Duzer T V and Turner C W 1998 *Superconductive Devices and Circuits* 2nd edn (Englewood Cliffs, NJ: Prentice Hall)
[121] Ueda K *et al* 1982 Magnetic remanences in migratory birds *J. Yamashina Instit. Ornithology* **14** 166–70
[122] Hong J, Bekyarova E, Liang P, de Heer W A, Haddon R C and Khizroev S 2012 Room-temperature magnetic ordering in functionalized graphene *Sci. Rep.* **2** 624
[123] Sueoka K, Subagyo A, Hosoi H and Mukasa K 2004 Magnetic imaging with scanning probe microscopy *Nanotechnology* **15** 10
[124] Gor'kov L P 1959 Microscopic derivation of the Ginzburg–Landau equations in the theory of superconductivity *Sov. Phys. JETP* **9** 1364
[125] Finkler A *et al* 2010 Self-aligned nanoscale SQUID on a tip *Nano Lett.* **10** 1046–9
[126] Cao Y *et al* 2018 Unconventional superconductivity in magic-angle graphene superlattices *Nature* **556** 43–50
[127] Cleuziou J-P, Wernsdorfer W, Bouchiat V, Ondarçuhu T and Monthioux M 2006 Carbon nanotube superconducting quantum interference device *Nat. Nanotechnol.* **1** 53–9
[128] Cardarelli F 2000 *Materials Handbook: A Concise Desktop Reference* 2nd edn (Berlin: Springer) pp 500–1

[129] Hodgkin A L and Katz B 1949 The effect of temperature on the electrical activity of the giant axon of the SQUID *J. Physiol.* **109** 240
[130] Gammel P L, Schneemeyer L F and Bishop D J 1991 SQUID picovoltometry of $YBa_2Cu_3O_7$ single crystals: evidence for a finite-temperature phase transition in the high-field vortex state *Phys. Rev. Lett.* **66** 953
[131] Park B G *et al* 2011 A spin-valve-like magnetoresistance of an antiferromagnet-based tunnel junction *Nat. Mater.* **10** 347
[132] Li L, Nishimura K, Hutchison W D, Qian Z, Huo D and NamiKi T 2012 Giant reversible magnetocaloric effect in $ErMn_2Si_2$ compound with a second order magnetic phase transition *Appl. Phys. Lett.* **100** 152403
[133] Matsumoto Y *et al* 2012 Room-temperature ferromagnetism in transparent transition metal-doped titanium dioxide *Science* **291** 854
[134] Tokmachev A M *et al* 2018 Emerging two-dimensional ferromagnetism in silicene materials *Nat. Commun.* **9** 854
[135] Kurter C, Finck A D K, Hor Y S and Van Harlingen D J 2015 Evidence for an anomalous current–phase relation in topological insulator Josephson junctions *Nat. Commun.* **6** 7130

IOP Publishing

Silicene-Based Layered Materials
Essential properties
Shih-Yang Lin, Hsin-Yi Liu, Duy Khanh Nguyen, Ngoc Thanh Thuy Tran, Hai Duong Pham, Shen-Lin Chang, Chiun-Yan Lin and Ming-Fa Lin

Chapter 4

Monolayer C-, Si-, Ge-, Sn- and Pb-based materials

Group-IV layered systems, which consist of C [1, 2], Si [3, 4], Ge [5, 6], Sn [7, 8] and Pb [9, 10] atoms, have been produced successfully in experimental laboratories since the first discovery of few-layer graphene systems through mechanical exfoliation in 2004 [11]. The carbon-related 2D materials are synthesized using various growth methods [9, 10, 16, 18–29], while the others are produced using molecular beam epitaxial growth and mechanical methods. The former have been identified to present the AAA, ABA, ABC and AAB stacking configurations, with specific layer numbers, according to high-resolution TEM measurements [12–15]. Using STM (or in combination with LEED/RHEED, details in sections 3.1 and 3.2) graphene-like honeycomb lattices have been clearly demonstrated in group-IV 2D systems, such as 2D C-, Si-, Ge-, Sn- and Pb-adlayers with hexagonal symmetries on, respectively, SiO_2, Cu and Bi_2Se_3 [16–18], Ag(111), Ir(111) and ZrB_2 (0001) [19–24], Pt(111), Al(111) and Au(111) [25–27], Bi_2Te_3[111] [28, 29] and Au(111), AlO and Si(111) [9, 10]. It should be noted that surface reconstructions appear frequently at the adlayer and substrate (details in section 3.2). Among the different systems, only graphene systems exhibit a planar structure with the absence of two-sublattice buckling, owing mainly to the dominant sp^2 bonding. The other group-IV materials have been demonstrated to possess buckled structures arising from the height difference between the A and B sublattices [30, 31]. Up-to-date data show that the buckling height of each layer are 0, 0.46, 0.68, 0.84 and 0.89 Å, respectively. The buckling degree grows with the increase of atomic number, indicating clearly the enhanced strength of sp^3 bonding. Through experimental studies on the band properties near the Fermi level, semiconducting and semimetallic behaviors have been confirmed in layered graphenes using STS and ARPES measurements (details in sections 3.2 and 3.3). The latter are also utilized to examine the presence or absence of Dirac-cone-like energy bands in other group-IV systems, illustrating clearly the critical role of the adlayer–substrate chemical bonding. The semiconducting and metallic adlayers have become mainstream 2D materials, in which many fundamental properties are worthy of systematic investigation. In addition, few-layer graphene systems can exhibit rich and unique physical, chemical, and material phenomena, e.g. the best mechanical properties [32, 33], stacking- and layer-number-varied band properties, diverse optical and magneto-optical spectra [34–48], unusual quantum Hall conductivities [49–51], and various single- and many-particle Coulomb excitations [52, 53].

Group-IV 2D pristine materials have also attracted a lot of theoretical attention. In general, there exist three methods, the first-principles method, the tight-binding model and the effective-mass approximation, for exploring electronic properties. They are available for studying energy bands, subenvelope functions and densities of states. Only the first-principles method is suitable for investigating the optimal geometric structures and orbital hybridization in chemical bonds. According to current theoretical predictions, monolayer group-IV systems are direct- or indirect-gap semiconductors, in which the energy gap grows from zero with the increase of atomic number. Also noted that single-layer graphene is a gapless semiconductor because of the vanishing density of states at the Fermi level. In the low-lying energy bands, C-, Si- and Ge-related systems are predicted to be dominated by π bonding associated with the p_z orbitals, while Sn- and Pb-dependent systems present significant sp^3-orbital hybridization. Such results are determined mainly by the buckling degree, the distribution ranges of the outer orbitals and the spin–orbit coupling (discussed later in figure 4.2). The direct combination of the intrinsic critical factors, magnetic fields and gate voltages will create varied fundamental properties, e.g. the magneto-electronic and optical properties in few-layer graphene [43–48], silicene [54–58], germanene [59] and tinene [60]. In addition, the generalized tight-binding model has been developed well to achieve a full understanding of the diverse magnetic quantization phenomena of layered graphene systems with various dimensions [61–66], geometric symmetries [2, 31, 67–69], stacking configurations [70–72], layer numbers [61, 73] and external fields [74]. However, the effective-mass method, which includes the perturbation concept, cannot deal with the magnetically quantized states from multi-pair, unusual, or complicated valence and conduction bands [43, 75], e.g. the Landau levels in Sn- and Pb-few-layer systems.

This chapter utilizes sensitive first-principles calculations to explore thoroughly the fundamental properties in group-IV monolayer materials, in particular to obtain a full understanding of the critical features. Through careful analyses, the lattice constants, buckling degrees, ground state energies, direct and indirect band gaps, band structures, spatial charge densities, orbital- and spin-decomposed densities of states, magnetic moments and spin configurations are investigated in detail. The strongly cooperative and competitive relations between the buckled structure, the multi- and single-orbital hybridization, and the spin–orbit coupling for specific materials can be deduced from the developed theoretical framework. The important differences between the C-, Si-, Ge-, Sn- and Pb-related layered systems will be discussed. Also, specific tight-binding models [56, 76–79], which are reliable in evaluating the low-lying energy bands, are available to support the concise physical pictures. The theoretical predictions agree roughly with up-to-date experimental measurements [8, 70, 80–86].

4.1 The geometric, electronic and spin properties of monolayer group-IV materials

Monolayer group-IV systems, as shown in figure 4.1, present stable honeycomb lattices in the isolated case. Such materials can provide full information about the π, sp^2 and sp^3 bonding due to, respectively, the p_z, (s, p_x, p_y) and (s, p_x, p_y, p_z) orbital hybridization. The main features, planar and buckled structures, ground state energies, lattice constants, and heights of the A and B sublattices, are consistent with the significant chemical bonding of the outermost s, p_x, p_y and p_z orbitals. Only graphene consists of normal hexagons (figure 4.1(a)), while the other four monoelements possess different heights in the A and B sublattices (height differences of 0, 0.48, 0.69, 0.84 and 0.89 Å for C, Si, Ge, Sn and Pb), shown in table 4.1). Most importantly, these two sublattices are equivalent on the (x, y)-plane for any fundamental properties of all the group-IV materials. This result illustrates clearly that the dominance of sp^2 bonding is reduced or enhanced gradually with the increase of atomic number, while the opposite is true for the importance of sp^3 bonding. Furthermore, the sp^2-bonding strength is much stronger than that of π bonding. As a result, C-related systems have the highest ground state energy and lowest lattice constant, and Pb-related systems have the lowest ground state energy and the highest lattice constant (see the calculated values in table 4.1), where the latter is defined as $(\sqrt{3}b)$ and b is the X–X bond length. In other words, layered graphenes are the most stable of the group-IV systems, indicating clearly the best mechanical features [33, 59, 87]. Host-atom-induced spin configurations have been included in the VASP calculations of the total ground state energies, in which the

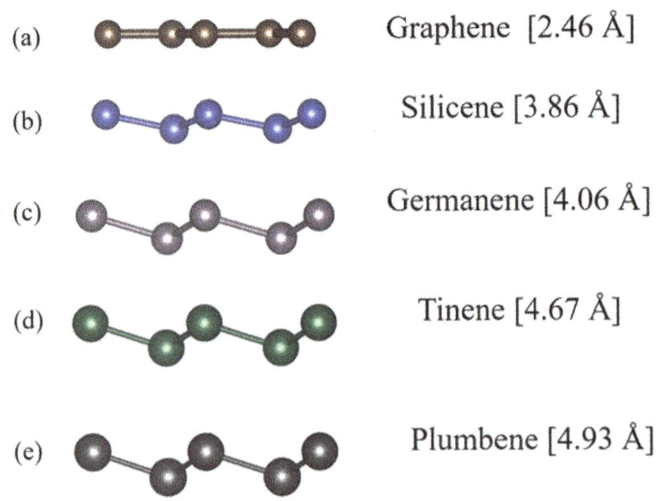

Figure 4.1. The optimal honeycomb lattices of single-layer group-IV systems: (a) graphene, (b) silicene, (c) germanene, (d) tinene and (e) plumbene, with different buckled structures and lattice constants.

Table 4.1. The significant geometric parameters, total ground state energy per unit cell, energy gap, lattice constant, and height difference between the A and B sublattices for monolayer group-IV materials.

Group-IV material	Lattice constant (Å)	Δz (Å)	b(X–X) (Å)	Ground state energy (eV/unit cell)
Graphene	2.46	0	1.42	−18.45
Silicene	3.86	0.48	2.28	−9.57
Germanene	4.06	0.69	2.44	−8.02
Tinene	4.67	0.84	2.83	−6.78
Plumbene	4.93	0.89	2.99	−6.00

spin–orbit interactions play a critical role in the geometric and electronic properties [5, 88–90]. In short, multi-orbital hybridization dominates the optimal geometric structures fully and thus enrich the fundamental properties greatly.

Monolayer group-IV systems exhibit varied valence and conduction bands, as illustrated in figures 4.2(a)–(e). All the electronic states, which consist of the spin-up- and spin-down-combined states with double degeneracy, can be classified into three groups according to the effects of sp^3 bonding and spin–orbit coupling. First, graphene has low-lying π and π^* bands in the energy range -3 eV $\leqslant E^{c.,v} \leqslant 3$ eV, and is asymmetric about the Fermi level ($E_F = 0$) as a result of the warping effect (the non-orthonormal property of the A- and B-sublattice $2p_z$-orbital tight-binding function [91]). The linear energy dispersions only intersect at the Dirac point (the K/K′ point) as a result of hexagonal symmetry [50, 51]. Furthermore, this Fermi-moment state only makes a zero contribution to the density of states (see figure 4.4(a) below). Graphene is a zero-gap semiconductor, as verified directly from STS measurements [36, 92–97]. The low-energy fundamental properties are dominated by the π bonding of $2p_z$ orbitals (see figure 4.4(a) below). When the valence state energies become deeper, the concave-downward Dirac cone changes into parabolic dispersions along the direction of KM, and when the conduction state energies become higher the concave-upward Dirac cone forms a saddle structure centered at the M point and then reaches the extreme Γ point (figure 4.2(a)). However, in the direction of KΓ, it is transformed uniformly into the parabolic Γ valley. The K and Γ valleys are very stable in initiating the fundamental physical properties, e.g. the initial Landau levels, with completely different behaviors [39, 98]. Specifically, the K, M and Γ points belong to the distinct critical points in the energy–wave-vector space; therefore, they will induce different Van Hove singularities in the density of states. In addition, the π-electronic state energy at the M point (~ -2.5 eV) is helpful in determining the strength of the nearest hopping integral for the tight-binding model [93, 94]. The $2p_x$ and $2p_y$ half-occupied orbitals create two σ parabolic bands which are initiated from the Γ valley at $E^v \sim -3.02$ eV. Such deeper electronic states arise from the stronger σ bonding, compared to the zero-energy states of the π bonding. The almost direct crossing between the σ and π bands occur between the Γ and K valleys along any direction, further illustrating the orthogonal feature of their chemical bonding. It should be noted that the specific σ band

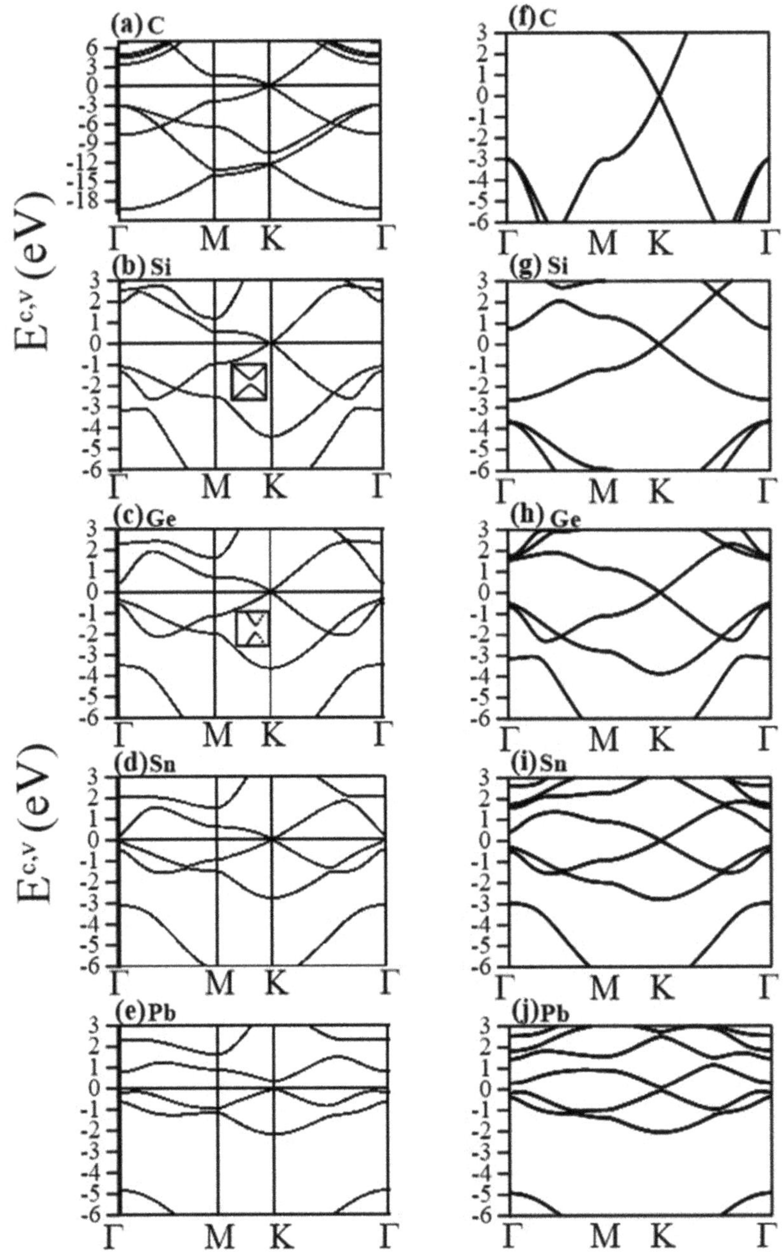

Figure 4.2. The band structures of monolayer (a) C-, (b) Si-, (c) Ge-, (d) Sn- and (e) Pb-related systems using the first-principles method, and the band structures of (f) C, (g) Si and (h) Ge for only π bonding, and (i) Sn and (j) Pb for sp^3 bonding using the tight-binding model.

associated with the 2s orbitals arise in a very deep-energy range ($E^v < -6$ eV) under the largest ionization energy (the lowest one-site energy in the tight-binding model [99]).

There exist certain important similarities and dissimilarities between silicene and germanene (figures 4.2(b) and (c)) and graphene (figure 4.2(a)). The low-lying valence and conduction bands of silicene and germanene are dominated by the π bonding of $3p_z$ and $4p_z$ orbitals. However, the low-energy ranges of $|E^{c,v}|$ without σ bands are, respectively, ~3.0, 1.0 and 0.5 eV for C-, Si- and Ge-related materials, respectively. This indicates reduced π bonding under larger lattice constants and wider orbitals. The Dirac-cone structure presents a slight difference in the Si- and Ge-related systems (the insets in figures 4.2(b) and (c)). The narrow band gaps of $E_g \sim 1.55$ meV and 24.0 meV for silicene and germanene are attributed to enhanced spin–orbit coupling in the buckled honeycomb lattice. This is consistent with the tight-binding model with nearest-neighbor hopping integrals and next-nearest-neighbor spin–orbit interactions (λ, $E_g = 2\lambda$ in [31, 100]). It can be deduced that single-p_z-orbital hybridization is roughly suitable for the low-energy physics of monolayer graphene, silicene and germanene (figures 4.2(f)–(h)). On the other hand, the energy spacing between the initial σ bands and the Fermi level at the Γ valley becomes small with the increase of atomic number, owing mainly to the enhanced sp^3 bonding and spin–orbit coupling. Furthermore, direct mixing (anti-crossing behavior) of π and σ bands might occur frequently along any direction related to the Γ point, e.g. those along MΓ and KΓ.

For Sn- and Pb-related monolayer systems (figures 4.2(d) and (e)), the significant sp^3 bonding and spin–orbit coupling play critical roles in the low-lying energy bands. Two σ bands and one π band appear to arise near the Fermi level simultaneously, initiated from, respectively, the Γ and K valleys. The former present splitting phenomena attributed to a sufficiently strong spin–orbit interaction (understood from the tight-binding model in section 4.2). The semiconducting or semimetallic behaviors are co-dominated by sp^3 bonding and spin–orbit coupling. Moreover, there exists more serious band mixing along MΓ and KΓ. The above-mentioned features illustrate clearly that single-orbital π bonding is not suitable for the low-energy fundamental properties of monolayer Sn and Pb systems, and that sp^3 orbital hybridization needs to be included in the phenomenological models (figures 4.2(i) and (j)). In short, the different strengths of sp^3 and sp^2 bonding and spin–orbit coupling produce varied band structures and thus other varied fundamental properties.

High-resolution ARPES measurements are the only sensitive way to test the theoretical predictions for the occupied electronic states of 2D group-IV materials efficiently (details in section 3.3), in particular for the diverse wave-vector dependence cases. ARPES has verified the linear energy dispersions of the K/K′ valleys for monolayer graphene [74], and even under n- or p-type doping (the blue or red shift of the Fermi level relative to the Dirac point [74]). Various band structures have also been identified in stacking- and layer-number-dependent few-layer graphene systems [13, 34, 35, 72, 75, 91]. The separated valence and conduction Dirac cones, the multi-Dirac-cone structure, the vanishing graphene-like energy bands and the quasi-hole semimetallic energy spectra have been found in, respectively, silicene–Ag(111) and silicene–Ag(110) [101, 102], germanene–Au(111) [26, 103], silicene–ZrB_2 (0001) [104, 105], and tinene–Bi_2Te_3 (111) and Pb–Au(111) [106–108]. Experimental studies of the π band at the K and M points, the Γ-valley σ bands with double degeneracy, and their obvious mixing along KΓ and MΓ (figures 4.2(a)–(e)) can provide sufficient information about the dominance of sp^2 or sp^3 bonding, the significance of spin–orbit coupling and their relationships to the buckled and planar honeycomb lattices.

Importantly, spatial charge distributions are available for identifying and examining multi- and single-orbital hybridization due to various chemical bonds [100, 109–111]. Both the total charge density distribution and its variation, ρ and $\Delta\rho$ in figures 4.3(a)–(e) with (x, z)-plane projection, are calculated with care from the first-principles calculations, in which the latter is obtained by subtracting the superposition of that in an isolated atom. Monolayer graphene presents π and σ chemical bonding (the black and red rectangles in figure 4.3(a)), respectively, arising from the $2p_z$ and (2s, $2p_x$, $2p_y$) orbitals. Such results are further illustrated by the charge variation after crystal formation (figure 4.3(b)): the charge density is enhanced between two carbon atoms (in red, associated with σ bonding), while it is reduced above and below each carbon atom (in dark blue, associated with π bonding). The highest charge densities appear to arise between two identical carbon atoms on the (x, y)-, (x, z)- and (y, z)-planes, which is responsible for the creation of σ bonding. Specifically, graphene has the largest valence electron density with the shortest C–C bond length, compared to the other group-IV systems (figures 4.3(c)–(j)). The carbon honeycomb lattice displays the best mechanical response with the lightest weight [112], therefore it has been very successful in material engineering applications [112]. The very strong σ bonding is reflected directly in their collective excitation modes; that is, the σ-electronic plasmon modes have higher oscillation frequencies, e.g. σ optical plasmons of graphene have frequencies higher than 10 eV in EELS and optical measurements [113, 114]. In addition to the (x, y)-plane three-orbital carriers, the electronic valence states, which arise from the $2p_z$ orbitals, are extended along the z-direction to generate the effective distribution width. W_z is estimated to be ~0.35 Å from the (x, z)- and (y, z)-plane projections [115]. The free extension of the π bonding on the (x, z)- and (y, z)-planes is the critical picture for the low-energy fundamental properties of graphene systems, according to consistent theoretical and experimental studies, e.g. the gapless and semiconducting Dirac-cone band structure [116], doping-induced Landau damping and 2D plasmons [117, 118], unusual magneto-electronic Landau levels [43], specific magneto-optical selection rules [43, 44, 119, 120], rich magnetoplasmon modes [121, 122] and unique quantum Hall conductivities [123, 124].

Both π and σ chemical bonding, which survive in the Si-, Ge-, Sn- and Pb-related monolayer materials, become non-orthogonal to each other, as observed clearly in figures 4.3(c)–(j). Their bonding strengths decline as the atomic number of the group-IV materials grows. This leads to longer bond lengths (table 4.1). However, the opposite is true for the observable or significant sp^3 bonding in buckled systems. The effective distribution widths related to the $3p_z$, $4p_z$, $5p_z$ and $6p_z$ orbitals are estimated to be, respectively, about 0.37, 0.39, 0.41 and 0.45 Å, as obtained from the

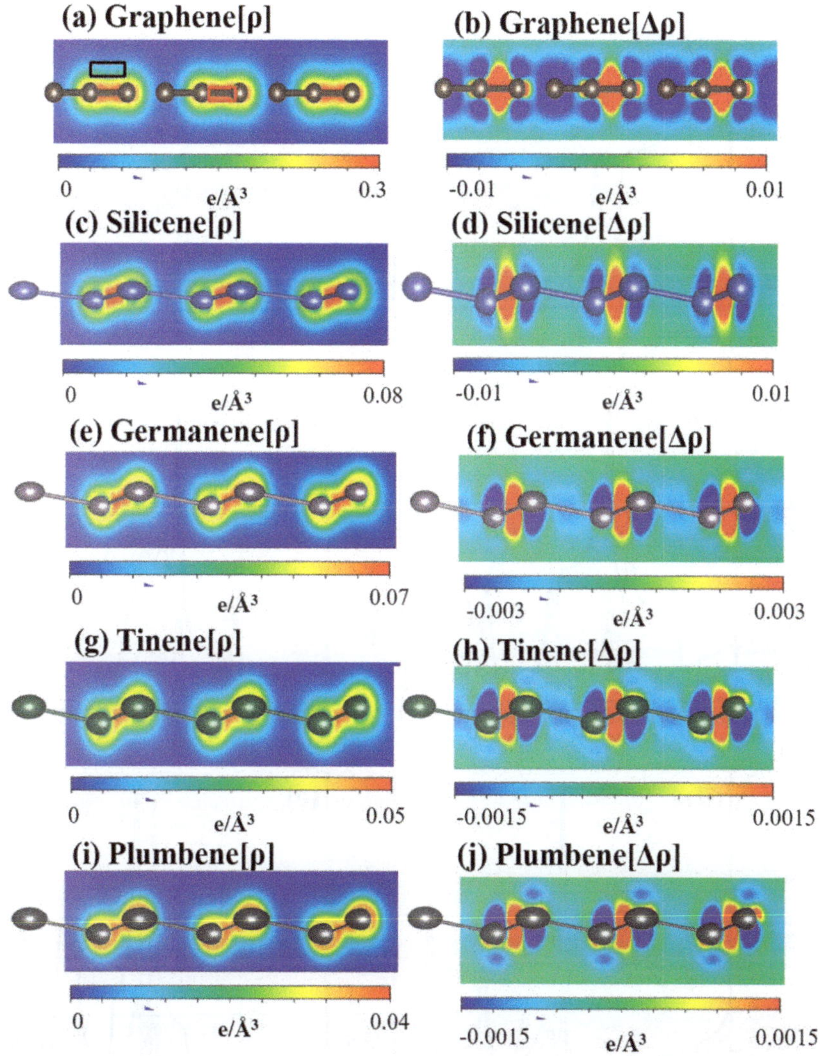

Figure 4.3. The spatial charge distributions (ρ) and their variation ($\Delta\rho$) for, respectively, monolayer (a) and (b) carbon-, (c) and (d) silicon-, (e) and (f) germanium-, (g) and (h) tin-, and (i) and (j) lead-related systems with (x, z)-plane projections.

(x, z)- and (y, z)-plane projections. They are roughly consistent with the probability distributions of isolated atomic orbitals. The total charge density can provide information on the A- and B-sublattice height difference (table 4.1), which is enhanced with an increase in atomic number. Large buckling indicates that it is relatively difficult to synthesize few-layer group-IV materials with heavy masses. To comprehend the fundamental properties fully, non-negligible sp^3 multi-orbital hybridization and spin–orbit coupling might have to be included in the tight-binding model simultaneously [31, 56, 78, 79].

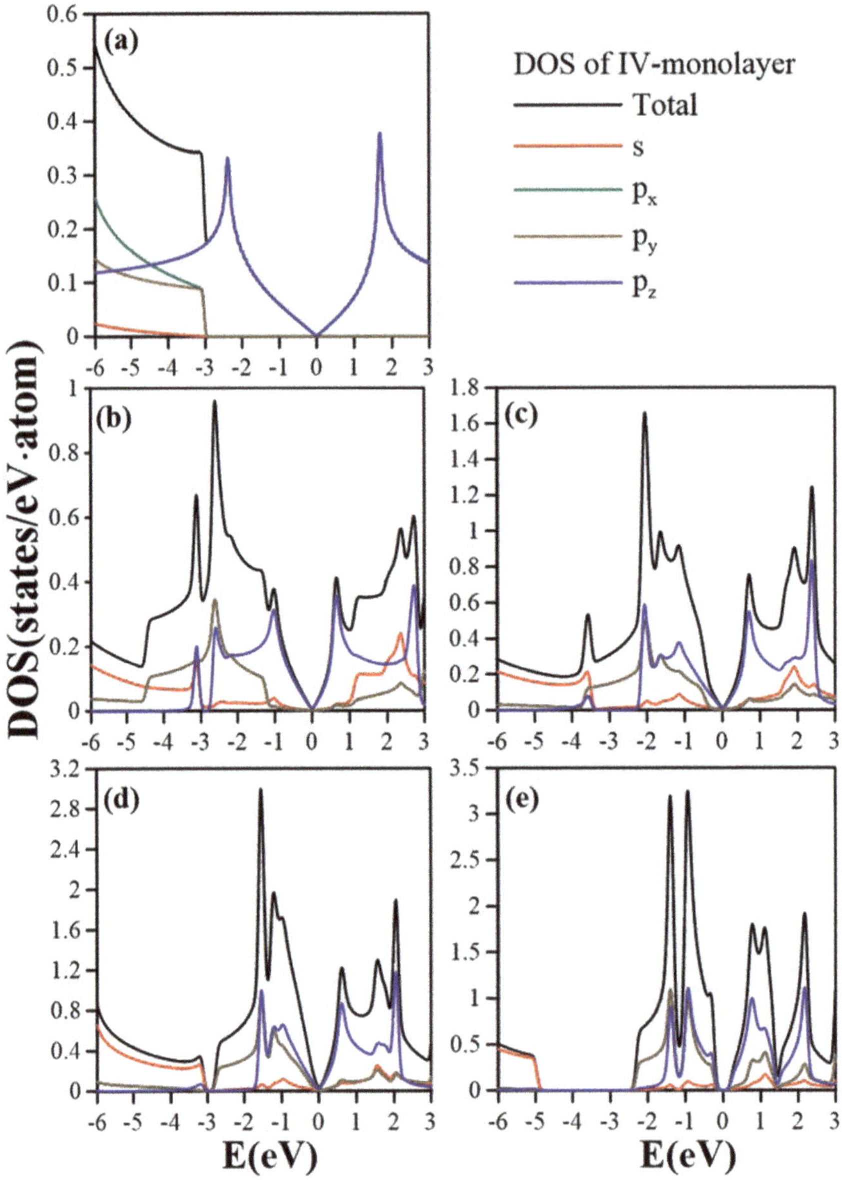

Figure 4.4. The four-orbital-projected density of states in monolayer (a) C-, (b) Si-, (c) Ge-, (d) Sn- and (e) Pb-generated systems.

The four-orbital-decomposed density of states is very useful in understanding the strong effects of multi-orbital hybridization on fundamental properties. The various band-edge states of electronic energy spectra, as illustrated clearly by the density of states in figures 4.4(a)–(e), can create diverse 2D van Hove singularities. Generally, the critical points in energy–wave-vector space, which appear in the energy bands of figures 4.2(a)–(e), include the Dirac-cone bottoms, the extreme states of parabolic dispersions, the saddle points and the constant-energy loops. Such electronic states generate, respectively, V-shaped structures, shoulders, logarithmically symmetric peaks and square-root-form asymmetric structures. For pristine monolayer graphene (figure 4.4(a)), its unusual density of states presents a linear energy dependence that is symmetric around the Fermi level, the ~−2.5 eV π peak and the ~1.90 eV π^* peak, and the ~−3.02 eV shoulder arising from the $2p_x$ and $2p_y$ orbitals. The energies of the symmetric π peak and the σ shoulder decline as the atomic number grows (figures 4.4(a)–(e)). The critical mechanisms are the reduced π bonding and the enhanced dominance of sp^3 bonding and spin–orbit coupling. The strong coupling of chemical and physical interactions further leads to the dramatic transformation of one into two shoulders in Sn and Pb (figures 4.4(d) and (e)). There exists an asymmetric peak after the formation of the σ bands, indicating clearly the mixing of orbital-dominated energy bands, such as the ~−1.52, −1.4, −1.3 and −1.2 eV peak structures for Si, Ge, Sn and Pb, respectively. This indicates the existence of significant sp^3 bonding. Studies using STS measurements on monolayer group-IV materials have been able to identify the main features for the π, sp^2 and sp^3 bonding, as well as the spin–orbit interaction.

4.2 Significant differences between monolayer group-IV systems

Monolayer group-IV materials, as indicated in the previous section, exhibit diverse low-lying energy bands because of buckled or planar honeycomb lattices, multi- or single-orbital hybridization, and significant or negligible spin–orbit coupling. The generalized tight-binding model [31, 42, 56, 68, 76–79], with reliable parameters in the presence of electric and magnetic fields, could be further developed to explore the varied fundamental properties fully through direct combination with the single- [52] and many-particle theories [53], such as optical absorption spectra and selection rules [43, 44], magneto-electronic properties [119, 120], Coulomb excitations and decay rates [5, 35, 52], and quantum Hall effects [50, 51, 123, 124].

Using the tight-binding model and the dynamic Kubo formula under the gradient approximation [56], monolayer tinene has been investigated for its optical properties, which are very useful in understanding the close relationships between the sp^3 bonding, spin–orbit coupling and perpendicular electric fields (gate voltages). Tinene (monolayer Sn) differs from germanene, silicene and graphene in its band structures and thus in other physical properties. Its vertical excitation spectra behave accordingly, owing mainly to the different strengths of orbital hybridization and spin–orbit interactions. The main features of the energy bands lie at the critical points (the band-edge states) and the orbital, spin and gate-voltage dependence. The critical points, which include the extreme and saddle points and the constant-energy loops, have unusual densities of states and van Hove singularities, which exhibit shoulders, and symmetric and asymmetric peaks in the density of states, respectively. The K/K′ valleys and the first pair of saddle points come mainly from the $5p_z$ orbitals (the π chemical bonding). $5p_x$, $5p_y$ and $5p_z$ co-dominate the constant-energy loops. The other critical points are related closely to the $5p_x$ and $5p_y$ orbitals. The spin-up- and spin-down-dominated energy bands in the deformed Dirac cones are split by an external electric field, leading to a non-uniform dependence of the band gap on the gate voltage. All the critical points create a lot of special absorption structures in optical spectra under the vertical transitions of intra- and inter-pairs of energy bands. The specific excitation structures, which are due to the first pair of energy bands, include the first and second shoulders, a symmetric peak in the logarithmic form and a square-root asymmetric peak, corresponding to the stable K and Γ valleys, M-saddle points and constant-energy loops, respectively. The Γ-induced low-frequency shoulder and the asymmetric peak are absent in graphene, silicene and germanene. The first shoulder becomes two splitting shoulders in the presence of a gate voltage, since the destruction of mirror symmetry about the (z = 0)-plane is combined directly with spin–orbit coupling. Moreover, there exist many absorption structures above the middle frequency, which are associated with the three pairs of energy bands. In addition, germanene and silicene exhibit gate-voltage-dependent splitting behavior, and all the group-IV systems have prominent π-electronic absorption peaks. The experimental measurements of energy bands [50, 80, 104], densities of states [60, 69, 125] and optical spectra [41, 42, 54, 59, 60, 75] can examine the orbital hybridization, the strength of spin–orbit coupling and the electric-field effects in 2D group-IV systems.

How to solve the various Hamiltonians with significant interactions is one of the basic topics in physics, in particular for magnetic quantization under phenomenological models (details in section 2.2). This is very interesting for comprehending fully the rich and unique physical phenomena of emergent 2D materials. Generally, such Hamiltonians include the strong coupling effects arising from multi-orbital hybridization, spin–orbit interactions, interlayer and intralayer hopping integrals, buckled and/or curved structures, layer numbers, stacking configurations, dimensions, magnetic fields and gate voltages, in which the external fields might be non-uniform and composite [61–66]. The critical mechanisms, which are related closely to lattice symmetries, intrinsic interactions, and external uniform or modulated fields, have been taken into account simultaneously through the generalized tight-binding model (details in section 2.2 [31, 68]). For example, systematic studies have been completed for 3D bulk graphite systems [65, 66], 2D few-layer graphene materials [72, 75, 91], and 1D carbon nanotubes [53] and graphene nanoribbons [63, 109, 126], clearly demonstrating varied magnetic quantization phenomena due to unusual Landau levels and magneto-optical selection rules. The generalized tight-binding model should be very helpful for the further development of various Hamiltonians in mainstream materials, such as 2D group-IV and group-V materials [56, 60, 120], topological insulators [127, 128] and MoS_2-related systems [129, 130].

Also, the low-energy effective-mass approximation (the perturbed form of the tight-binding model) is utilized frequently by theoretical research groups [31, 99, 100]. In general, the perturbation of electronic states around the high symmetry point is suitable only for condensed-matter systems with simple electronic structures, e.g. monolayer graphene, silicene and germanene [31, 99, 100], and AA- and AB-stacked graphene systems [52, 131, 132]. However, this method might be too coarse to handle the magnetically quantized electronic states from complicated band structures. For example, the low-lying valence and conduction bands, which have been demonstrated clearly in monolayer tinene (discussed below) [60], bilayer AA- and AB-bt silicene systems (discussed in detail in section 5.1) [54, 55, 133–136] and ABC- [39, 48] and AAB-stacked graphene materials [35], have multi-constant-energy loops and/or they originate from multi-orbital hybridization and significant spin–orbit coupling. The diverse quantization phenomena in such materials can be explored thoroughly using the generalized tight-binding model as well as the Kubo formulas and modified random-phase approximation, such as the rich and unique magneto-electronic states [43, 54–58], magneto-optical absorption spectra [119, 120], quantum Hall conductivities [123, 124], and inter-Landau-level excitations and magnetoplasmon modes [121, 122].

According to the calculated results in section 4.1, a buckled tinene honeycomb lattice possesses strong sp^3 chemical bonding and thus creates multi-degenerate energy bands (figure 4.2(d)). These intrinsic interactions, gate voltages (electric fields) and magnetic fields can be included fully in the simultaneous calculations of the generalized tight-binding model [61–66]. The magnetically quantized energy spectra and the sublattice-, orbital- and spin-dependent wave functions (subenvelope functions) could be computed very efficiently through the exact diagonalization method even for a rather large complex Hamiltonian matrix. Furthermore, the chemical and physical pictures are responsible for the varied magneto-electronic properties (and also for the magneto-optical absorption spectra and selection rules [43, 44, 119, 120]). Recently, calculated results have shown that tinene presents unusual magnetic quantization [60] which is absent in graphene, silicene and germanene. The multi-valley electronic structure, which correspond to the low-energy π and σ energy bands, is capable of generating two groups of low-lying Landau levels (LLs), with different orbital components, spin configurations, localization centers, state degeneracies, and magnetic- and electric-field dependence. The first and second groups are dominated by the $5p_z$ and ($5p_x$, $5p_y$) orbitals, respectively. The LL splittings in the first and second groups are due to the coupling effects of the perpendicular electric and magnetic fields, respectively. Specifically, LL anti-crossing appears only in the first group during variation of the electric field. The inter-atomic spin–orbit coupling between the $5p_z$ and ($5p_x$,$5p_y$) orbitals and the Coulomb potential difference between the A and B sublattices results in a dramatic probability redistribution of the opposite spin components and thus frequent LL anti-crossing behaviors. Similar diverse magneto-electronic phenomena are expected to be revealed in monolayer Pb. On the other hand, monolayer graphene, silicene and germanene present completely different phenomena, owing mainly to the π-bonding-dominated low-energy physics. There is only one group of well-behaved valence and conduction LLs, in which each localized oscillation mode has a specific zero-point number. When buckled silicene and gemanene are in the presence of a gate voltage, the electronic states and magneto-electronic states are split through co-operation of the spin–orbit coupling and the sublattice-dependent Coulomb potentials, as observed in the tinene and Pb systems. There exist valley-split or spin-split LL energy spectra, which are very sensitive to the strength of the electric and magnetic fields. The theoretical predictions for the diverse magneto-electronic properties of 2D group-IV materials can be verified using high-resolution measurements of magnetic van Hove singularities [93, 94], quantum Hall transport [49] and magneto-optical absorption spectra [43, 44, 119, 120]. In short, Sn-, Pb-, Si-, Ge- and C-related monolayer materials have different electronic properties, and the other fundamental properties also differ. Furthermore, their composite bilayer systems, with weak but significant van der Waals interactions, will induce unusual moiré-superlattice effects (non-uniform chemical and physical environments), as investigated thoroughly in chapter 12.

References

[1] Shao Y *et al* 2018 Epitaxial growth of flat antimonene monolayer: a new honeycomb analogue of graphene *Nano Lett.* **18** 2133

[2] Pham K D and Nguyen C V 2018 First principles calculations of the geometric structures and electronic properties of van der Waals heterostructure based on graphene, hexagonal boron nitride and molybdenum diselenide *Diam. Relat. Mater.* **88** 151

[3] Yaokawa R, Ohsuna T, Morishita T, Hayasaka Y, Spencer M J S and Nakano H 2016 Monolayer-to-bilayer transformation of silicenes and their structural analysis *Nat. Commun.* **7** 10657

[4] Shu H, Cao D, Liang P, Wang X, Chen X and Lu W 2014 Two-dimensional silicene nucleation on a Ag(111) surface: structural evolution and the role of surface diffusion *Phys. Chem. Chem. Phys.* **16** 304

[5] Shih P H, Chiu Y H, Wu J Y, Shyu F L and Lin M F 2017 Coulomb excitations of monolayer germanene *Sci. Rep.* **7** 40600

[6] Dávila M E and Lay G L 2016 Few layer epitaxial germanene: a novel two-dimensional Dirac material *Sci. Rep.* **6** 20714

[7] Zhu F F *et al* 2015 Epitaxial growth of two-dimensional stanene *Nat. Mater.* **14** 1020

[8] Liao M, Zang Y, Guan Z, Li H, Gong Y and Xue Q K 2018 Superconductivity in few-layer stanene *Nat. Photonics* **14** 344

[9] Ganz E, Hwang I S, Xiong F, Theiss S K and Golovchenko J 1991 Growth and morphology of Pb on Si(111) *Surf. Sci.* **257** 259

[10] Yakes M and Tringides M C 2011 Probing the buried Pb/Si(111) interface with SPA LEED and STM on Si(111)–Pb $\propto \sqrt{3} \times \sqrt{3}$ *J. Phys. Chem.* A **115** 7096

[11] Novoselov K S *et al* 2004 Electric field effect in atomically thin carbon film *Science* **306** 666

[12] Zhou L *et al* 2018 Surface structure of few layer graphene *Carbon* **136** 255

[13] Ping J and Fuhrer M S 2012 Layer number and stacking sequence imaging of by transmission electron microscopy *Nano Lett.* **12** 4635

[14] Warner J H, Mukai M and Kirkland A I 2012 Atomic structure of ABC rhombohedral stacked trilayer graphene *ACS Nano* **6** 5680

[15] Zhang F, Sahu B, Min H and MacDonald A H 2010 Band structure of ABC-stacked graphene trilayers *Phys. Rev.* B **82** 035409

[16] Vishwakarma R *et al* 2017 Transfer free graphene growth on SiO_2 substrate at 250 °C *Sci. Rep.* **7** 43756

[17] Li X *et al* 2010 Graphene films with large domain size by a two-step chemical vapor deposition process *Nano Lett.* **10** 4328

[18] Andzane J *et al* 2019 Effect of graphene substrate type on formation of Bi_2Se_3 nanoplates *Sci. Rep.* **9** 4791

[19] Osborne D A, Morishita T, Tawfik S A, Yayamab T and Spencer M J S 2019 Adsorption of toxic gases on silicene/Ag(111) *Phys. Chem. Chem. Phys.* **10** 1039

[20] Arafune R *et al* 2013 Structural transition of silicene on Ag(111) *Surf. Sci.* **608** 297

[21] Meng L *et al* 2013 Buckled silicene formation on Ir(111) *Nano Lett.* **13** 685

[22] Gill T G *et al* 2017 Metallic atomically-thin layered silicon epitaxially grown on silicene/ZrB_2 *2D Mater.* **4** 021015

[23] Wiggers F B, Bui H V, Friedlein R, Yamada-Takamura Y, Schmitz J and Jong M P 2016 A nitride-based epitaxial surface layer formed by ammonia treatment of silicene-terminated ZrB_2 *J. Chem. Phys.* **144** 134703

[24] Fleurence A and Yamada-Takamura Y 2017 Insights into the spontaneous formation of silicene sheet on diboride thin films *Appl. Phys. Lett.* **110** 041601

[25] Li L *et al* 2014 Buckled germanene formation on Pt(111) *Adv. Mater.* **26** 4820

[26] Cantero E D *et al* 2017 Growth of germanium on Au(111): formation of germanene or intermixing of Au and Ge atoms? *Phys. Chem. Chem. Phys.* **19** 18580

[27] Martinez E A, Fuhr J D, Grizzi O, Sánchez E A and Cantero E D 2019 Growth of germanene on Al(111) hindered by surface alloy formation *J. Phys. Chem.* C **123** 12910

[28] Chen S W, Wu C Y, Wu H J and Chiu W T 2014 Interfacial reactions in Sn/Bi_2Te_3, Sn/Bi_2Se_3 and Sn/Bi_2 $(Te_{1-x}Se_x)_3$ couples *J. Alloy Compd.* **611** 313

[29] Kim K C *et al* 2015 Sn doping in thermoelectric Bi_2Te_3 films by metal–organic chemical vapor deposition *Appl. Surf. Sci.* **353** 232

[30] Tsai W, Huang C, Chang T, Lin H, Jeng H and Bansil A 2013 Gated silicene as a tunable source of nearly 100% spin-polarized electrons *Nat. Commun.* **4** 1500

[31] Kaloni T P, Modarresi M, Tahir M and Roknabadi M R 2015 Electrically engineered band gap in two-dimensional Ge, Sn, and Pb: a first-principles and tight-binding approach *J. Phys. Chem.* C **119** 11896

[32] Guinea F, Katsnelson M and Geim A 2010 Energy gaps and a zero-field quantum Hall effect in graphene by strain engineering *Nat. Phys.* **6** 30

[33] Wong J H, Wu B R and Lin M F 2012 Strain effect on the electronic properties of single layer and bilayer graphene *J. Phys. Chem.* C **116** 8271

[34] Kamal C, Chakrabarti A, Banerjee A and Deb S K 2013 Silicene beyond mono-layers? Different stacking configurations and their properties *J. Phys. Condens. Matter* **25** 085508

[35] Lin C Y, Huang B L, Ho C H, Gumbs G and Lin M F 2018 Geometry-diversified Coulomb excitations in trilayer AAB stacking graphene *Phys. Rev.* B **98** 195442

[36] Que Y, Xiao W, Chen H, Wang D, Du S and Gao H J 2015 Stacking-dependent electronic property of trilayer graphene epitaxially grown on Ru (0001) *Appl. Phys. Lett.* **107** 263101

[37] Padiha J E, Lima M P, da Silva A J R and Fazzio A 2011 Bilayer graphene dual-gate nanodevice: an *ab initio* simulation *Phys. Rev.* B **84** 113412

[38] Wang Y P, Li X G, Fry J N and Cheng H P 2016 First-principles studies of electric field effects on the electronic structure of trilayer graphene *Phys. Rev.* B **94** 165428

[39] Yuan S, Roldn R and Katsnelson M I 2011 Landau level spectrum of ABA- and ABC-stacked trilayer graphene *Phys. Rev.* B **84** 125455

[40] Pamuk B, Baima J, Mauri F and Calandra M 2017 Magnetic gap opening in rhombohedral-stacked multilayer graphene from first principles *Phys. Rev.* B **95** 075422

[41] Bonaccorso F, Sun Z, Hasan T and Ferrari A C 2010 Graphene photonics and optoelectronics *Nat. Photonics* **4** 611

[42] Bravo S, Correa J, Chico L and Pacheco M 2018 Tight-binding model for opto-electronic properties of penta-graphene nanostructures *Sci. Rep.* **8** 11070
[43] Lin C Y, Wu J Y, Chiu Y H and Lin M F 2015 Magneto-electronic properties of multilayer graphenes *Phys. Chem. Chem. Phys.* **17** 26008
[44] Do T N, Shih P H, Chang C P, Lin C Y and Lin M F 2016 Rich magneto-absorption spectra in AAB-stacked trilayer graphene *Phys. Chem. Chem. Phys.* **18** 17597
[45] Koshino M 2013 Stacking-dependent optical absorption in multilayer graphene *New J. Phys.* **15** 015010
[46] Chiu C W and Chen R B 2016 Influence of electric fields on absorption spectra of AAB-stacked trilayer graphene *Appl. Phys. Express* **9** 065103
[47] Chen S C, Chiu C W, Wu C L and Lin M F 2014 Shift- enriched optical properties in bilayer graphene *RSC Adv.* **4** 63779
[48] Chiu C W, Huang Y C, Shyu F L and Lin M F 2011 Excitation spectra of ABC-stacked graphene superlattice *Appl. Phys. Lett.* **98** 261920
[49] Bao W *et al* 2011 Stacking-dependent band gap and quantum transport in trilayer graphene *Nat. Phys.* **7** 948
[50] Zhang Y, Tan Y W, Stormer H L and Kim P 2005 Experimental observation of the quantum Hall effect and Berry's phase in graphene *Nature* **438** 201
[51] Zhang L, Zhang Y, Camacho J, Khodas M and Zaliznyak I 2011 The experimental observation of quantum Hall effect of $l = 3$ chiral quasiparticles in trilayer graphene *Nat. Phys.* **7** 953
[52] Ho J H, Lu C L, Hwang C C, Chang C P and Lin M F 2006 Coulomb excitations in AA- and AB-stacked bilayer graphites *Phys. Rev.* B **74** 085406
[53] Shyu F L and Lin M F 2000 Loss spectra of graphite-related systems: a multiwall carbon nanotube, a single-wall carbon nanotube bundle, and graphite layers *Phys. Rev.* B **62** 8508
[54] Huang B, Deng H X, Lee H K, Yoon M, Sumpter B G and Wei S H 2014 Exceptional optoelectronic properties of hydrogenated bilayer silicene *Phys. Rev.* X **4** 021029
[55] Wang R, Wang S and Wu X 2014 The formation and electronic properties of hydrogenated bilayer silicene from first-principles *J. Appl. Phys.* **116** 024303
[56] Do T N *et al* 2019 Peculiar optical properties of bilayer silicene under the influence of external electric and magnetic fields *Sci. Rep.* **9** 36547
[57] Wang X and Wu Z 2017 Intrinsic magnetism and spontaneous band gap opening in bilayer silicene and germanene *Phys. Chem. Chem. Phys.* **19** 2148
[58] Do T N, Shih P H, Gumbs G, Huang D, Chiu C W and Lin M F 2018 Diverse magnetic quantization in bilayer silicene *Phys. Rev.* B **97** 125416
[59] Chen X *et al* 2016 The electronic and optical properties of novel germanene and antimonene heterostructures *J. Mater. Chem.* C **4** 5434
[60] Chen R B, Chen S C, Chiu C W and Lin M F 2017 Optical properties of monolayer tinene in electric fields *Sci. Rep.* **7** 1849
[61] Ohta T, Bostwick A, McChesney J L, Seyller T, Horn K and Rotenberg E 2007 Interlayer interaction and electronic screening in multilayer graphene investigated with angle-resolved photoemission spectroscopy *Phys. Rev. Lett.* **98** 206802
[62] Siegel D A, Regan W, Fedorov A V, Zettl A and Lanzara A 2013 Charge-carrier screening in single-layer graphene *Phys. Rev. Lett.* **110** 146802
[63] Bostwick A, Ohta T, Seyller T, Horn K and Rotenberg E 2007 Quasiparticle dynamics in graphene *Nat. Phys.* **3** 36

[64] Ruffieux P *et al* 2012 Electronic structure of atomically precise graphene nanoribbons *ACS Nano* **6** 6930

[65] Sugawara K, Sato T, Souma S, Takahashi T and Suematsu H 2006 Fermi surface and edge-localized states in graphite studied by high-resolution angle-resolved photoemission spectroscopy *Phys. Rev.* B **73** 045124

[66] Grüneis A *et al* 2008 Electron–electron correlation in graphite: a combined angle-resolved photoemission and first principles study *Phys. Rev. Lett.* **100** 037601

[67] Yu D, Lupton E M, Gao H J, Zhang C and Liu F 2008 A unified geometric rule for designing nanomagnetism in graphene *Nano Res.* **1** 497

[68] Kahl K and Kintscher N 2018 Geometric multigrid for the tight-binding Hamiltonian of graphene *SIAM J. Numer. Anal.* **56** 499

[69] Castro Neto A H, Guinea F, Peres N M R, Novoselov K S and Geim A K 2009 The electronic properties of graphene *Rev. Mod. Phys.* **81** 109

[70] Ohta T, Bostwick A, Seyller T, Horn K and Rotenberg E 2006 Controlling the electronic structure of bilayer graphene *Science* **313** 951

[71] Kim K S *et al* 2013 Coexisting massive and massless Dirac fermions in symmetry-broken bilayer graphene *Nat. Mater.* **12** 887892

[72] Sutter P, Hybertsen M S, Sadowski J T and Sutter E 2009 Electronic structure of few layer epitaxial graphene on Ru (0001) *Nano Lett.* **9** 2654

[73] Coletti C *et al* 2013 Revealing the electronic band structure of trilayer graphene on SiC: an angle-resolved photoemission study *Phys. Rev.* B **88** 155439

[74] Huang Y K, Chen S C, Ho Y H, Lin C Y and Lin M F 2014 Feature-rich magnetic quantization in sliding bilayer graphenes *Sci. Rep.* **4** 7509

[75] Lu C L, Chang C P, Huang Y C, Chen R B and Lin M F 2006 Influence of an electric field on the optical properties of few-layer graphene with AB stacking *Phys. Rev.* B **73** 144427

[76] Chiu Y H, Lai Y H, Ho J H, Chuu S and Lin M F 2008 Electronic structure of a two-dimensional graphene monolayer in a spatially modulated magnetic field: Peierls tight-binding model *Phys. Rev.* B **77** 045407

[77] Konschuh S, Gmitra M and Fabian J 2010 Tight-binding theory of the spin–orbit coupling in graphene *Phys. Rev.* B **82** 245412

[78] Fan Y, Liu X, Wang J, Ai H and Zhao M 2018 Silicene and germanene on InSe substrates: structures and tunable electronic properties *Phys. Chem. Chem. Phys.* **20** 11369

[79] Jiang L *et al* 2017 A tight binding and $\overleftarrow{k} \cdot \overleftarrow{p}$ study of monolayer stanene *Sci. Rep.* **7** 12069

[80] Sadeddine S *et al* 2017 Compelling experimental evidence of a Dirac cone in the electronic structure of a 2D silicene layer *Sci. Rep.* **7** 44400

[81] Balog R *et al* 2010 Bandgap opening in graphene induced by patterned hydrogen adsorption *Nat. Mater.* **9** 315

[82] Grassi R, Low T and Lundstrom M 2011 Scaling of the energy gap in pattern-hydrogenated graphene *Nano Lett.* **11** 4574

[83] Wang X, Song J and Qu J 2019 Antimonene: from experimental preparation to practical application *Angew. Chem. Int. Ed.* **58** 1574

[84] Peters J L *et al* 2018 Mono- and multilayer silicene-type honeycomb lattices by oriented attachment of PbSe nanocrystals: synthesis, structural characterization, and analysis of the disorder *Chem. Mater.* **30** 4831

[85] Du Y, Zhuang J, Wang J, Li Z, Liu H and Dou S X 2016 Quasi-freestanding epitaxial silicene on Ag(111) by oxygen intercalation *Sci. Adv.* **2** 1600067

[86] Deng J *et al* 2018 Epitaxial growth of ultraflat stanene with topological band inversion *Nat. Mater.* **17** 1081

[87] Molle A, Grazianetti C, Tao L, Taneja D, Alam M H and Akinwande D 2018 Silicene, silicene derivatives, and their device applications *Chem. Soc. Rev.* **47** 6370

[88] Singh R 2018 Spin–orbit coupling in graphene, silicene and germanene: dependence on the configuration of full hydrogenation and fluorination *Bull. Mater. Sci.* **41** 158

[89] Acun A *et al* 2015 Germanene: the germanium analogue of graphene *J. Phys. Condens. Matter* **27** 443002

[90] Xu Y *et al* 2013 Large-gap quantum spin Hall insulators in tin films *Phys. Rev. Lett.* **111** 136804

[91] Grüneis A *et al* 2008 Tight-binding description of the quasiparticle dispersion of graphite and few-layer graphene *Phys. Rev.* B **78** 205425

[92] Luican A *et al* 2011 Single-layer behavior and its breakdown in twisted graphene layers *Phys. Rev. Lett.* **106** 126802

[93] Li G *et al* 2010 Observation of van Hove singularities in twisted graphene layers *Nat. Phys.* **6** 109

[94] Cherkez V, de Laissardiere G T, Mallet P and Veuillen J Y 2015 van Hove singularities in doped twisted graphene bilayers studied by scanning tunneling spectroscopy *Phys. Rev.* B **95** 155428

[95] Lauffer P *et al* 2008 Atomic and electronic structure of few-layer graphene on SiC (0001) studied with scanning tunneling microscopy and spectroscopy *Phys. Rev.* B **77** 155426

[96] Yankowitz M, Wang F, Lau C N and LeRoy B J 2013 Local spectroscopy of the electrically tunable band gap in trilayer graphene *Phys. Rev.* B **87** 165102

[97] Pierucci D *et al* 2015 Evidence for flat bands near the Fermi level in epitaxial rhombohedral multilayer graphene *ACS Nano* **9** 5432

[98] Zhou X Y *et al* 2015 Landau levels and magneto-transport property of monolayer phosphorene *Sci. Rep.* **5** 12295

[99] Bena C and Montambaux G 2009 Remarks on the tight-binding model of graphene *New J. Phys.* **11** 095003

[100] Lin S Y, Chang S L, Thuy Tran N T, Yang P H and Lin M F 2015 H–Si bonding-induced unusual electronic properties of silicene: a method to identify hydrogen concentration *Phys. Chem. Chem. Phys.* **17** 26443

[101] Wang X, Liu H and Tu S T 2015 Study of formaldehyde adsorption on silicene with point defects by DFT method *RSC Adv.* **5** 65255

[102] Gao N, Lu G Y, Wen Z and Jiang Q 2017 Electronic structure of silicene: effects of the organic molecular adsorption and substrate *J. Mater. Chem.* C **5** 627

[103] Ni Z, Minamitani E, Andoa Y and Watanabe S 2015 The electronic structure of quasi-free-standing germanene on monolayer MX (M = Ga, In; X = S, Se, Te) *Phys. Chem. Chem. Phys.* **17** 19039

[104] Lee C C, Fleurence A, Yamada-Takamura Y, Ozaki T and Friedlein R 2014 Band structure of silicene on zirconium diboride (0001) thin-film surface: convergence of experiment and calculations in the one-Si-atom Brillouin zone *Phys. Rev.* B **90** 075422

[105] Lee C C, Fleurence A, Friedlein R, Yamada-Takamura Y and Ozaki T 2013 First-principles study on competing phases of silicene: effect of substrate and strain *Phys. Rev.* B **88** 165404

[106] Michiardi M *et al* 2014 Bulk band structure of Bi_2Te_3 *Phys. Rev.* B **90** 075 105

[107] Li A H *et al* 2010 Electronic structure and thermoelectric properties of Bi_2Te_3 crystals and graphene-doped Bi_2Te_3 *Thin Solid Films* **518** 57
[108] Crepaldi A *et al* 2013 Combined ARPES and STM study of Pb/Au(111) moiré structure: one overlayer, two symmetries *Phys. Rev.* B **87** 115138
[109] Nguyen D K, Lin Y T, Lin S Y, Chiu Y H, Thuy Trana N T and Lin M F 2017 Fluorination-enriched electronic and magnetic properties in graphene nanoribbons *Phys. Chem. Chem. Phys.* **19** 20667
[110] Lin S Y, Lin Y T, Thuy Tran N T, Su W P and Lin M F 2017 Feature-rich electronic properties of aluminum-adsorbed graphenes *Carbon* **120** 209
[111] Thuy Tran N T, Lin S Y, Lin Y T and Lin M F 2016 Chemical bonding-induced rich electronic properties of oxygen adsorbed few-layer graphenes *Phys. Chem. Chem. Phys.* **18** 4000
[112] Zhang X, Rajaraman B R S, Liu H and Ramakrishna S 2014 Graphene's potential in materials science and engineering *RSC Adv.* **4** 28987
[113] Johari P and Shenoy V B 2011 Modulating optical properties of graphene oxide: role of prominent functional groups *ACS Nano* **5** 7640
[114] Lundie M, Sljivancanin Z and Tomic S 2015 Electronic and optical properties of reduced graphene oxide *J. Mater. Chem.* C **3** 7632
[115] Wang J, Ma F and Sun M 2017 Graphene, hexagonal boron nitride, and their heterostructures: properties and applications *RSC Adv.* **7** 16801
[116] Novoselov K S *et al* 2007 Electronic properties of graphene *Phys. Status Solidus* **244** 4106
[117] Langer T, Baringhaus J, Pfnür H, Schumacher H W and Tegenkamp C 2010 Plasmon damping below the Landau regime: the role of defects in epitaxial graphene *New J. Phys.* **12** 033017
[118] Tegenkamp C, Pfnür H, Langer T, Baringhaus J and Schumacher H W 2015 Plasmon electron–hole resonance in epitaxial graphene *J. Phys. Condens. Mat.* **23** 012001
[119] Ho Y H, Chiu Y H, Lin D H, Chang C P and Lin M F 2010 Magneto-optical selection rules in bilayer Bernal graphene *ACS Nano* **4** 1465
[120] Wu J Y, Chen S C, Do T N, Su W P, Gumbs G and Lin M F 2018 The diverse magneto-optical selection rules in bilayer black phosphorus *Sci. Rep.* **8** 13303
[121] Politano A and Chiarello G 2014 Plasmon modes in graphene: status and prospect *Nanoscale* **6** 10927
[122] Jiao N, Kang S, Han K, Shen X and Wang W 2019 Shape effects on graphene magneto-plasmons *Phys. Rev.* B **99** 195447
[123] Kiryu T and Koshino M 2019 Quantum Hall effect in three-dimensional graphene *Phys. Rev.* B **99** 085443
[124] Cysne T P, Garcia J H, Rocha A R and Rappoport T G 2018 Quantum Hall effect in graphene with interface-induced spin–orbit coupling *Phys. Rev.* B **97** 085413
[125] Ali M, Pi X, Liu Y and Yang D 2017 Electronic and magnetic properties of graphene, silicene and germanene with varying vacancy concentration *AIP Adv.* **7** 045308
[126] Chung H C, Su W P and Lin M F 2013 Electric-field-induced destruction of quasi-Landau levels in bilayer graphene nanoribbons *Phys. Chem. Chem. Phys.* **15** 868
[127] Song K, Soriano D, Cummings A W, Robles R, Ordejón P and Roche S 2018 Spin proximity effects in graphene/topological insulator heterostructures *Nano Lett.* **18** 2033
[128] Ezawa M 2015 Monolayer topological insulators: silicene, germanene and stanene *J. Phys. Soc. Jpn* **84** 121003

[129] Ridolfi E, Le D, Rahman T S, Mucciolo E R and Lewenkopf C H 2015 A tight-binding model for MoS_2 monolayers *J. Phys. Condens. Mat.* **23** 365501
[130] Zahid F, Liu L, Zhu Y, Wang J and Guo H 2013 A generic tight-binding model for monolayer, bilayer and bulk MoS_2 *AIP Adv.* **3** 052111
[131] Chiu C W, Lee S H, Chen S C, Shyu F L and Lin M F 2010 Absorption spectra of AA-stacked graphite *New J. Phys.* **12** 083060
[132] Lu C L, Chang C P, Huang Y C, Ho J H, Hwang C C and Lin M F 2007 Electronic properties of AA- and ABC-stacked few-layer graphites *J. Phys. Soc. Jpn* **76** 024701
[133] Mohan B, Kumar A and Ahluwalia P K 2013 A first principles calculation of electronic and dielectric properties of electrically gated low-buckled mono and bilayer silicene *Physica* E **53** 233
[134] Fu H, Zhang J, Ding Z, Li H and Meng S 2014 Stacking-dependent electronic structure of bilayer silicene *Appl. Phys. Lett.* **104** 131904
[135] Liu Y, Shu H, Liang P, Cao D, Chen X and Lu W 2013 Structural, electronic, and optical properties of hydrogenated few-layer silicene: size and stacking effects *J. Appl. Phys.* **114** 094308
[136] Liu J and Zhang W 2013 Bilayer silicene with an electrically-tunable wide band gap *RSC Adv.* **3** 21943

Shih-Yang Lin, Hsin-Yi Liu, Duy Khanh Nguyen, Ngoc Thanh Thuy Tran, Hai Duong Pham, Shen-Lin Chang, Chiun-Yan Lin and Ming-Fa Lin

Chapter 5

Stacking-configuration-enriched fundamental properties in bilayer silicenes

The emergent two-dimensional materials have attracted a lot of experimental and theoretical attention since the first discovery of few-layer graphene systems using mechanical exfoliation in 2004 [1]. To date, the mono-element group-IV and group-V systems include silicene [2], germanene [3], tinene [4], phosphorene [5], antimonene [6, 7] and bismuthene [7, 8]. Such layered systems are highly suitable for studying diverse physical, chemical and material phenomena, owing mainly to their various Hamiltonians with rich and unique intrinsic geometric symmetries and atomic interactions. Furthermore, they are expected to have great potential for applications such as nano-electronics [9–17], optoelectronics [8, 18–24] and energy storage [25–38]. How to determine and comprehend the fundamental properties thoroughly in atom-dependent layered materials has become a mainstream topic. The critical factors, based on important concepts, appear to originate mainly from the planar and non-planar lattices, stacking configurations, layer numbers, intralayer and interlayer multi- and single-orbital hybridization, and spin–orbit coupling. The focus of the current work is determining the diverse fundamental properties in bilayer silicene systems through the relative shifts between two layers, for which first-principles calculations with the spin configurations are available for carrying out thorough investigations.

Recently, few-layer germanene, silicene, stanene (tinene) and plumbene have been synthesized successfully on various substrate surfaces, e.g. Si on Ag(111), Ir(111) and ZrB_2 surfaces [39–45], Ge on Pt(111), Au(111) and Al(111) surfaces [46–48], Sn on Bi_2Te_3 surfaces [49, 50] and Pb on Si(111) surfaces [51, 52]. Such group-IV layered materials possess observable buckled honeycomb lattices, as observed in images obtained using high-resolution tunneling electron microscopy [2, 53, 54]. This indicates significant mixing of sp^2 and sp^3 orbital hybridization, since the relatively weak chemical bonding between larger atoms cannot maintain a planar structure.

doi:10.1088/978-0-7503-3299-6ch5

On the other hand, graphene systems present planar geometries because of their dominant σ bonding [55]. In addition to layer numbers, stacking configurations will also play a critical role in the various fundamental properties. To date, AA [56], AB [57], ABC [58–60], AAB [61] and twisted [62] stacking has been identified in 2D graphenes. The other group-IV systems are expected to have more complicated stacking symmetry structures when using experimental growth methods as a result of buckling effects, as shown in the detailed analysis of both the AB-bt and AB-bb configurations in bilayer silicene systems using high-angle annular dark-field scanning transmission electron microscopy [2, 54].

Theoretical models are utilized extensively to explore fully the diverse phenomena in emergent group-IV materials, e.g. the first-principles method, tight-binding model and effective-mass approximation for the geometric [63–75], electronic [76–83], optical [18, 22, 82–85] and magnetic properties [62, 72, 85–89]. Interestingly, monolayer graphene is a zero-gap semiconductor with a linear Dirac-cone structure across the Fermi level [89], as verified using angle-resolved photoemission spectroscopy (ARPES) [90]. However, the other group-IV single-layer systems are predicted to exhibit finite-gap semiconducting behaviors under modified and separated valence and conduction cones [74]. The main mechanisms underlying this are the significant spin–orbit coupling and sp^3 bonding, in particular the spin–orbit coupling. The energy gaps grow with the increase of atomic number because of the stronger spin–orbit interactions. Within the theoretical predictions, there exist a lot of meta-stable stacking configurations in few-layer systems. For example, the sliding bilayer silicenes and graphenes [91, 92] will exhibit important differences in terms of chemical bonding and spin configurations, leading to orbital-varied band properties (finite-gap semiconductors, zero-gap semiconductors, semimetals and metals) and magnetic configurations (non-magnetism, ferromagnetism and anti-ferromagnetism).

A theoretical framework based on first-principles calculations has been constructed successfully for the fundamental properties of 2D layered graphenes [58, 61, 93–95] and 1D graphene nanoribbons [15, 72, 96, 97]. The critical mechanisms, layer numbers, stacking configurations, planar, curved and folded surfaces, edge structures, quantum confinement and adatom-enriched chemisorption are explored fully using sensitive analyses. Concise physical and chemical pictures are proposed to account for the optimal geometries, rich electronic properties and diverse magnetic configurations. These pictures include multi- and single-orbital hybridization in chemical bonds, and host- and/or guest-atom-induced spin distributions. These theoretical models are further developed for the emergent silicene-related systems. For example, the significant differences between the sliding bilayer silicene and graphene systems [91, 92] are worthy of systematic investigation, and are due mainly to the intrinsic geometric structures, intralayer and interlayer hopping integrals, and spin–orbit interactions.

The diverse fundamental properties in sliding bilayer silicene systems are investigated thoroughly using first-principles calculations. The evaluated results include the significant dependence of the ground state energy, lattice constant, inter-sublattice distances, buckling angles, electronic energy spectra, energy gaps, wave functions, spatial charge densities, densities of states, magnetic moments and spin density distributions on the relative shift between the upper and lower layers. Furthermore, the layer-, sublattice-, orbital- and spin-decomposed physical quantities are very useful in obtaining the critical orbital hybridization and the magnetic configurations. Specifically, zero-gap semiconductors, finite-gap semiconductors and semimetals will be identified clearly in the 3D band structures and the specific densities of states at the Fermi level. The non-magnetic, ferromagnetic and anti-ferromagnetic configurations are also examined fully using the spin-dependent energy bands, moments, densities and densities of states. Specifically, the valley structures, which are initiated from the various band-edge states (the critical points) in the energy–wave-vector space, are proposed to comprehend the main features of the valence and conduction bands. A detailed comparison of the sliding bilayer silicenes and graphenes is made. The theoretical predictions for the geometric structures, valence bands and densities of states can be verified using, respectively, high-resolution measurements using scanning tunneling microscopy (STM) and tunneling electron microscopy (TEM) [2, 98–101], ARPES [76, 90, 102–106], and scanning tunneling spectroscopy (STS) [107–115].

5.1 The geometric, electronic and magnetic properties of sliding bilayer silicenes

Sliding bilayer silicene materials are created through a relative shift between two subsystems, where δ is in units of the lattice constant ($\sqrt{3}\,b = 2.32$ Å and $b = 2.48$ Å). The typical sliding types are chosen as follows (figure 5.1): (a) $\delta_a = 0$ (AA-bb), (b) $\delta_a = 1/8$, (c) $\delta_a = 3/8$, (d) $\delta_a = 4/8$, (e) $\delta_a = 6/8$, (f) $\delta_a = 8/8$ (AB-bt), (g) $\delta_a = 11/8$ and (h) $\delta_a = 12/8$ along the armchair direction, and (i) $\delta_z = 1/8$ and (j) $\delta_z = 3/8$ along the zigzag direction. There are four silicon atoms in the unit cell of each stacking configuration. They have lower ground state energies (−20.30 to −20.05 eV per unit cell in table 5.1), compared to that of monolayer silicene (–9.57 eV per unit cell; table 4.1). This indicates significant attractive forces between the two monolayer subsystems. Among the different stacking configurations, the AA $\delta_a = 3/8$ configuration shows the lowest total energy, suggesting the most stable geometric structure [81]. The height difference between the intralayer A and B sublattices is enhanced greatly, from 0.48 Å in the monolayer (table 4.1) to 0.66–0.88 Å in the bilayer. Moreover, the interlayer distance of ~2.43–2.72 Å is shorter than the 3.52 and 3.26 Å for the AA- and AB-stacked bilayer graphenes, respectively. The buckled angle, which is defined as that between the position vector of the nearest A and B sublattices $\overleftarrow{R_{AB}}$ and the x-axis, is in the range $15.43 \leqslant \theta < 22.0$. Of all the buckled honeycomb lattices, the AB-bt and AA-bb stacking configurations present, respectively, the lowest and highest buckling. In short, the increased buckling and the reduced interlayer distance in bilayer silicene systems are expected to induce more prominent multi-orbital hybridization of Si-3s, -3p$_x$, -3p$_y$ and -3p$_z$.

Monolayer silicene exhibits an unusual band structure, as shown in figure 4.2(b). The occupied valence bands are asymmetric to the unoccupied conduction bands around the Fermi level (E_F). The first pair of energy bands, which is measured from E_F, possesses almost linear energy dispersions with a small gap ~1.5 meV. That is,

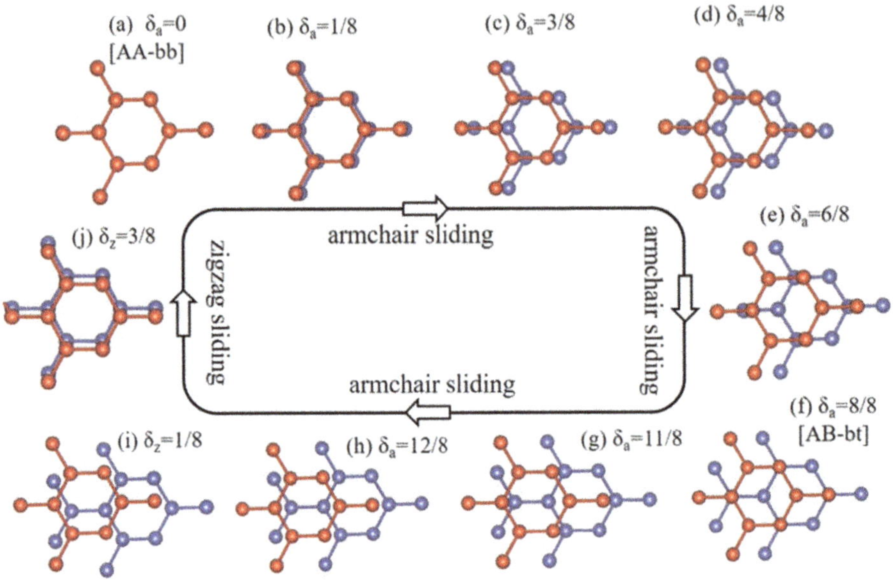

Figure 5.1. The relative shifts between two buckled silicene systems along the armchair and zigzag directions: (a) $\delta_a = 0$ (in units of lattice constant; AA-bb), (b) $\delta_a = 1/8$, (c) $\delta_a = 3/8$, (d) $\delta_a = 4/8$, (e) $\delta_a = 6/8$, (f) $\delta_a = 8/8$ (AB-bt), (g) $\delta_a = 11/8$ and (h) $\delta_a = 12/8$ (armchair), and (i) $\delta_z = 1/8$ and (j) $\delta_z = 3/8$ (zigzag).

Table 5.1. The geometric properties of sliding bilayer silicenes: lattice constant, interlayer distance (Δd), height difference (Δz), between the intralayer A and B sublattices, buckling angles (θ), Si–Si bond lengths (b) and ground state energies per unit cell.

Bilayer silicene	Lattice constant (Å)	Δd (Å)	Δz (Å)	θ (°)	b (Si–Si) (Å)	Ground state energy (eV/unit cell)
$\delta_a = 0$ (AA-bb)	3.79	2.72	0.88	21.89	2.36	–20.19
$\delta_a = 1/8$	3.84	2.69	0.84	20.85	2.36	–20.19
$\delta_a = 3/8$	3.84	2.44	0.78	19.30	2.36	–20.30
$\delta_a = 4/8$	3.84	2.43	0.73	18.10	2.35	–20.27
$\delta_a = 6/8$	3.84	2.48	0.68	17.04	2.32	–20.24
$\delta_a = 8/8$ (AB-bt)	3.84	2.48	0.66	15.43	2.48	–20.25
$\delta_a = 11/8$	3.84	2.52	0.69	17.23	2.33	–20.15
$\delta_a = 12/8$	3.84	2.57	0.70	17.33	2.35	–20.05
$\delta_z = 1/8$	3.84	2.53	0.72	17.84	2.35	–20.10
$\delta_z = 3/8$	3.84	2.60	0.84	20.67	2.38	–20.24

the bands have slightly separated Dirac-cone structures (stable upward and downward Dirac-cone valleys from the K/K′ point). The low-lying electronic states, which are related closely to the fundamental physical properties, are identified to arise from the π bonding of $3p_z$ orbitals (see figure 5.4(a) for the density of states). This further confirms the dominant sp^2 bonding in monolayer buckled silicene. An anisotropic energy spectrum is revealed clearly with the increase of state energy. For example, distinct state energies of $|E^{c,v}|$ are revealed along MK and KΓ, in which the saddle valence M-point has $E^v \sim -1.01$ eV, corresponding to the nearest hopping integral in the tight-binding model [67]. The total π band width within the composite MKΓ structure is about 1.20 eV, which is much smaller than those (>6.5 eV) in any graphene system [116]. Interestingly, the other valence and conduction bands are initiated from the Γ valleys, in particular for the second and third valence bands. These occupied valence states are predicted to be dominated by the $3p_x$, $3p_y$ and 3s orbitals (figure 4.2(b)). It should be noted that the degenerate initial energy ($E^v \sim -1.20$ eV) of the second and third valence bands is very close to the saddle-point energy of the first valence band. Furthermore, the σ-band widths are ~3.21–3.41 eV for the energy spectrum along the ΓMK direction. The above-mentioned band structure shows that most band-edge states appear at K, M and Γ, while some of them arise between the high-symmetry points due to band mixing. For example, the band-edge states, which occur at $E = -2.7$ and -3.2 eV, can form constant-energy loops arising from π–σ band mixing, or they are able to create quasi-1D parabolic energy dispersions. Such electronic states belong to the critical points in the energy–wave-vector space and might be revealed as the local maxima, minima, saddle points and constant-energy loops. Diverse van Hove singularities are thus expected to exist in the density of states.

Bilayer silicene systems are completely different to monolayer systems, as illustrated clearly by the detailed comparison in figures 5.2(b)–(l) and 5.2(a). The important differences originate mainly from the greatly enhanced buckling, thus leading to giant interlayer hopping integrals, more prominent sp^3 bonding, and even layer- and sublattice-dependent spin–orbit coupling in some configurations (the critical mechanisms using the phenomenological tight-binding model in [87]). Most importantly, the strongly modified electronic structures include the stacking-enriched valence and conduction bands, the semiconducting and semimetallic behaviors, the obvious electron–hole asymmetric energy spectra around the Fermi level, more oscillatory (non-uniform) valence and conduction bands even close to E_F, frequent band crossing and anti-crossing, and many band-edge states and valley structures. As a result, it is very difficult to characterize the stable valleys due to the initial and final π and σ bands; that is, the separation of π and σ bonding would become meaningless. Pure π bonding, which comes from the $3p_z$ orbital, might be not suitable for any low-energy fundamental properties in the phenomenological models.

The entire band structures are varied greatly by the relative shift between two silicene sublayers. The $\delta_a = 0$ system, as shown in figure 5.2(b), presents six valence and four conduction bands, which is double those of monolayer silicene (figure 5.2(a)). The small overlaps of the valence and conduction bands indicate semimetallic behavior, e.g. the overlaps at the Γ point. The valence states above the Fermi level possess free holes while the conduction states below the Fermi level possess electrons. Three valence bands and one conduction band, with oscillatory energy dispersions, are very close to E_F. This result illustrates clearly the significant

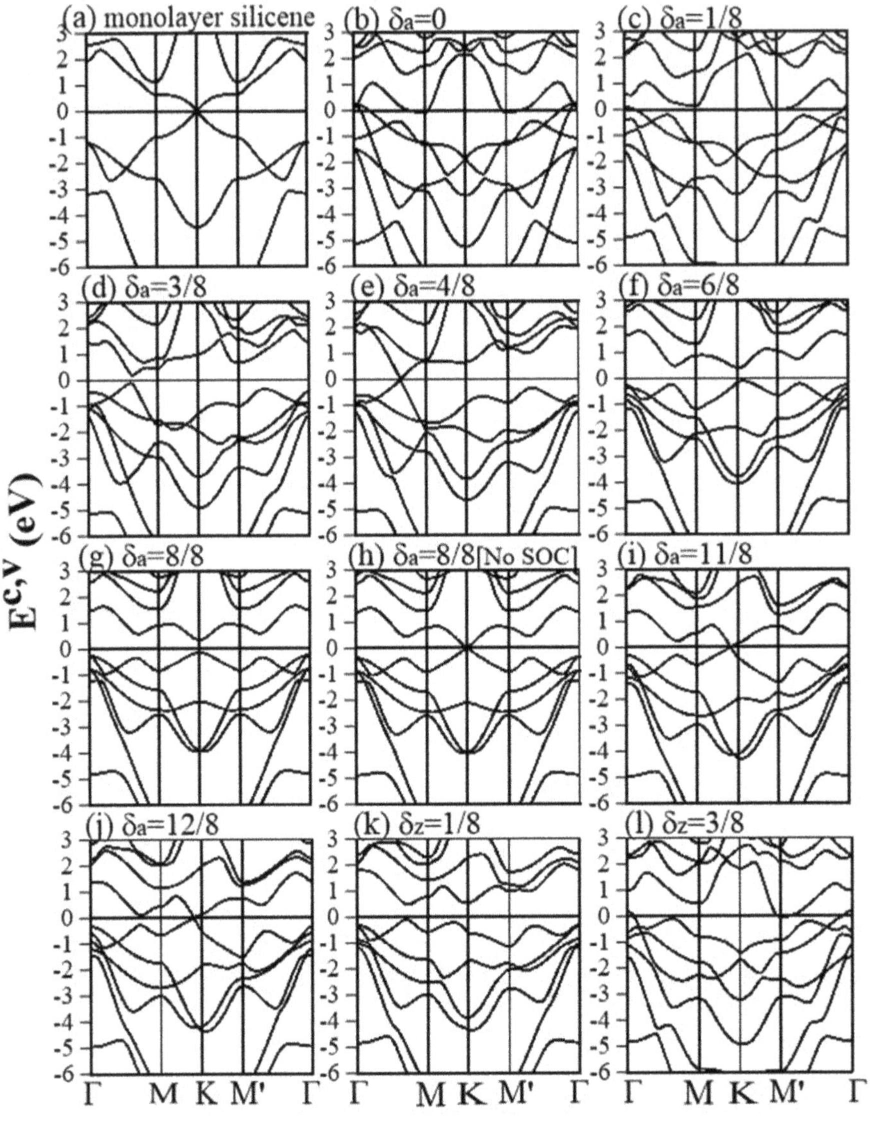

Figure 5.2. Electronic energy spectra of sliding bilayer silicene systems for the stacking configurations: (b) $\delta_a = 0$, (c) $\delta_a = 1/8$, (d) $\delta_a = 3/8$, (e) $\delta_a = 4/8$, (f) $\delta_a = 6/8$, (g) $\delta_a = 8/8$, (i) $\delta_a = 11/8$ and (j) $\delta_a = 12/8$ along the armchair direction, and (k) $\delta_z = 1/8$ and (l) $\delta_z = 3/8$ along the zigzag direction. Electronic energy spectra of (a) monolayer silicene and (h) AB-bt without spin-dependent interactions.

contributions of π and σ electrons to the low-energy fundamental properties. Specifically, the first conduction band presents a partially flat energy dispersion near the M and M′ points, with a high density of states (see figure 5.4(b)). Constant-energy loops, which are associated with band mixing, might arise along ΓM/ΓM′ and KM/KM′. Saddle M-point structures are revealed in all energy bands, thus leading to prominent van Hove singularities. The Γ and K valleys are parabolic structures, and may be meaningless in characterizing the band widths of π and σ bonding. The composite valley structures, similar to those of monolayer silicene (the KMΓ and ΓMK valleys due to, respectively, the π and σ bands in figure 4.2(b)), cannot survive in this bilayer system. The outer four orbitals of silicon atoms need to be taken into account simultaneously for multi-orbital hybridization. The above-mentioned electronic energy spectra are doubly degenerate in the absence of spin configurations. The spin-dependent interactions are fully negligible in bilayer AA-bb stacking, owing mainly to the highest-symmetry stacking and the very strong interlayer interactions. In addition, another AA-bt stacking presents a completely different band structure, in which there exist Dirac-cone structures along the MK and M Γ directions based on π-band evidence [60, 79].

The electronic energy spectra are very sensitive to δ_a and δ_z. For $\delta_a = 1/8$ in figure 5.2(c), the bilayer system retains the semimetallic property near the Fermi level, while the main features of the electronic structures present diverse behaviors. The band overlap of the valence and conduction states occurs at different wave vectors, not only the Γ point. Specifically, the state degeneracy of the high-symmetry M and M′ points is destroyed completely through variation of the stacking configuration. That is to say, the obvious change from a six-fold to a two-fold rotational symmetry results in their energy splitting with a wide range of ~0.5 eV. Only the saddle M- and M′-point structures are capable of accumulating a lot of electronic states (the density of states in figure 5.4(c)). Therefore, they are expected to exhibit very prominent optical absorption peaks. This phenomenon can also be found in carbon-related sp^2-bonding systems, such as layered graphenes [117–122], carbon nanotubes [123–125] and graphite [126–131]. It should be noted that the electronic states in some energy bands create parabolic valleys near the M/M′ points. All the band-edge states appear at the Γ, M, M′ and K points and between them. Similar semimetallic behavior is revealed as a weak band overlap under $\delta_a = 11/8$, $\delta_a = 12/8$, $\delta_z = 1/8$ and $\delta_z = 3/8$ (figures 5.2(i), (j), (k) and (l), respectively). The other stacking configurations belong to the zero-, direct- and indirect-gap semiconductors corresponding to, respectively, $\delta_a = 4/8$, 6/8 and 8/8 without the spin configuration, and $\delta_a = 3/8$ and $\delta_a = 8/8$ with the spin configuration (figures 5.2(e)–(g)). Also note that the former present modified Dirac-cone structures (figures 5.2(e), (f) and (h)). Specifically, AB-bt bilayer silicene, as shown in figures 5.2(g) and (h), exhibits unusual energy bands. This system has six-fold rotational symmetry even under enhanced buckling, creating degenerate M and M′ points. In the energy range of $|E^{c,v}| < 0.5$ eV, the band structures, with and without spin-dependent interactions, are quite different from each other. These results suggest clearly that spin–orbit coupling is enhanced greatly by the AB-bt stacking configuration. Its strength is estimated to be ~0.5 eV. There exists an indirect band gap of $E_g \sim 290$ meV which corresponds to the lowest unoccupied conduction state at the K point and the highest occupied valence state along the KΓ direction. The first pair of energy bands appears to present non-uniform dispersion relations. To simulate such unique energy spectra, very strong interlayer hopping integrals (larger than the intralayer integral) and layer-dependent spin–orbit interactions need to be included in the $3p_z$-orbital tight-binding model (details in [85]). In addition, the weak overlap of the first pair of energy bands occurs near the K point in the absence of spin–orbit coupling (figure 5.2(h)). These two bands are deduced to be dominated by the π-electronic states; that is, they are almost independent of σ orbitals. Their low-energy physical properties depend little on the sp^3 bonding, e.g. the magnetic quantization [87] and magneto-optical selection rules [85]. However, the opposite is true for the other stacking configurations.

High-resolution ARPES is the only method available for identifying the strong wave-vector dependence of the occupied electronic energy spectra and quasiparticle decay rates (details in section 3.3). The study of theoretical predictions for layered graphene systems has shown much progress, such as the linear valence Dirac cone in monolayer systems [57, 58], n-type doping by alkali absorption [132, 133], two parabolic valence bands in bilayer AB stacking [57], coexistent linear and parabolic dispersions in twisted bilayer systems [62], linear and parabolic bands in tri-layer ABA stacking [60], partially flat, sombrero-shaped and linear dispersions in tri-layer ABC stacking [58, 59], semimetal–semiconductor transitions in adatom- and molecule-adsorbed graphenes [132, 133], and substrate-modified low-lying energy bands in bilayer AB stacking [112] and few-layer ABC stacking [113]. Moreover, the experimental verification of the other group-IV layered materials includes the modified Dirac-cone structure with $E_g \sim 0.6$ eV for Si-adlayer–Ag [111] and [110] [39–41], its thorough absence in silicene–ZrB_2 (0001) [43, 44], the multi-Dirac cones of few-layer germanene on Au[111], and the metallic and semimetallic behaviors of bilayer-Sn–Bi_2Te_3 [111] and monolayer-Pb–Au(111). The significant interactions with the coexistent substrate play a critical role in varying low-lying energy spectra. The completely modified band structures in sliding bilayer silicene systems (figures 5.2(a)–(l)) are worthy of systematic APRES examination: the vanishing Dirac-cone structures, the semimetals and direct- or indirect-gap semiconductors, the many band-edge states, the multi-constant loops, the parabolic (Γ, K, M) valleys, the M- and M′-related split energy bands except for high-symmetry AA and AB stacking, the negligible or important spin-configuration effects for non-AB and AB stacking, and the meaningless π and σ band widths. Such identifications could provide very useful conclusions about the cooperative and competitive relationships between π and σ bonding, and the presence or absence of spin–orbit coupling.

The spatial charge density distributions present dramatic transformations after including interlayer atomic interactions. The bilayer silicene systems are completely different from monolayer systems, as shown in figures 5.3(a)–(j) and 4.3(c). The charge density, which is distributed between two neighboring silicon atoms on the same layer (the red regions), is reduced or transferred greatly, compared to isolated silicene. The lower part of the π bonding in the first layer and the upper part in the second layer show an obvious and dramatic transformation; that is, it is completely absent. Also, more charge densities come to exist in the interlayer spacing (the orange or yellow regions). These illustrate clearly that electron carriers are transferred from the first and second layers to the interlayer spacing and there exists important four-orbital hybridization (the (3s, $3p_x$, $3p_y$, $3p_z$)–(3s, $3p_x$, $3p_y$, $3p_z$) bonding). Furthermore, the orbital-created intralayer and interlayer hopping integrals are very sensitive to the stacking configurations. As a result, diverse

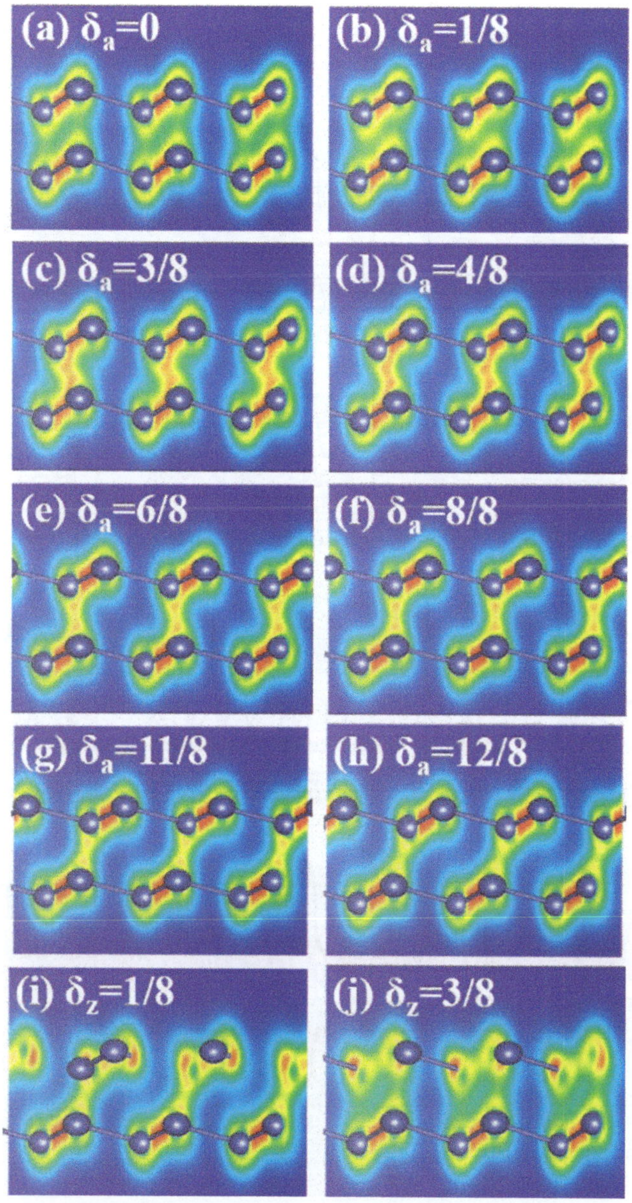

Figure 5.3. (a)–(j) The spatial charge densities for sliding bilayer silicene systems with the same stacking configurations as in figure 5.1, shown on an (x, z)-plane projection.

electronic properties are revealed directly in sliding bilayer silicenes, e.g. their band structures (figures 5.2(b)–(l)) and van Hove singularities. Of course, it is almost impossible to simulate the stacking-dependent electronic energy spectra through the sp^3-based tight-binding model [87]. For example, to date, the theoretical simulation has only been performed for the first pair of valence and conduction bands nearest to

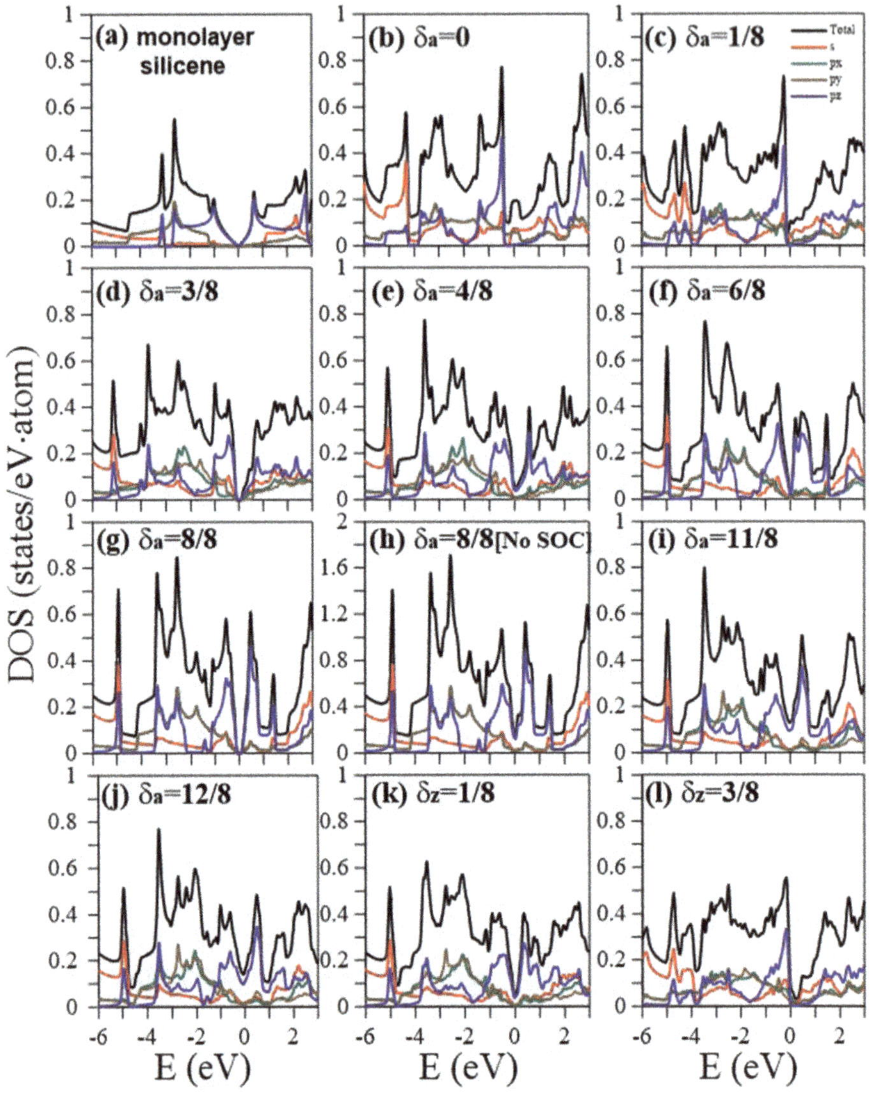

Figure 5.4. (a)–(l) The orbital-projected densities of states for sliding bilayer silicenes corresponding to the band structures in figures 5.2(a)–(l).

the Fermi level in AB-bt bilayer silicene [87]. Similar chemical bonding and difficulties in phenomenological models are expected to be revealed in more complicated few-layer silicene systems, e.g. tri- and four-layer systems [53, 54, 60, 101, 134].

The 2D van Hove singularities, shown in figures 5.4(a)–(l), reflect directly the primary features of the electronic energy spectra (figures 5.2(a)–(l)). Furthermore, the orbital-decomposed densities of states are very useful in exploring fully the significant contributions due to the 3s, $3p_x$, $3p_y$ and $3p_z$ orbitals of silicene-based materials and thus distinguishing the important differences between the π, σ and sp^3 bonding. For pristine monolayer silicene, the density of states is vanishing at the Fermi level (the black curve in figure 5.4(a)), illustrating obvious semiconducting behavior. The $3p_z$-orbital contributions (the blue curve) present a linear E-dependence at low energy, a symmetric peak in the logarithmic divergent form at $E = -1.03$ eV and $E = 0.63$ eV (a π and π^* peak, respectively), and then vanishes after $E < -3.20$ eV. These features, which arise from, respectively, a very narrow gap, the almost isotropic linear bands initiated from the K valley, the saddle M-point structures of the π valence and π^* conduction bands, and the parabolic Γ valley, are determined mainly by the π bonding (the valence and conduction band-edge states in the KMΓ composite valleys in figure 5.2(a)). In addition, the π and π^* peaks are contributed partially by the 3s, $3p_x$ and $3p_y$ orbitals (the red, green and brown curves, respectively). The $3p_x$- and $3p_y$-dominated valence van Hove singularities include an initial shoulder structure at $E = -1.2$ eV, a prominent symmetric peak in the logarithmic form at $E = -2.70$ eV and a terminated shoulder peak at $E = -4.50$ eV. They correspond to the first valence band along the ΓMK direction. The π- and σ-orbital hybridization appears to be characterized by two very strong asymmetric peaks at $E = -2.21$ eV and $E = -2.72$ eV. Moreover, most 3s-orbital contributions arise in the ranges $E < -3.0$ eV and $E > 1.0$ eV. The above-mentioned results have shown both the separation and significant hybridization of the π and σ bands in different energy ranges.

The main features in the densities of states, as illustrated in figures 5.4(b)–(l), become very complicated under any interlayer coupling in bilayer silicene systems. The semimetallic and semiconducting bilayer silicene systems, which are characterized by their values near the Fermi level, are presented in, respectively, figures 5.4(b), (c) and (i)–(l) and figures 5.4(d)–(h). The finite densities of states in the former are accompanied frequently by strong special structures, owing mainly to the weak energy dispersions across $E_F = 0$ in figures 5.2(b), (c) and (i)–(l). However, the vanishing density-of-state ranges of the latter might be close to zero or observable where the asymmetric V-shapes in figures 5.4(e), (f) and (h) across E_F correspond to the modified Dirac-cone structures (figures 5.2(e), (f) and (h)). Whether the AB-bt system has an intrinsic spin–orbit coupling can be determined from the band gap of ~0.31 eV (figure 5.4(g)) or the Dirac point at the Fermi level (figure 5.4(h)). Within the low-energy range of $|E| < 0.5$ eV, only a few stacking configurations are dominated by the $3p_z$ orbitals such as $\delta_a = 6/8$ and 8/8 in figures 5.4(f) and (g), respectively. Generally, it is very difficult to distinguish the initial, middle (saddle-point structure) and final π and σ bands; therefore, bilayer and monolayer silicenes (figures 5.4(b)–(l) and (a)) are completely different from each other. There are a lot of symmetric peaks, antisymmetric peaks and shoulders in the whole energy range, owing mainly to multi-orbital hybridization, oscillatory energy dispersions and frequent anti-crossing.

It is well known that the STS measurements are able to identify band properties and band-edges states (details in section 3.2). Much experimental progress has been achieved on graphene-related systems, but not other group-IV layered materials. Current experimental studies of the rich van Hove singularities in graphene-related systems include a symmetric V-shaped energy dependence vanishing at the Dirac point in monolayer graphene [116, 135], a blue shift of E_F in alkali-adsorbed graphenes [116, 132, 133], the twisting-induced symmetric peak in the logarithmic form for bilayer graphene [62], an electric-field-created band gap in bilayer AB stacking and tri-layer ABC stacking [57–59], a prominent peak centered around the Fermi level due to the partially flat bands in tri-layer and penta-layer ABC stacking [114, 115], a sharp dip structure at E_F accompanied by a pair of asymmetric peaks in tri-layer AAB stacking (a narrow-gap semiconductor with low-lying constant-energy loops [114]), a finite density of states at E_F for Bernal graphite, and its split π and π^* prominent peaks at deeper and higher state energies [136, 137]. Further experimental studies are required to identify the unusual van Hove singularities in silicene-based materials: (i) for monolayer silicene, the V-shaped structure vanishing at the Fermi level, the well-defined π and σ band widths through the initial cusp or shoulder, logarithmic peak and final shoulder, as well as the two square-root asymmetric peaks of the band mixing, and (ii) for sliding bilayer systems, the finite or zero density of states at E_F or a gap region, the many stacking-enriched shoulders, symmetric and asymmetric peaks, and the undefined π and σ electronic structures. Such information would be very useful in distinguishing the π- or sp^3-bonding-dominated low-energy fundamental properties.

5.2 Significant differences between bilayer silicenes and graphenes

The optimal geometric structures and electronic properties of configuration-dependent bilayer graphene systems are explored thoroughly using the first-principles method [57, 92, 93, 121]; also see the tight-binding model in [67, 84, 89, 138]. The geometric structures for sliding bilayer graphene systems are illustrated under a specific path in figures 5.5(a) $\delta = 0$ (AA), (b) $\delta = 1/8$, (c) $\delta = 4/8$, (d) $\delta = 8/8$ (AB), (e) $\delta = 11/8$ and (f) $\delta = 12/8$ (AA′) along the armchair direction, and (g) $\delta = 1/8$ and (h) $\delta = 3/8$ along the zigzag direction. In addition, all carbon atoms possess an identical environment for AA′ stacking (figure 5.5(f)), while their (x, y)-projections are different on the first and second layers. Any stacking configurations present planar honeycomb lattices, illustrating clearly that the dominant σ bonding on the (x, y)-plane is affected little by the stacking configuration. Clearly, the interlayer distance, as indicated in table 5.2, declines and then grows through variation of the relative shift between two layers. Furthermore, the interlayer distance is shortest for the AB stacking (3.26 Å), but longest for the AA stacking (3.52 Å). This result is consistent with the TEM measurements (3.35 Å and 3.55 Å, respectively, for AB and AA stacking [99]) and other theoretical predictions [139, 140]. The total ground state energy is affected strongly by the relative shift (table 5.2). As a result, the shortest interlayer distance creates the strongest van der Waals interactions among the $2p_z$ orbitals. The AB stacking is thus expected to be the most stable configuration of bilayer graphene systems, as demonstrated frequently for experimental samples [99, 141]. The stacking-induced difference of ground state energy is about 20 meV, which is one order smaller than that in bilayer silicene systems (table 5.1). Such a result

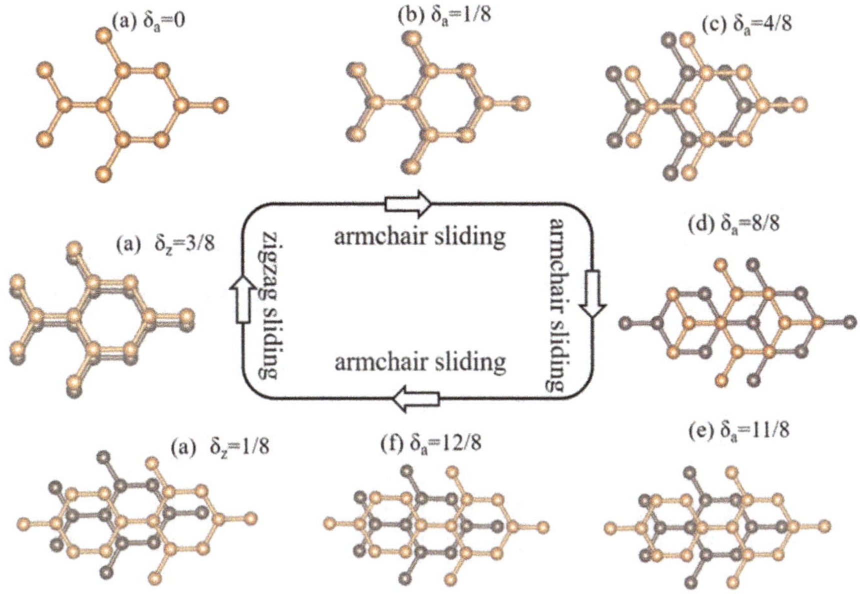

Figure 5.5. The geometric structures for sliding bilayer graphene systems under a specific path: (a) $\delta_a = 0$, (b) $\delta_a = 1/8$, (c) $\delta_a = 4/8$, (d) $\delta_a = 8/8$, (e) $\delta_a = 11/8$ and (f) $\delta_a = 12/8$ along the armchair direction, and (g) $\delta_z = 1/8$ and (h) $\delta_z = 3/8$ along the zigzag direction.

Table 5.2. Similar to table 5.1, but shown for sliding bilayer graphene systems with planar honeycomb lattices.

Bilayer graphene	Lattice constant (Å)	Δd (Å)	b (C–C) (Å)	Ground state energy (eV/unit cell)
$\delta_a = 0$ (AA)	2.47	3.53	1.42	–37.19
$\delta_a = 1/8$	2.47	3.51	1.42	–37.19
$\delta_a = 4/8$	2.47	3.41	1.42	–37.20
$\delta_a = 8/8$ (AB)	2.47	3.26	1.42	–37.22
$\delta_a = 11/8$	2.47	3.36	1.42	–37.21
$\delta_a = 12/8$	2.47	3.36	1.42	–37.21
$\delta_z = 1/8$	2.47	3.38	1.42	–37.21
$\delta_z = 3/8$	2.47	3.46	1.42	–37.19

illustrates clearly the presence of more strong interlayer chemical bonding in the latter.

The calculated results, as illustrated in figures 5.6–5.8, appear to demonstrate dramatic transformations of electronic properties, in particular for the sensitive dependence of low-lying energy bands on the magnitude and direction of the relative shift between two graphene sublayers. The π-electronic energy bands, which are generated through the AA $\longrightarrow$ AB $\longrightarrow$ AA′$\longrightarrow$ AA path (figure 5.5), present

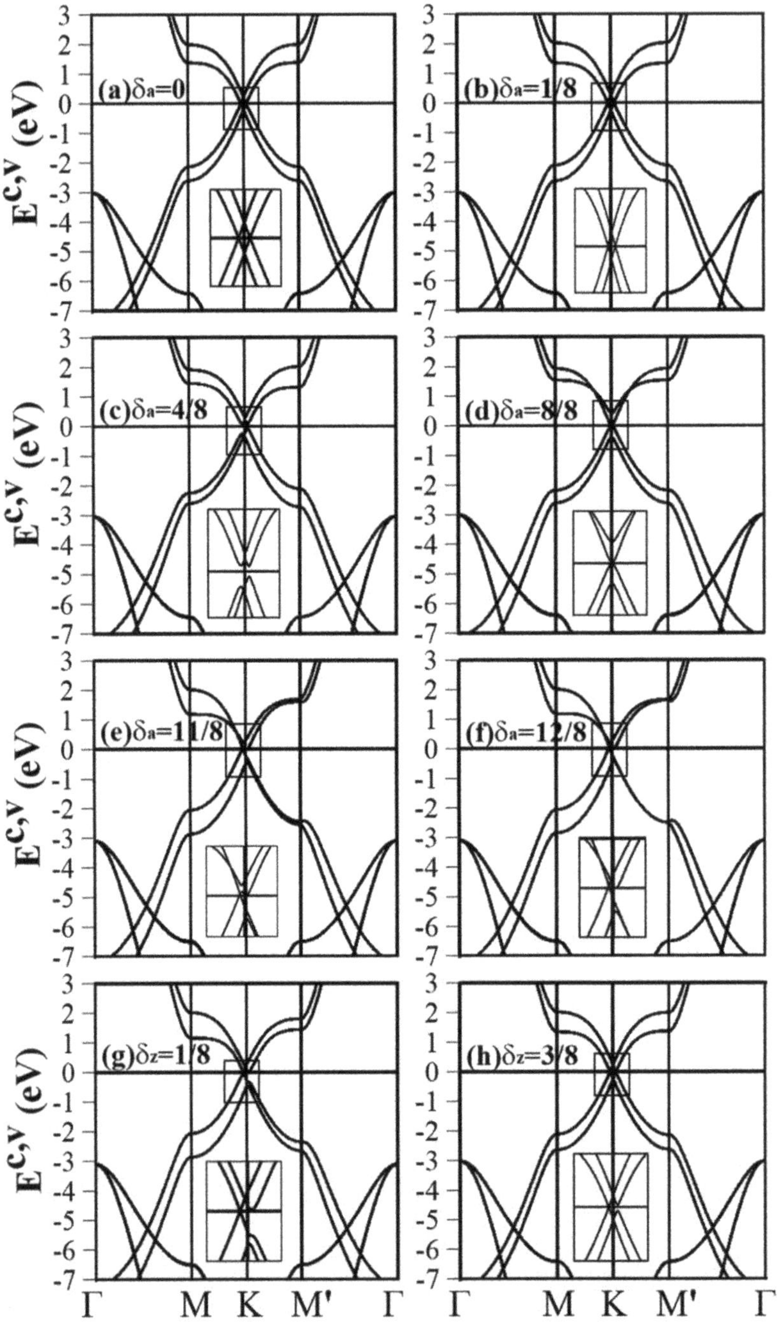

Figure 5.6. As in figure 5.2, but shown for the energy bands of sliding bilayer graphene systems.

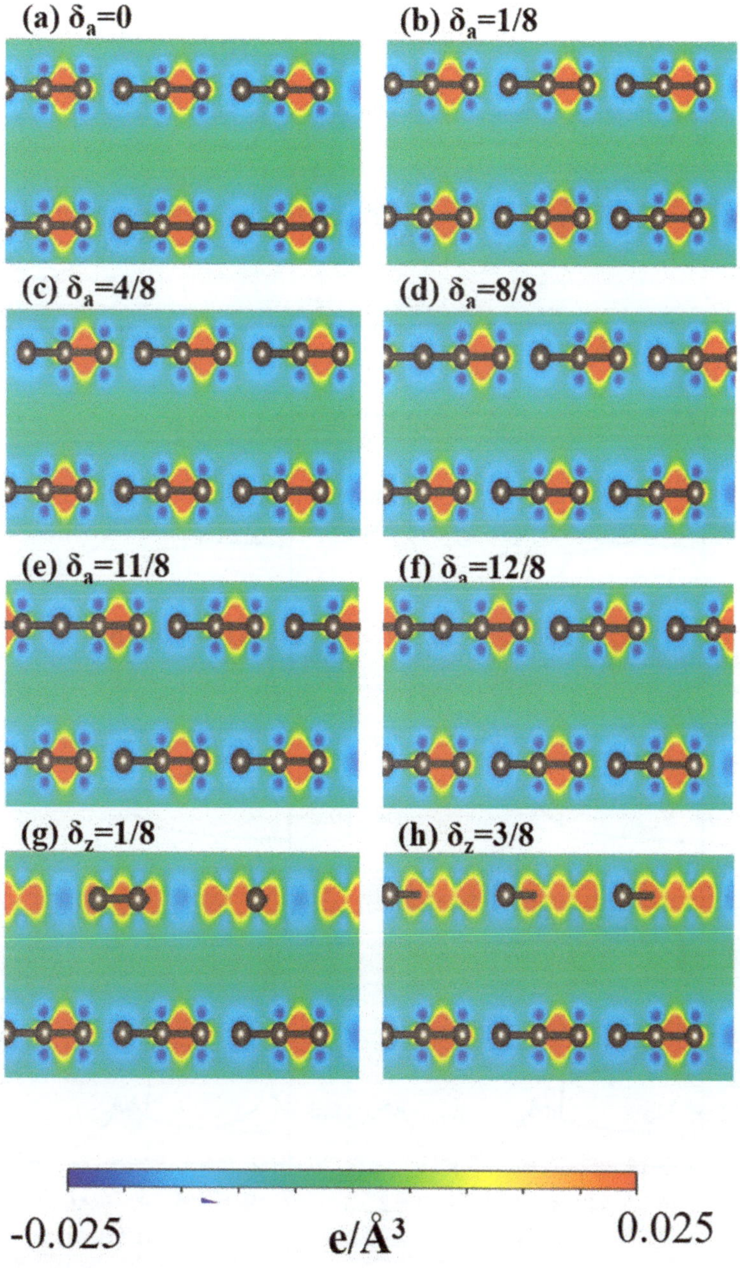

Figure 5.7. The variation of spatial charge densities ($\Delta\rho$) under the different stacking configurations in figure 5.5, illustrated on the (x, z)-planes.

stacking-varied phenomena. Specifically, two pairs of vertical Dirac-cone structures (two Dirac points at the K and K′ valleys in figure 5.6(a) for $\delta_a = 0$) are transformed dramatically into parabolic bands (figure 5.6(d) for $\delta_a = 8.8$), non-vertical Dirac cones (figure 5.6(f) at $\delta_a = 12/8$) and non-well-behaved electronic energy spectra.

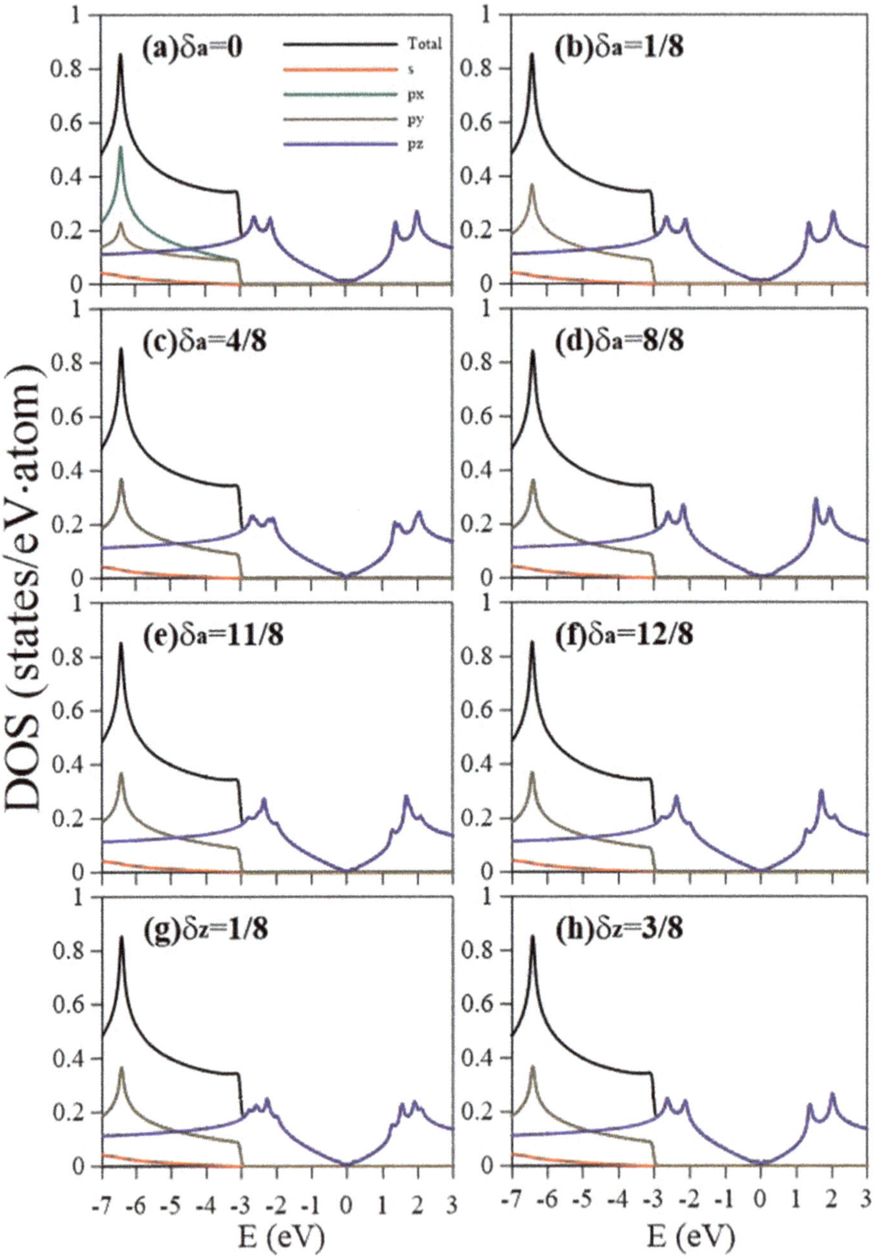

Figure 5.8. The 2s-, $2p_x$-, $2p_y$- and $2p_z$-decomposed van Hove singularities in sliding bilayer graphene materials arising from the electronic energy spectra in figure 5.6.

The latter present the creation of an arc-shaped stateless region near the Fermi level (figures 5.6(b), (c), (e), (g) and (h)) at, respectively, $\delta_a = 1/8$, 4/8 and 11/8 (the armchair direction) and $\delta_z = 1/8$ and 3/8 (the zigzag direction), strongly distorted energy dispersions, significant overlaps of the valence and conduction bands, extra low-energy saddle points and the splitting of saddle M-point states (also in the AA′ stacking of figure 5.6(f)). The π-band widths, which are associated with the KMΓ/KM′Γ composite structures, can reach the range of ~7 − 8 eV for any stacking configurations. They have been extended by interlayer hopping integrals, compared to that of monolayer graphene [66]. The σ bands due to the $2p_x$ and $2p_y$ orbitals are initiated from the stable Γ valleys and their state degeneracy becomes double. Apparently, the interlayer atomic interactions depend little on the C-2s, $-2p_x$ and $-2p_y$ orbitals in two graphene subsystems. Such degenerate energy dispersions also present a saddle M-/M′-point structure and terminate the stable K/K′ valleys; therefore, their widths are greater than 4 eV. Among all the bilayer graphene systems, the AA stacking has the highest free carrier density, in which the band overlap even reaches ~1 eV (the largest density of states at the Fermi level) [92]. In short, the feature-rich band structures of bilayer graphene systems, which include the highly anisotropic energy dispersions, the dramatic transformations and drastic changes of Dirac-cone structures near the stable K/K′ valleys, the extra low-lying saddle points, the eye-shaped stateless region, the split saddle M and M′ points in non-AA and non-AB stacking, the degeneracy-enhanced σ bands, and the direct crossing of the π and σ bands, can be studied directly using high-resolution ARPES measurements (details in section 3.3). Furthermore, the M and M′ saddle points are expected to cause very strong absorption peaks at about 4–5 eV due to the high concentration of π-electronic states.

The spatial charge densities in sliding bilayer graphene systems present observable and significant variation, owing mainly to the interlayer van der Waals interactions. Various $\Delta\rho$ are shown in figures 5.7(a)–(h), depending on various stacking configurations. Apparently, the π and σ chemical bonding remain well-defined forms; that is, they are identified easily from the carrier distributions between two neighboring carbon atoms. They are orthogonal to each other on the planar honeycomb lattices. The σ bonding remains almost the same after bilayer coupling; therefore, the σ valence bands exhibit only a rigid shift (figures 5.6(a)–(h)). Only the π bonding becomes somewhat modified under attractive interlayer atomic interactions. The effective distribution width of $2p_z$ orbitals becomes wider, compared to that in monolayer systems (table 5.2). The interlayer hopping integrals related to them are responsible for diverse physical phenomena, such as the dramatic transformation of band structures and van Hove singularities during the variation of bilayer stacking configurations (figures 5.6(a)–(h) and 5.8(a)–(h), respectively). Of course, their values are sensitive to the stacking symmetries. The clear separation of π and σ chemical bonding, as well as the dominance of the former in the low-energy properties, can be found in all sp^2-based carbon materials, e.g. AA- [56], AB- [57] and ABC-stacked [58] materials, bulk graphites [56], layered graphene systems with AA, AB [56, 57, 93], ABC [80, 94] and AAB stacking [61], achiral and chiral carbon nanotubes [123–125], and graphene nanoribbons [15, 96, 97]. The current tight-binding model, which is constructed on the $2p_z$-orbital periodical functions [65, 67], is capable of fitting the first-principles low-lying energy bands. Thus it is very useful in creating and understanding the varied fundamental properties in graphene-related emergent materials [116, 132, 133, 135, 139–143], such as the rich and unique magnetic quantization phenomena with strong dependence on stacking configurations [90, 144–146], layer numbers [102, 103], dimensions [103–105, 147–149], field modulations [150], composite fields [151, 152] and free carrier doping [136, 153–157].

The densities of states in sliding bilayer graphenes exhibit a lot of unusual structures due to the rich band-edge states, as shown in figures 5.8(a)–(h). The van Hove singularity at the Fermi level presents a plateau (figures 5.8(a), (e), (f) and (g)) or a dip structure (figures 5.8(b)–(d) and (h)), respectively, associated with the Dirac-cone-like (figures 5.6(a), (e), (f) and (g)) and parabolic bands (figures 5.6(b)–(d) and (h)). Both structures possess finite values at E_F, illustrating semimetallic behaviors in all bilayer graphene systems. The free carrier density shows a non-uniform dependence on the relative shift. The critical mechanisms, the stacking-enriched interlayer hopping of C-2p_z orbitals, create significant overlaps of the low-lying valence and conduction bands. The neighboring special structures of the valance and the conduction band belong to, respectively, cusps (figures 5.8(a), (e)–(g)) and shoulders or symmetric peaks (figures 5.8(d), (h)). The low-lying logarithmic peaks arise from the extra saddle points of greatly modified energy dispersions, and their number is very sensitive to the stacking configuration (figure 5.6). Moreover, the middle-energy prominent peaks from the M and M′ points are revealed as single- and twin-peak structures. The intensity and energy of the prominent peaks depend strongly on their splitting and merging as the relative shift occurs. Induced saddle points, and the splitting of the middle-energy peaks and free carriers at E_F can be verified using STS experiments (details in section 3.2). The low- and middle-energy saddle points and the derived prominent peaks are related closely to the stacking configurations; therefore, experimental measurements are available for resolving the subangström misalignment stacking of bilayer graphene systems [121, 144].

The sliding bilayer silicene and graphene systems are in sharp contrast to each other for any fundamental property. According to the current theoretical predictions of first-principles calculations and phenomenological models, the important differences in geometric, electronic, magnetic and transport properties include the buckled or planar honeycomb lattices [2, 21, 57, 78, 79, 81, 83, 85–87, 90, 92, 132, 136, 144, 150, 157], the band properties near the Fermi level [74, 92, 116], the absence or presence of two pairs of valence and conduction Dirac cones, the prominent or weak electron–hole asymmetry around E_F, the strong mixing or good separation of π and σ bonding [74, 92, 116], the disordering or whole valley structures associated with the K-, M- and Γ-symmetry points, the sublattice-, layer- and even the spin-dependent magnetic subenvelope functions [158], the magnetic- and electric-field-induced Landau levels, the magneto-optical selection rules [159, 160], and the normal or regular Hall conductivities [161].

For bilayer graphene systems, the interlayer attractive forces are identified to be clearly due to the van der Waals interactions of pure carbon atoms [63, 162]. The σ bonding, which is generated by 2s, 2p_x and 2p_y orbitals, lies only on the (x, y)-plane honeycomb lattice. Their valence energy spectra, with $E^v < -3.0$ eV, are barely affected by bilayer coupling, and accordingly the higher- or deeper-energy physical properties. The weak, but significant interlayer atomic interactions arise mainly from the π bonding of C-2p_z orbitals. The interlayer hopping integrals are much smaller than the intralayer ones, while the low-energy physical properties are related closely to their stacking dependence. That is, the single $2p_z$-orbital Hamiltonian [138, 163] is sufficient and reliable for stacking-enriched phenomena in few-layer graphene systems. The phenomenological models are very successful in exploring the other fundamental properties, such as the rich optical properties [84, 119, 121], Coulomb excitations [61, 164] and inelastic scattering rates [165]. On the other hand, the sliding bilayer silicenes have larger height differences between the intralayer A and B sublattices and smaller interlayer distances, compared to monolayer silicene and bilayer graphenes. These two critical factors will enhance sp^3 bonding and enlarge the interlayer hopping integrals greatly, where the latter even dominate the kinetic energies of the intrinsic Hamiltonian [67]. Consequently, the π and σ bands are difficult to distinguish from complicated electronic energy spectra. To date, the few existing calculations of the tight-binding model focused on specific AA-bt and AB-bt bilayer silicenes [87], have only been conducted on the $3p_z$-orbital interactions. As a result, they are useful in understanding unusual magnetic quantization phenomena and optical absorption spectra due to the first pair of valence and conduction bands. Even for well-defined stacking configurations in bilayer silicenes, it is very difficult to simulate the low-lying electronic structures with strong wave-vector dependence completely using first-principles calculations. This has created significant challenges for future studies of phenomenological models for few-layer silicene, germanene and tinene systems [3, 53, 54, 60, 73, 82, 101, 166].

According to first-principles calculations [63, 94, 95], bilayer graphene systems possess low-energy π-electronic states under any stacking configuration. Their band structures can also be understood from the tight-binding model with well-fitted intralayer and interlayer hopping integrals from the empirical interaction formula [138]. The generalized tight-binding model, which is based on the subenvelope functions of specific sublattices, is developed successfully to explore the rich magnetic quantization phenomena. The relative shift between two graphene layers induces a dramatic transformation between the vertical and non-vertical Dirac cones and the parabolic energy dispersions, thus leading to the dramatic transformation of Landau levels in the spatial symmetry, initial formation energy, intergroup anti-crossing and crossing, state degeneracy and semiconductor–semimetal transition. There exist three kinds of magneto-electronic states; that is, the well-behaved, perturbed and undefined Landau levels are characterized by, respectively, a specific mode, a main mode plus side modes and a disordered mode. Their spatial distribution symmetries are reflected directly in the diverse magneto-optical selection rules, inert-Landau–Coulomb excitations and magnetoplasmon modes [159, 160]. Specifically, the undefined Landau levels exhibit frequent intergroup anti-crossing in field-dependent energy spectra, and show a large number of absorption peaks without specific inter-Landau-level vertical transitions. In general, the carbon-$2p_z$ orbitals in all the sp^2-dominated systems are sufficient for characterizing the rich and unique phenomena, e.g. the π and π^* bands in the AA- [56], AB- and ABC-stacked graphites [57–60], few-layered graphene systems with AAA, ABA, ABC and AAB stacking [61, 98, 100, 112, 113, 142], achiral and chiral [167] carbon nanotubes, 1D graphene nanoribbons with open or passivated edges [159, 162], carbon onions

[168, 169] and C_{60}-related fullerenes [170, 171]. This fact is confirmed by both theoretical predictions (the numerical and phenomenological methods) and experimental studies on band structures [172, 173], magneto-electronic states [144], absorption spectra [174, 175], quantum transport [176], and Coulomb excitations [177, 178] and decay rates [179]. In short, the pure few-layer graphene systems can be well understood through the calculations of phenomenological models, while chemical modifications due to absorption, substitution and defects might create structure- and bonding-dependent non-uniform environments and thus the open issues of the near future [98].

References

[1] Novoselov K S *et al* 2004 Electric field effect in atomically thin carbon film *Science* **306** 666

[2] Yaokawa R, Ohsuna T, Morishita T, Hayasaka Y, Spencer M J S and Nakano H 2016 Monolayer-to-bilayer transformation of silicenes and their structural analysis *Nat. Commun.* **7** 10657

[3] Dávila M E and Lay G L 2016 Few layer epitaxial germanene: a novel two-dimensional Dirac material *Sci. Rep.* **6** 20714

[4] Zhu F F, Chen W J, Xu Y, Gao C I, Guan D D and Jia J F 2015 Epitaxial growth of two-dimensional stanene *Nat. Mater.* **14** 2020

[5] Liu H *et al* 2014 Phosphorene: an unexplored 2D semiconductor with a high hole mobility *ACS Nano* **8** 4033

[6] Shao Y *et al* 2018 Epitaxial growth of flat antimonene monolayer: a new honeycomb analogue of graphene *Nano Lett.* **18** 2133

[7] Pumera M and Sofer Z 2017 2D monoelemental arsenene, antimonene, and bismuthene: beyond black phosphorus *Adv. Mater.* **29** 1605299

[8] Yang Q Q *et al* 2018 Phosphorene: 2D bismuthene fabricated via acid-intercalated exfoliation showing strong nonlinear near-infrared responses for mode-locking lasers *Nanoscale* **10** 21106

[9] Zhang P *et al* 2018 Two-dimensional penta-Sn_3H_2 monolayer for nanoelectronics and photocatalytic water splitting: a first-principles study *RSC Adv.* **8** 11799

[10] Ferdous N, Islam M S, Park J and Hashimoto A 2019 Tunable electronic properties in stanene and two dimensional silicon-carbide heterobilayer: a first principles investigation *AIP Adv.* **9** 025120

[11] Sato S 2015 Graphene for nanoelectronics *Jpn J. Appl. Phys.* **54** 040102

[12] Bishnoi B and Ghosh B 2013 Spin transport in silicene and germanene *RSC Adv.* **3** 26153

[13] Pan F *et al* 2015 Silicene nanomesh *Sci. Rep.* **5** 09075

[14] Ji J *et al* 2016 Two-dimensional antimonene single crystals grown by van der Waals epitaxy *Nat. Commun.* **7** 13352

[15] Wang Y and Ding Y 2015 Electronic structure and carrier mobilities of arsenene and antimonene nanoribbons: a first-principle study *Nanoscale Res. Lett.* **10** 254

[16] Kistanov A A, Khadiullin S K, Zhou K, Dmitriev S V and Korznikova E A 2019 Environmental stability of bismuthene: oxidation mechanism and structural stability of 2D pnictogens *J. Mater. Chem.* C **7** 9195

[17] Mayorga-Martinez C C, Gusmão R, Sofer Z and Pumera M 2019 Pnictogen-based enzymatic phenol biosensors: phosphorene, arsenene, antimonene, and bismuthene *Angew. Chem. Int. Ed.* **58** 134

[18] Jiang S, Butler S, Bianco E, Restrepo O D, Windl W and Goldberger J E 2014 Improving the stability and optical properties of germanane via one-step covalent methyl-termination *Nat. Commun.* **5** 3389
[19] Cheng J, Wang C, Zou X and Liao L 2019 Recent advances in optoelectronic devices based on 2D materials and their heterostructures *Adv. Optical Mater.* **7** 1800441
[20] Bonaccorso F, Sun Z, Hasan T and Ferrari A C 2010 Graphene photonics and optoelectronics *Nat. Photonics* **4** 611
[21] Huang B, Deng H X, Lee H K, Yoon M, Sumpter B G and Wei S H 2014 Exceptional optoelectronic properties of hydrogenated bilayer silicene *Phys. Rev.* X **4** 021029
[22] Chen X *et al* 2016 The electronic and optical properties of novel germanene and antimonene heterostructures *J. Mater. Chem.* C **4** 5434
[23] Lu L *et al* 2018 Few-layer bismuthene: sonochemical exfoliation, nonlinear optics and applications for ultrafast photonics with enhanced stability *Laser Photonics Rev.* **12** 1700221
[24] Garg P, Choudhuri I, Mahata A and Pathak B 2017 Band gap opening in stanene induced by patterned B–N doping *Phys. Chem. Chem. Phys.* **19** 3660
[25] Li X and Zhi L 2018 Graphene hybridization for energy storage applications *Chem. Soc. Rev.* **47** 3189
[26] Mao J, Iocozzia J, Huang J, Meng K, Lai Y and Lin Z 2018 Graphene aerogels for efficient energy storage and conversion *Energy Environ. Sci.* **11** 772
[27] Molle A, Grazianetti C, Tao L, Taneja D, Alam M H and Akinwande D 2018 Silicene, silicene derivatives, and their device applications *Chem. Soc. Rev.* **47** 6370
[28] Xu Q, Yang G M, Fan X F and Zheng W T 2019 Adsorption of metal atoms on silicene: stability and quantum capacitance of silicene-based electrode materials *Phys. Chem. Chem. Phys.* **21** 4276
[29] Derivaz M *et al* 2015 Continuous germanene layer on Al(111) *Nano Lett.* **15** 2510
[30] Zhu J, Chroneos A and Schwingenschlögl U 2016 Silicene/germanene on MgX_2 (X = Cl, Br, and I) for Li-ion battery applications *Nanoscale* **8** 7272
[31] Ma J, Gu J, Li B and Yang S 2019 Facile fabrication of 2D stanene nanosheets via a dealloying strategy for potassium storage *Chem. Commun.* **55** 3983
[32] Tao Y *et al* 2019 Few-layer phosphorene: an emerging electrode material for electrochemical energy storage *Appl. Mater. Today* **15** 18
[33] Qiu M, Sun Z T, Sang D K, Han X G, Zhang H and Niu C M 2017 Current progress in black phosphorus materials and their applications in electrochemical energy storage *Nanoscale* **9** 13384
[34] Gao Y *et al* 2019 Tailoring natural layered β-phase antimony into few layer antimonene for Li storage with high rate capabilities *J. Mater. Chem. A.* **7** 3238
[35] Martinez-Perinán E, Down P, Gibaja C, Lorenzo E, Zamora F and Banks C E 2018 Antimonene: a novel 2D nanomaterial for supercapacitor applications *Adv. Energy Mater.* **8** 1702606
[36] Wang X, Song J and Qu J 2019 Antimonene: from experimental preparation to practical application *Angew. Chem. Int. Ed.* **58** 1574
[37] Pan H *et al* 2018 Giant energy density and high efficiency achieved in bismuth ferrite-based film capacitors via domain engineering *Nat. Commun.* **9** 1813
[38] Li F, Jiang T, Zhai J, Shen B and Zeng H 2018 Exploring novel bismuth-based materials for energy storage applications *J. Mater. Chem.* C **6** 7976

[39] Osborne D A, Morishita T, Tawfik S A, Yayamab T and Spencer M J S 2019 Adsorption of toxic gases on silicene/Ag(111) *Phys. Chem. Chem. Phys.* **10** 1039
[40] Shu H, Cao D, Liang P, Wang X, Chen X and Lu W 2014 Two-dimensional silicene nucleation on a Ag(111) surface: structural evolution and the role of surface diffusion *Phys. Chem. Chem. Phys.* **16** 304
[41] Arafune R *et al* 2013 Structural transition of silicene on Ag(111) *Surf. Sci.* **608** 297
[42] Meng L *et al* 2013 Buckled silicene formation on Ir(111) *Nano Lett.* **13** 685
[43] Gill T G *et al* 2017 Metallic atomically-thin layered silicon epitaxially grown on silicene/ ZrB_2 *2D Mater.* **4** 021015
[44] Wiggers F B, Bui H V, Friedlein R, Yamada-Takamura Y, Schmitz J and Jong M P 2016 A nitride-based epitaxial surface layer formed by ammonia treatment of silicene-terminated ZrB_2 *J. Chem. Phys.* **144** 134703
[45] Fleurence A and Yamada-Takamura Y 2017 Insights into the spontaneous formation of silicene sheet on diboride thin films *Appl. Phys. Lett.* **110** 041601
[46] Li L *et al* 2014 Buckled germanene formation on Pt(111) *Adv. Mater.* **26** 4820
[47] Cantero E D *et al* 2017 Growth of germanium on Au(111): formation of germanene or intermixing of Au and Ge atoms? *Phys. Chem. Chem. Phys.* **19** 18580
[48] Martinez E A, Fuhr J D, Grizzi O, Sánchez E A and Cantero E D 2019 Growth of germanene on Al(111) hindered by surface alloy formation *J. Phys. Chem.* C **123** 12910
[49] Chen S W, Wu C Y, Wu H J and Chiu W T 2014 Interfacial reactions in Sn/Bi_2Te_3, Sn/ Bi_2Se_3 and Sn/Bi_2 $(Te_{1-x}Se_x)_3$ couples *J. Alloy Compd.* **611** 313
[50] Kim K C *et al* 2015 Sn doping in thermoelectric Bi_2Te_3 films by metal–organic chemical vapor deposition *Appl. Surf. Sci.* **353** 232
[51] Ganz E, Hwang I S, Xiong F, Theiss S K and Golovchenko J 1991 Growth and morphology of Pb on Si(111) *Surf. Sci.* **257** 259
[52] Yakes M and Tringides M C 2011 Probing the buried Pb/Si(111) interface with SPA LEED and STM on Si(111)-Pb $\propto \sqrt{3} \times \sqrt{3}$ *J. Phys. Chem.* A **115** 7096
[53] Peters J L *et al* 2018 Mono- and multilayer silicene-type honeycomb lattices by oriented attachment of PbSe nanocrystals: synthesis, structural characterization, and analysis of the disorder *Chem. Mater.* **30** 4831
[54] Li H, Fu H X and Meng S 2015 Silicene: from monolayer to multilayer—a concise review *Chin. Phys.* B **24** 086102
[55] Johns J E and Hersam M C 2013 Atomic covalent functionalization of graphene *Acc. Chem. Res.* **46** 77
[56] Song M K, do Hong S and No K T 2001 The structure of lithium intercalated graphite using an effective atomic charge of lithium *J. Electrochem. Soc.* **148** 1159
[57] McCann E, Abergel D S and Fal'ko V I 2007 The low energy electronic band structure of bilayer graphene *Eur. Phys. J.: Spec. Top.* **148** 91
[58] Warner J H, Mukai M and Kirkland A I 2012 Atomic structure of ABC rhombohedral stacked trilayer graphene *ACS Nano* **6** 5680
[59] Zhang F, Sahu B, Min H and MacDonald A H 2010 Band structure of ABC-stacked graphene trilayers *Phys. Rev.* B **82** 035409
[60] Kamal C, Chakrabarti A, Banerjee A and Deb S K 2013 Silicene beyond mono-layers: different stacking configurations and their properties *J. Phys. Condens. Matter* **25** 085508
[61] Lin C Y, Huang B L, Ho C H, Gumbs G and Lin M F 2018 Geometry-diversified Coulomb excitations in trilayer AAB stacking graphene *Phys. Rev.* B **98** 195442

[62] Sharpe A L, Fox E J, Barnard A W, Finney J, Watanabe K and Kastner M A 2019 Emergent ferromagnetism near three-quarters filling in twisted bilayer graphene *Science* **10** 1126

[63] Pham K D and Nguyen C V 2018 First principles calculations of the geometric structures and electronic properties of van der Waals heterostructure based on graphene, hexagonal boron nitride and molybdenum diselenide *Diam. Relat. Mater.* **88** 151

[64] Yu D, Lupton E M, Gao H J, Zhang C and Liu F 2008 A unified geometric rule for designing nanomagnetism in graphene *Nano Res.* **1** 497

[65] Kaloni T P, Modarresi M, Tahir M and Roknabadi M R 2015 Electrically engineered band gap in two-dimensional Ge, Sn, and Pb: a first-principles and tight-binding approach *J. Phys. Chem.* C **119** 11896

[66] Li X, Wu S, Zhou S and Zhu Z 2014 Structural and electronic properties of germanene/MoS_2 monolayer and silicene/MoS_2 monolayer superlattices *Nanoscale Res. Lett.* **9** 110

[67] Kahl K and Kintscher N 2018 Geometric multigrid for the tight-binding Hamiltonian of graphene *SIAM J. Numer. Anal.* **56** 499

[68] Castro Neto A H, Guinea F, Peres N M R, Novoselov K S and Geim A K 2009 The electronic properties of graphene *Rev. Mod. Phys.* **81** 109

[69] Fan Y, Liu X, Wang J, Ai H and Zhao M 2018 Silicene and germanene on InSe substrates: structures and tunable electronic properties *Phys. Chem. Chem. Phys.* **20** 11369

[70] Quhe R *et al* 2012 Tunable and sizable band gap in silicone by surface adsorption *Sci. Rep.* **2** 853

[71] Qin R, Wang C H, Zhu W and Zhang Y 2012 First-principles calculations of mechanical and electronic properties of silicene under strain *AIP Adv.* **2** 022159

[72] Zhang X, Zhang D, Xie F, Zheng X, Wang H and Long M 2017 First-principles study on the magnetic and electronic properties of Al or P doped armchair silicene nanoribbons *Phys. Lett.* A **381** 2097

[73] Aierken Y, Leenaertsa O and Peeters F M 2016 A first-principles study of stable few-layer penta-silicene *Phys. Chem. Chem. Phys.* **18** 18486

[74] Lin S Y, Chang S L, Thuy Tran N T, Yang P H and Lin M F 2015 H–Si bonding-induced unusual electronic properties of silicene: a method to identify hydrogen concentration *Phys. Chem. Chem. Phys.* **17** 26443

[75] Ni Z, Minamitani E, Ando Y and Watanabe S 2017 Germanene and stanene on two-dimensional substrates: Dirac cone and Z_2 invariant *Phys. Rev.* B **96** 075427

[76] Sadeddine S *et al* 2017 Compelling experimental evidence of a Dirac cone in the electronic structure of a 2D silicene layer *Sci. Rep.* **7** 44400

[77] Chen L *et al* 2012 Evidence for Dirac fermions in a honeycomb lattice based on silicon *Phys. Rev. Lett.* **109** 056804

[78] Wang R, Wang S and Wu X 2014 The formation and electronic properties of hydrogenated bilayer silicene from first-principles *J. Appl. Phys.* **116** 024303

[79] Mohan B, Kumar A and Ahluwalia P K 2013 A first principles calculation of electronic and dielectric properties of electrically gated low-buckled mono and bilayer silicene *Physica* E **53** 233

[80] Yuan S, Roldn R and Katsnelson M I 2011 Landau level spectrum of ABA- and ABC-stacked trilayer graphene *Phys. Rev.* B **84** 125455

[81] Fu H, Zhang J, Ding Z, Li H and Meng S 2014 Stacking-dependent electronic structure of bilayer silicene *Appl. Phys. Lett.* **104** 131904

[82] Liu Y *et al* 2013 Structural, electronic, and optical properties of hydrogenated few-layer silicene: size and stacking effects *J. Appl. Phys.* **114** 094308
[83] Liu J and Zhang W 2013 Bilayer silicene with an electrically-tunable wide band gap *RSC Adv.* **3** 21943
[84] Bravo S, Correa J, Chico L and Pacheco M 2018 Tight-binding model for opto-electronic properties of penta-graphene nanostructures *Sci. Rep.* **8** 11070
[85] Do T N *et al* 2019 Peculiar optical properties of bilayer silicene under the influence of external electric and magnetic fields *Sci. Rep.* **9** 36547
[86] Wang X and Wu Z 2017 Intrinsic magnetism and spontaneous band gap opening in bilayer silicene and germanene *Phys. Chem. Chem. Phys.* **19** 2148
[87] Do T N, Shih P H, Gumbs G, Huang D, Chiu C W and Lin M F 2018 Diverse magnetic quantization in bilayer silicene *Phys. Rev.* B **97** 125416
[88] Zhou X Y *et al* 2015 Landau levels and magneto-transport property of monolayer phosphorene *Sci. Rep.* **5** 12295
[89] Chiu Y H, Lai Y H, Ho J H, Chuu S and Lin M F 2008 Electronic structure of a two-dimensional graphene monolayer in a spatially modulated magnetic field: Peierls tight-binding model *Phys. Rev.* B **77** 045407
[90] Ohta T, Bostwick A, Seyller T, Horn K and Rotenberg E 2006 Controlling the electronic structure of bilayer graphene *Science* **313** 951
[91] Padiha J E and Pontes R B 2015 Free-standing bilayer silicene: the effect of stacking order on the structural, electronic, and transport properties *J. Phys. Chem.* C **119** 3818
[92] Thuy Tran N T, Lin S Y, Glukhova O E and Lin M F 2015 Configuration-induced rich electronic properties of bilayer graphene *J. Phys. Chem.* C **119** 10623
[93] Padiha J E, Lima M P, da Silva A J R and Fazzio A 2011 Bilayer graphene dual-gate nanodevice: an *ab initio* simulation *Phys. Rev.* B **84** 113412
[94] Wang Y P, Li X G, Fry J N and Cheng H P 2016 First-principles studies of electric field effects on the electronic structure of trilayer graphene *Phys. Rev.* B **94** 165428
[95] Pamuk B, Baima J, Mauri F and Calandra M 2017 Magnetic gap opening in rhombohedral-stacked multilayer graphene from first principles *Phys. Rev.* B **95** 075422
[96] Yeo P S E, Sullivan M B, Lohb K P and Gan C K 2013 First-principles study of the thermoelectric properties of strained graphene nanoribbons *J. Mater. Chem.* A **1** 10762
[97] Jippo H and Ohfuchi M 2013 First-principles study of edge-modified armchair graphene nanoribbons *J. Appl. Phys.* **113** 183715
[98] Zhou L *et al* 2018 Surface structure of few layer graphene *Carbon* **136** 255
[99] Lee J K, Lee S C, Ahn J P, Kim S C, Wilson J I and John P 2008 The growth of AA graphite on (111) diamond *J. Chem. Phys.* **129** 234709
[100] Ping J and Fuhrer M S 2012 Layer number and stacking sequence imaging of few-layer graphene by transmission electron microscopy *Nano Lett.* **12** 4635
[101] Tsai H S, Hsiao C H, Chen C W, Ouyang H and Liang J H 2016 Synthesis of nonepitaxial multilayer silicene assisted by ion implantation *Nanoscale* **8** 9488
[102] Coletti C *et al* 2013 Revealing the electronic band structure of trilayer graphene on SiC: an angle-resolved photoemission study *Phys. Rev.* B **88** 155439
[103] Ohta T, Bostwick A, McChesney J L, Seyller T, Horn K and Rotenberg E 2007 Interlayer interaction and electronic screening in multilayer graphene investigated with angle-resolved photoemission spectroscopy *Phys. Rev. Lett.* **98** 206802

[104] Siegel D A, Regan W, Fedorov A V, Zettl A and Lanzara A 2013 Charge-carrier screening in single-layer graphene *Phys. Rev. Lett.* **110** 146802

[105] Bostwick A, Ohta T, Seyller T, Horn K and Rotenberg E 2007 Quasiparticle dynamics in graphene *Nat. Phys.* **3** 36

[106] Du Y, Zhuang J, Wang J, Li Z, Liu H and Dou S X 2016 Quasi-freestanding epitaxial silicene on Ag(111) by oxygen intercalation *Sci. Adv.* **2** 1600067

[107] Podsiadly-Paszkowskaa A and Krawiec M 2015 Dirac fermions in silicene on Pb(111) surface *Phys. Chem. Chem. Phys.* **17** 2246

[108] Xu X *et al* 2014 Effects of oxygen adsorption on the surface state of epitaxial silicene on Ag (111) *Sci. Rep.* **4** 7543

[109] Luican A *et al* 2011 Single-layer behavior and its breakdown in twisted graphene layers *Phys. Rev. Lett.* **106** 126802

[110] Li G *et al* 2010 Observation of van Hove singularities in twisted graphene layers *Nat. Phys.* **6** 109

[111] Cherkez V, de Laissardiere G T, Mallet P and Veuillen J Y 2015 van Hove singularities in doped twisted graphene bilayers studied by scanning tunneling spectroscopy *Phys. Rev.* B **95** 155428

[112] Lauffer P *et al* 2008 Atomic and electronic structure of few-layer graphene on SiC (0001) studied with scanning tunneling microscopy and spectroscopy *Phys. Rev.* B **77** 155426

[113] Yankowitz M, Wang F, Lau C N and LeRoy B J 2013 Local spectroscopy of the electrically tunable band gap in trilayer graphene *Phys. Rev.* B **87** 165102

[114] Que Y, Xiao W, Chen H, Wang D, Du S and Gao H J 2015 Stacking-dependent electronic property of trilayer graphene epitaxially grown on Ru (0001) *Appl. Phys. Lett.* **107** 263101

[115] Pierucci D *et al* 2015 Evidence for flat bands near the Fermi level in epitaxial rhombohedral multilayer graphene *ACS Nano* **9** 5432

[116] Lin S Y, Lin Y T, Thuy Tran N T, Su W P and Lin M F 2017 Feature-rich electronic properties of aluminum-adsorbed graphenes *Carbon* **120** 209

[117] Lin C Y, Wu J Y, Chiu Y H and Lin M F 2015 Magneto-electronic properties of multilayer graphenes *Phys. Chem. Chem. Phys.* **17** 26008

[118] Do T N, Shih P H, Chang C P, Lin C Y and Lin M F 2016 Rich magneto-absorption spectra in AAB-stacked trilayer graphene *Phys. Chem. Chem. Phys.* **18** 17597

[119] Koshino M 2013 Stacking-dependent optical absorption in multilayer graphene *New J. Phys.* **15** 015010

[120] Chiu C W and Chen R B 2016 Influence of electric fields on absorption spectra of AAB-stacked trilayer graphene *Appl. Phys. Express* **9** 065103

[121] Chen S C, Chiu C W, Wu C L and Lin M F 2014 Shift-enriched optical properties in bilayer graphene *RSC Adv.* **4** 63779

[122] Chiu C W, Huang Y C, Shyu F L and Lin M F 2011 Excitation spectra of ABC-stacked graphene superlattice *Appl. Phys. Lett.* **98** 2631920

[123] Shyu F L and Lin M F 2002 Electronic and optical properties of narrow-gap carbon nanotubes *J. Phys. Soc. Jpn* **71** 1820

[124] Lin M F and Shung K W K 1994 Plasmons and optical properties of carbon nanotubes *Phys. Rev.* B **50** 17744

[125] Ruzicka B *et al* 2000 Optical and dc conductivity study of potassium-doped single-walled carbon nanotube films *Phys. Rev.* B **61** 2468

[126] Djurisic A B and Li E H 1999 Optical properties of graphite *J. Appl. Phys.* **85** 7404
[127] Chiu C W, Lee S H, Chen S C, Shyu F L and Lin M F 2010 Absorption spectra of AA-stacked graphite *New J. Phys.* **12** 083060
[128] Taft E A and Philipp H R 1965 Optical properties of graphite *Phys. Rev.* **138** 197
[129] Ichikawa Y H and Kobayashi K 1996 Optical properties of graphite in the infrared region *Carbon* **3** 401
[130] Dovbeshko G I *et al* 2015 Optical properties of pyrolytic carbon films versus graphite and graphene *Nanoscale Res. Lett.* **10** 234
[131] Ergun S, Yasinsky J B and Townsend J R 1967 Transverse and longitudinal optical properties of graphite *Carbon* **5** 403
[132] Praveen C S, Piccinin S and Fabris S 2015 Adsorption of alkali adatoms on graphene supported by the Au/Ni(111) surface *Phys. Rev.* B **92** 075403
[133] Hao J H, Wang Z J, Wang Y F, Yin Y H, Jiang R and Jin Q H 2015 Adsorption of alkali adatoms on graphene supported by the Au/Ni(111) surface *Solid State Sci.* **50** 69
[134] Padova P D *et al* 2016 Multilayer silicene: clear evidence *2D Mater.* **3** 049501
[135] Li G, Luican A and Andrei E Y 2009 Scanning tunneling spectroscopy of graphene on graphite *Phys. Rev. Lett.* **102** 176804
[136] Lai Y H, Ho J H, Chang C P and Lin M F 2008 Magnetoelectronic properties of bilayer Bernal graphene *Phys. Rev.* B **77** 085426
[137] Ho Y H, Chiu Y H, Lin D H, Chang C P and Lin M F 2010 Magneto-optical selection rules in bilayer Bernal graphene *ACS Nano* **4** 1465
[138] Konschuh S, Gmitra M and Fabian J 2010 Tight-binding theory of the spin–orbit coupling in graphene *Phys. Rev.* B **82** 245412
[139] Xu Y, Li X and Dong J 2010 Infrared and Raman spectra of AA stacking bilayer graphene *Nanotechnology* **21** 065711
[140] Wang Q, Wei L, Sullivan M, Yang S W and Chen Y 2013 Graphene layers on Cu and Ni (111) surfaces in layer controlled graphene growth *RSC. Adv.* **3** 3406
[141] Rong Z Y and Kuiper P 1993 Electronic effects in scanning tunneling microscopy: moiré pattern on a graphite surface *Phys. Rev.* B **48** 17427
[142] Thuy Tran N T, Lin S Y, Lin Y T and Lin M F 2016 Chemical bonding-induced rich electronic properties of oxygen adsorbed few-layer graphenes *Phys. Chem. Chem. Phys.* **18** 4000
[143] Nguyen D K, Lin Y T, Lin S Y, Chiu Y H, Thuy Trana N T and Lin M F 2017 Fluorination-enriched electronic and magnetic properties in graphene nanoribbons *Phys. Chem. Chem. Phys.* **19** 20667
[144] Huang Y K, Chen S C, Ho Y H, Lin C Y and Lin M F 2014 Feature-rich magnetic quantization in sliding bilayer graphenes *Sci. Rep.* **4** 7509
[145] Kim K S *et al* 2013 Coexisting massive and massless Dirac fermions in symmetry-broken bilayer graphene *Nat. Mater.* **12** 887892
[146] Sutter P, Hybertsen M S, Sadowski J T and Sutter E 2009 Electronic structure of fewlayer epitaxial graphene on Ru (0001) *Nano Lett.* **9** 2654
[147] Ruffieux P *et al* 2012 Electronic structure of atomically precise graphene nanoribbons *ACS Nano* **6** 6930
[148] Sugawara K, Sato T, Souma S, Takahashi T and Suematsu H 2006 Fermi surface and edge-localized states in graphite studied by high-resolution angle-resolved photoemission spectroscopy *Phys. Rev.* B **73** 045124

[149] Grüneis A *et al* 2008 Electron–electron correlation in graphite: a combined angle-resolved photoemission and first principles study *Phys. Rev. Lett.* **100** 037601
[150] Bhattacharyya S and Singh A K 2016 Lifshitz transition and modulation of electronic and transport properties of bilayer graphene by sliding and applied normal compressive strain *Carbon* **99** 432
[151] Chen S C, Wu C L, Wu J Y and Lin M F 2016 Magnetic quantization of sp^3 bonding in monolayer gray tin *Phys. Rev.* B **94** 045410
[152] Ho Y H, Chiu C W, Su W P and Lin M F 2014 Magneto-optical spectra of transition metal dichalcogenides: a comparative study *Appl. Phys. Lett.* **105** 222411
[153] Lu C L, Chang C P, Huang Y C, Chen R B and Lin M F 2006 Influence of an electric field on the optical properties of few-layer graphene with AB stacking *Phys. Rev.* B **73** 144427
[154] Lu C L, Chang C P, Huang Y C, Ho J H, Hwang C C and Lin M F 2007 Electronic properties of AA- and ABC-stacked few-layer graphites *J. Phys. Soc. Jpn* **76** 024701
[155] Yan J, Zhang Y, Kim P and Pinczuk A 2007 Electric field effect tuning of electron–phonon coupling in graphene *Phys. Rev. Lett.* **98** 166802
[156] Goerbig M 2011 Electronic properties of graphene in a strong magnetic field *Rev. Mod. Phys.* **83** 1193
[157] Castro E V *et al* 2007 Biased bilayer graphene: semiconductor with a gap tunable by the electric field effect *Phys. Rev. Lett.* **99** 216802
[158] Chung H C, Chiu C W and Lin M F 2019 Spin-polarized magneto-electronic properties in buckled monolayer GaAs *Sci. Rep.* **9** 2332
[159] Chung H C, Su W P and Lin M F 2013 Electric-field-induced destruction of quasi-Landau levels in bilayer graphene nanoribbons *Phys. Chem. Chem. Phys.* **15** 868
[160] Wu J Y, Chen S C, Do T N, Su W P, Gumbs G and Lin M F 2018 The diverse magneto-optical selection rules in bilayer black phosphorus *Sci. Rep.* **8** 13303
[161] Do T N, Chang C P, Shih P H, Wu J Y and Lin M F 2017 Stacking-enriched magneto-transport properties of few-layer graphenes *Phys. Chem. Chem. Phys.* **19** 29525
[162] Santos H, Ayuela A, Chico L and Artacho E 2012 van der Waals interaction in magnetic bilayer graphene nanoribbons *Phys. Rev.* B **85** 245430
[163] Konschuh S, Gmitra M, Kochan D and Fabian J 2011 Theory of spin–orbit coupling in bilayer graphene *Phys. Rev.* B **85** 115423
[164] Shih P H, Chiu Y H, Wu J Y, Shyu F L and Lin M F 2017 Coulomb excitations of monolayer germanene *Sci. Rep.* **7** 40600
[165] Schütt M, Ostrovsky P M, Gornyi I V and Mirlin A D 2011 Coulomb interaction in graphene: relaxation rates and transport *Phys. Rev.* B **83** 155441
[166] Liao M, Zang Y, Guan Z, Li H, Gong Y and Xue Q K 2018 Superconductivity in few-layer stanene *Nat. Photonics* **14** 344
[167] Rossato J, Baierle R J, Fazzio A and Mota R 2018 Vacancy formation process in carbon nanotubes: first-principles approach *Nano Lett.* **5** 197
[168] Kratschmer W, Lamb L D, Fostiropoulos K and Huffman D 1990 Solid C_{60}: a new form of carbon *Nature* **347** 27
[169] Chen X *et al* 2001 New method of carbon onion growth by radio-frequency plasma-enhanced chemical vapor deposition *Chem. Phys. Lett.* **68** 631
[170] Mintmire J, Dunlap B and White C 1992 Are fullerene tubules metallic? *Phys. Rev. Lett.* **68** 631
[171] Liu J *et al* 1998 Fullerene pipes *Science* **280** 1253

[172] Balog R *et al* 2010 Bandgap opening in graphene induced by patterned hydrogen adsorption *Nat. Mater.* **9** 315
[173] Grassi R, Low T and Lundstrom M 2011 Scaling of the energy gap in pattern-hydrogenated graphene *Nano Lett.* **11** 4574
[174] Ellis C T *et al* 2013 Magneto-optical fingerprints of distinct graphene multilayers using the giant infrared Kerr effect *Sci. Rep.* **3** 3143
[175] Chen J Y, Zhu J, Zhang D, Lattery D M, Li M and Wang X 2016 Time-resolved magneto-optical Kerr effect of magnetic thin films for ultrafast thermal characterization *J. Phys. Chem. Lett.* **7** 2328
[176] Bao W *et al* 2011 Stacking-dependent band gap and quantum transport in trilayer graphene *Nat. Phys.* **7** 948
[177] Ho J H, Lu C L, Hwang C C, Chang C P and Lin M F 2006 Coulomb excitations in AA- and AB-stacked bilayer graphites *Phys. Rev.* B **74** 085406
[178] Shyu F L and Lin M F 2000 Loss spectra of graphite-related systems: a multiwall carbon nanotube, a single-wall carbon nanotube bundle, and graphite layers *Phys. Rev.* B **62** 8508
[179] Levy G, Nettke W, Ludbrook B M, Veenstra C N and Damascelli A 2014 Deconstruction of resolution effects in angle-resolved photoemission *Phys. Rev.* B **90** 045150

IOP Publishing

Shih-Yang Lin, Hsin-Yi Liu, Duy Khanh Nguyen, Ngoc Thanh Thuy Tran, Hai Duong Pham, Shen-Lin Chang, Chiun-Yan Lin and Ming-Fa Lin

Chapter 6

Hydrogen-adsorbed silicene materials

6.1 Hydrogenated silicenes

Group IV materials with 2D structure have attracted considerable attention in the fields of chemistry, materials science and physics [1–6] due to their planar or buckled structures of sp^2 or sp^3 bonding formed by four valence electrons. The motivation for studies on the electronic structure [7–12], optical adsorption [13–16] and transport properties [17–20] of such materials is thus provided. Electronic properties with a band gap opening are an important focus for possible use in next-generation nanoelectronic devices [21–24]. Several methods have been realized experimentally, including chemical functionalization [25–31], which is particularly promising for tuning electronic structures. The hydrogenation of graphene has created a material, graphane, with a large band gap [32–35]. Its valence band structures have been verified using angle-resolved photoemission spectroscopy (ARPES) [33–35]. The conduction and valence band structure of graphane can be modified significantly through reversible hydrogenation [32], however, the corresponding concentration is difficult to estimate. Graphene still faces a lot of challenges in applications, such as the difficulty of synthesis, material toxicity, integrating it within current silicon-based industry, and the availability of large-area production. This chapter investigates the unique H–Si bonds with feature-rich geometric and electronic properties in hydrogenated silicene.

Two-dimensional hexagonal lattices of Si, namely silicene (a counterpart to graphene), have been fabricated on silver substrates [3, 36–38] and iridium substrates [39] using chemical vapor deposition. Chemical exfoliation methods have created suspended monolayer silicene doped with magnesium and covered with organic groups [40–42]. These experimental improvements can provide suitable environments for chemical functionalizations of single-sided and double-sided adsorption, and further tune the electronic properties of silicene. Silicene, with its diverse electronic properties, could enable the electronics industry to produce components without the

doi:10.1088/978-0-7503-3299-6ch6

need to retool, which is very different to graphene and other carbon-based nanostructures

Unlike graphene, silicene is not stable as a perfectly planar sheet. sp^3 hybridization is more stable in silicene than sp^2 hybridization, so its ground state is a low buckled structure. Such a structure has been confirmed using density functional theory (DFT) and phonon calculations [43]. The electronic structure of silicene has been verified to be a zero-gap semiconductor using scanning tunneling spectroscopy (STS) and ARPES measurements [3, 5, 44]. Theoretical works exploring silicene sheets terminated with hydrogen show an indirect middle band gap [45, 46]. Such results indicate that the electronic properties of hydrogenated silicene can be tuned through competition between hydrogenation and sp^3 hybridization. However, silicene sheets have only been realized in the last few years and thus the influence of hydrogen concentrations under various configurations on silicene has not yet been explored thoroughly.

In this chapter, the feature-rich electronic structures induced by competition between the H–Si bonds and the Si–Si σ and π bonds is studied using first principles. Two specific configurations are selected, including double-sided (the chair configuration in figure 6.1(a)) and single-sided hydrogenations (the top configuration in figure 6.1(b)). They are expected to reveal two basic types of electronic properties, which can be further tuned by various hydrogen concentrations. Many kinds of energy dispersions are revealed in this chapter, including the existence of a Dirac cone, partially flat bands induced by H–Si bonds and localized states near the Fermi level induced by broken mirror symmetry. All the predicted energy bands are further reflected in van Hove signatures in the density of states (DOS), such as strong peaks related to the hydrogen concentration. The hydrogenation effects on band structures and the DOS can be understood through the analysis of the π and σ bond charge distributions. The predicted band gaps and energy dispersions, and the van Hove structures of the DOS, can be further verified using ARPES and STS [3, 5, 47, 48], respectively. The latter method is very useful, and is a newly invented method to determine the hydrogen concentration from its relation to the intensities of peaks. The rich features of hydrogenated silicenes may promote their future application in energy storage materials and nanoelectronic devices.

The specific relationships between the geometric structures of hydrogenated silicene and various hydrogen configurations are obtained. Two different atomic arrangements are taken into consideration for fully hydrogenated silicene. In the chair-like configuration, as shown in figure 6.1(a), hydrogen atoms are located alternately on both sides of the silicene sheet. Hydrogen atoms are selected to be distributed uniformly, and each H atom is separated from its six nearest H atoms by the same distance (figure 6.1(c)). The different sites of hydrogen atoms (the green spheres) are set to be at the middle of three front-site hydrogen atoms (red spheres). The hydrogen–silicon ratios are reduced to 2:2, 2:8 and 2:32, where the corresponding hydrogen concentrations are 100%, 25% and 6.3%, respectively (figure 6.1(c)–(e)). For the other structure, the top configuration, the hydrogen atoms exist only on one side of the silicene sheet (figure 6.1(b)), that is, all the back-side hydrogen atoms are not present. Such a structure has a H concentration half that of the chair-like

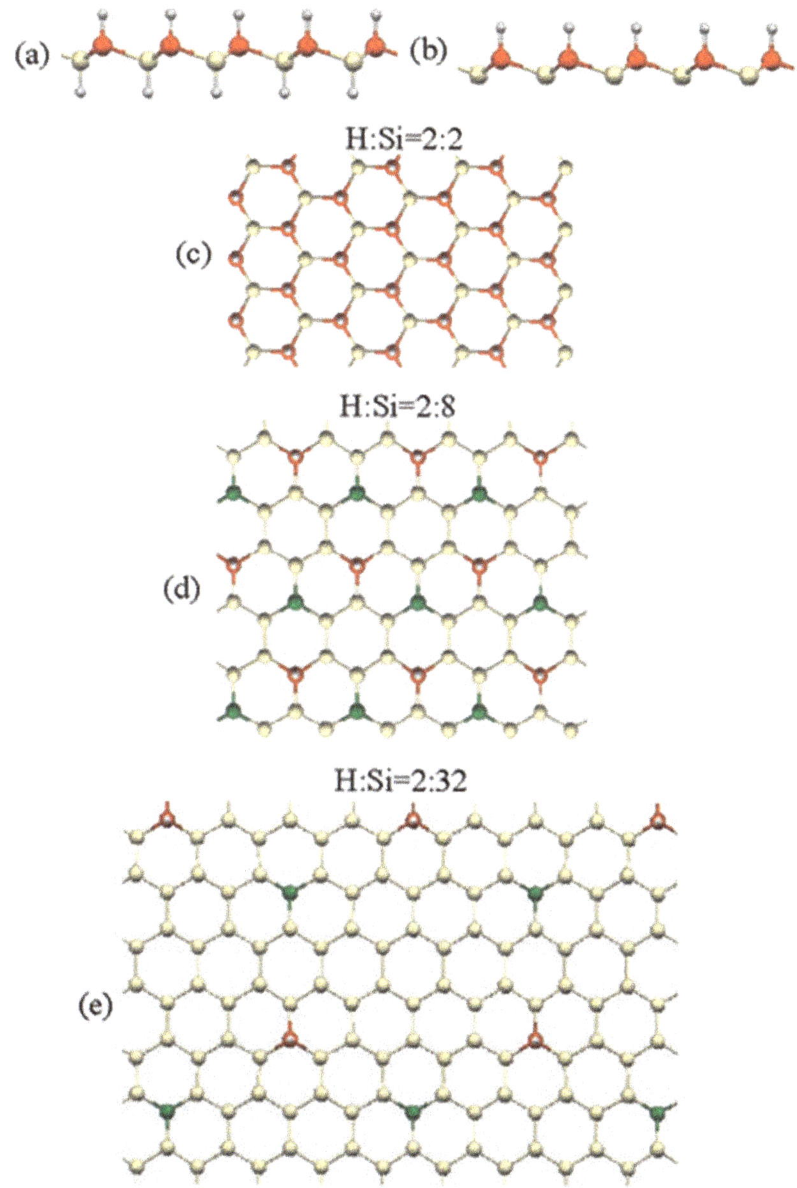

Figure 6.1. Side view atomic configurations of hydrogenated silicenes: (a) chair configuration, where the H atoms appear on both sides of the silicene plane, and (b) top configuration, where all H atoms are located on the same side of the plane. Top view for various concentrations, namely (c) 2:2, (d) 2:8 and (e) 2:32.

configuration. These two structures are shown to have similar geometric deformations to each other, although they may lead to diverse electronic structures due to the breaking of mirror symmetry.

The main features of the geometric structures are determined by the H concentration, including the Si–Si bond lengths, H–Si–Si angles and H–Si bond lengths.

The Si–Si bond lengths are non-uniform in a silicene sheet after hydrogenation, while they are all of the same value in pristine silicene, 2.25 Å. The H–Si–Si angles change significantly from 107.0° to 111.7° as the hydrogen concentration is decreased from 100% to 3.1%. The obvious attraction of electrons from Si atoms to H forms strong H–Si covalent bonds (discussed later). This charge transfer results in weaker nearest Si–Si bonds but stronger second-nearest bonds as viewed from the hydrogen atoms. Such bond lengths become longer and shorter in the ranges of 2.31~2.33 Å and 2.23~2.24 Å, respectively. It is noted that the σ bonding determines the planar Si–Si bond lengths due to the 3s, $3p_x$ and $3p_y$ orbitals. However, the H–Si bond lengths are insensitive to hydrogen concentration—they remain at 1.50 Å, owing to the stable and strong covalent bonding. The above-mentioned bonding schemes are related to the strong $3p_z$ and 1s orbital covalent bonds, and the hybridization of the four orbitals (3s, $3p_x$, $3p_y$ and $3p_z$) with $3p_z$ orbitals perpendicular to the plane (discussed later).

Hydrogenated silicene reveals special energy bands, which originate from the competition of weak sp^3 hybridization and strong H–Si bonding. Pristine silicene has an almost vanishing gap of ~1 meV, and the linear dispersions at low energy turn into parabolic dispersions with a saddle at the M point, as shown in figure 6.2(a). These bands are regarded as π bands, which come from the $3p_z$ orbital bonding of two nearest silicon atoms. The clear separation of the π bands and σ bands within ±2 eV is indicated by the weak sp^3 hybridization. The occupied and unoccupied σ bands are initiated at about −1.2 and 1 eV with parabolic energy dispersions, respectively. The lowest unoccupied states (LUSs) are not degenerate states at the M point, while the highest occupied states (HOSs) are doubly degenerate at the Γ point. The LUSs has energy bands related closely to the $3p_x$ and $3p_y$ orbitals with $3p_z$ orbital hybridization at lower energy (~−2.5 eV). The third σ band with an initial energy at −11.2 eV increases gradually with parabolic behavior, which is dominated by the 3s orbitals. However, this σ band becomes partially flat at −3.2 eV near the Γ point, where the $3p_z$ orbitals hybridize with 3s orbitals (discussed later). The above-mentioned main characteristics of silicene are changed after hydrogenation.

Chair-like systems are semiconductors and possess H concentration-dependent energy gaps. A linear Dirac cone does not exist in fully hydrogenated silicene, resulting from the strong H–Si bonding, as illustrated in figure 6.2(b). A large indirect band gap of $E_g = 2.1$ eV is present, which is determined by the HOS and the LUS, respectively. All of the $3p_z$ orbitals are saturated by 1s orbitals, and thus the low-lying π bands become the σ bands corresponding to Si atoms. The amount of contribution is indicated by the radius of the red circles in figure 6.2(b). A clear revelation is that the $3p_x$ and $3p_y$ orbitals form valence σ bands. On the other hand, the weakly dispersed band at about −3.5 eV arises mainly from one of the specific 1s orbitals of two H atoms (blue circles), and the other 1s orbitals contribute equally to the two partially flat bands. These electronic states can be treated roughly as localized electronic states. The fact that the 1s orbital has a much lower bound state energy than the other orbitals illustrates H-dependent localization behavior. Furthermore, the energy gap is determined by the σ bands, because the 1s and $3p_z$ orbital hybridizations push the π bands away from $E_F = 0$.

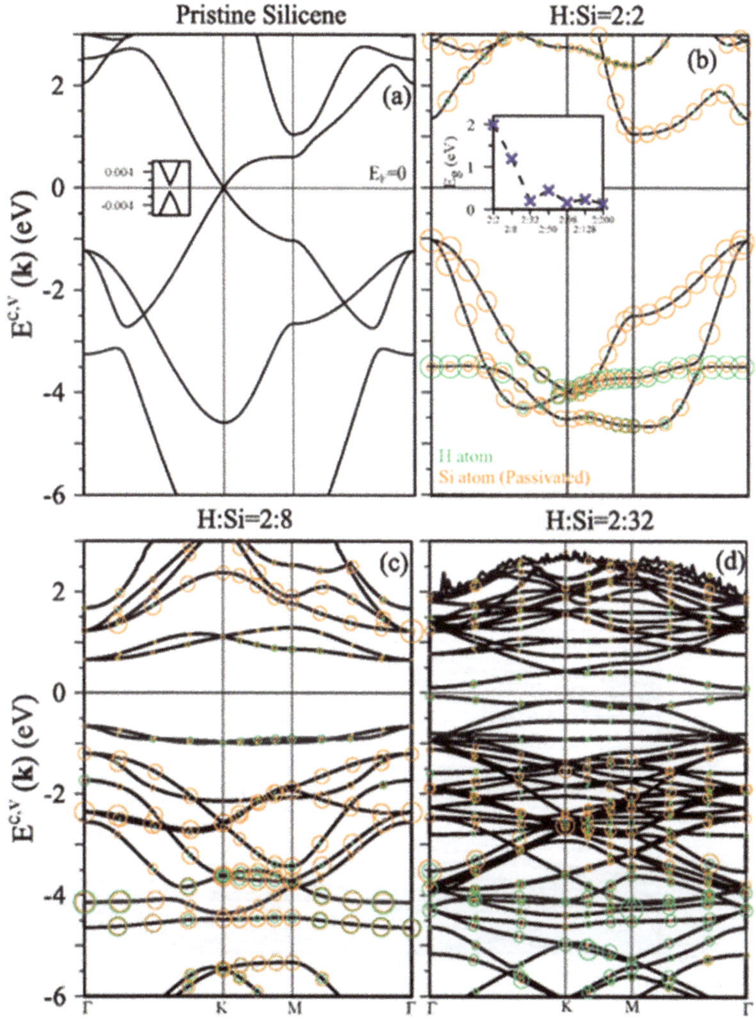

Figure 6.2. Energy bands of (a) pristine silicene, and hydrogenated silicenes with a chair configuration for (b) 2:2, (c) 2:8 and (d) 2:32 hydrogen–silicon ratios. The inset in (a) shows the low-lying energy bands and the inset in (b) the band gaps of various concentrations.

When the concentration decreases further, the H–Si bonds contribute much less and compete with the weak sp^3 hybridization to decide the energy bands. The 25% and 6.3% structures are intermediate configurations between fully hydrogenated and pristine silicene, and they possess 1.3 and 0.2 eV direct band gaps at the Γ point, respectively, as shown in figures 6.2(c) and (d). There are two pairs of energy bands nearest to $E_F = 0$, which exhibit relatively weak dispersions, and they originate from the non-passivated silicon atoms' $3p_z$ orbitals. Such features suggest that the low-energy σ bands are replaced gradually by π bands. In the middle-energy range (−1 to −3.5 eV), the σ bands originate from more silicon atoms, including passivated (red circles in figure 6.2(c)) and

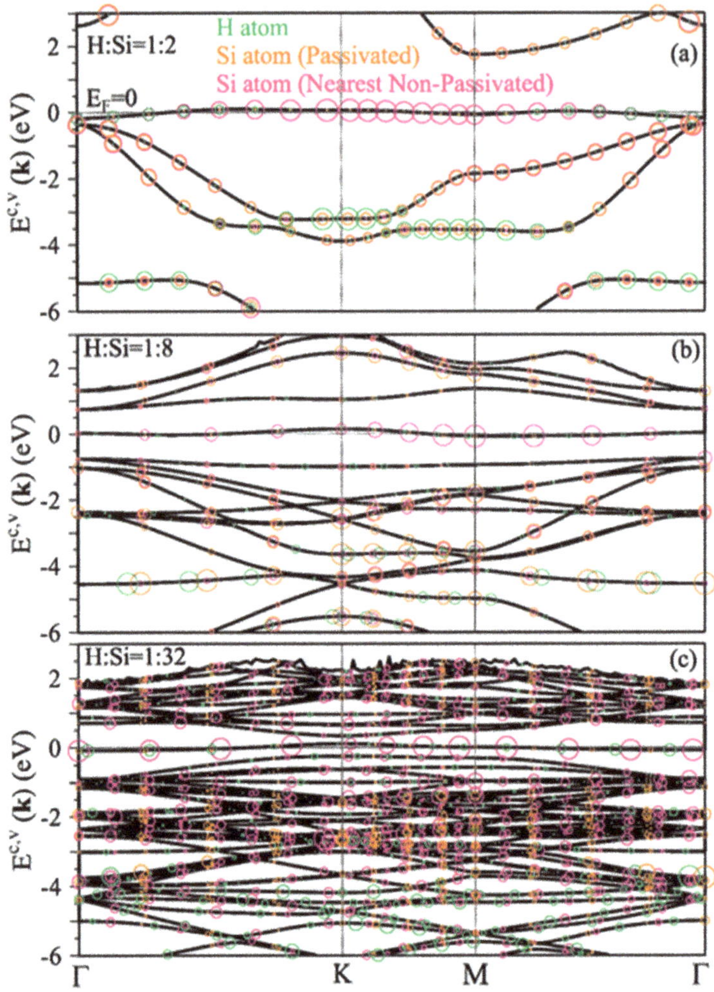

Figure 6.3. Energy bands of the top configuration for (a) 1:2, (b) 1:8 and (c) 1:32 concentrations.

non-passivated ones. The latter dominate mainly such bands in the 6.3% structure, which results from the higher ratio of such atoms. The hydrogen atoms contribute weakly to the dispersed band (blue circles), and they move to −4.0 and −4.2 eV, respectively, for the 25% and 6.3% structures. The other low concentrations have energy gaps that fluctuate and decline gradually. The tunable direct band gap in hydrogenated silicene may be applied in nanophotonic and nanoelectronic devices in the future.

The highly localized states at the Fermi level induced by broken mirror symmetry are observed for the top configuration. Fifty per cent of the structure has an almost flat band structure at $E_F = 0$ along different k-space directions (figure 6.3(a)). This originates mainly from the three non-passivated silicon atoms (green circles) nearest to the passivated silicon atom. The quasi-0D states are not influenced by the hydrogen concentration, as shown by the 12.5% and 3.1% structures (figures 6.3(b)

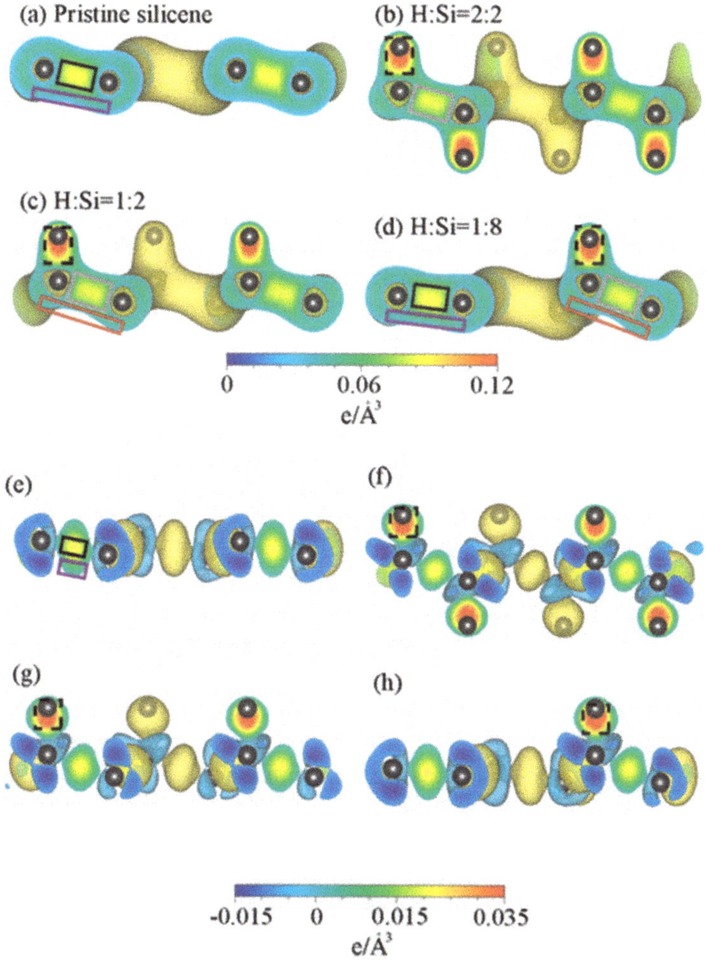

Figure 6.4. The carrier density ρ of (a) pristine silicene, and hydrogenated silicenes with (b) 2:2, (c) 1:2 and (d) 1:8 concentrations. The variation of carrier density $\Delta\rho$ of (e) pristine silicene, and hydrogenated silicenes with (f) 2:2, (g) 1:2 and (h) 1:8 concentrations.

and (c)). The π bonding between the non-passivated and passivated Si atoms is largely suppressed by the significant charge transfer between the Si atoms and the H atoms. Importantly, these systems are gapless, but are not good conductors due to the extremely low group velocities. The main difference between the top and chair configurations is in the intriguing localized states. Furthermore, the weakly dispersed band of the chair configurations is absent in top configuration structures due to the decreasing single effective 1s orbital.

The top and chair configurations share certain similarities, including the localized states created by the H–Si bond, the σ bands related to the Si–Si bond, and the recreation of π bands at low hydrogen concentrations. These two systems exhibit Si-atom-dominated low-energy bands as the H concentration decreases, and they both possess stronger dispersions with larger energy widths, as indicated in figures 6.2(d) and 6.3(c). These bands originate mostly from $3p_z$ orbitals with the π bonding reformed by the large amount of non-passivated Si atoms. The initial energy, band width and state degeneracy of the σ bands within −1 to −3.5 eV are very similar between the two systems (figures 6.2(a)–(c) and 6.3(a)–(c)) due to the barely affected planar σ bonds. In addition, the H–Si-bond-induced partially flat bands (the blue and red circles) are restrained in a similar energy range compared to their counterparts in the chair configuration. This is attributed to the much lower bound state energy of hydrogen. The above-mentioned similarities are affected little by the reduction of the mirror symmetry.

The carrier density (ρ) and the variation of carrier density ($\Delta\rho$) are used to understand the charge distribution in the bonding of critical orbitals, and thus are very useful in explaining the changes of low-energy bands. ρ exhibits the bond strength of the H–Si and Si–Si bonds directly, as shown in figures 6.4(a)–(d). Strong covalent σ bonds are present between two Si atoms in a pristine system or two non-passivated Si atoms, as illustrated by the black rectangles in figures 6.4(a), (c) and (d). These σ bonds become slightly weaker when the Si–H bonds are formed (the gray rectangles in figures 6.4(b)–(d)). The small change can be attributed to the similar σ bands between the two systems under different hydrogen concentrations. The carrier density is lower, while its distribution is extended between them away from the central line between two Si atoms (the purple rectangle). This can provide strong evidence of the π-electronic states, exhibiting the recovery of π bands at low hydrogen concentrations. However, barely extended states, which relate to the π-electron absent region, are discovered between the passivated and non-passivated atoms, as shown by the red rectangle in figure 6.4(d). A similar charge distribution is found in isolated Si atoms, demonstrating the disappearance of π bonding. On the other hand, the H–Si bonds reveal a much stronger charge density than those of all the Si–Si bonds (the dashed rectangle). Such strong bonds are not sensitive to the hydrogen concentration in all hydrogenated silicene systems; hence, the H-dominated bands appear within the range of −3.5 to −5.5 eV. The charge distributions can also explain the changes in geometric structures, i.e. only the nearest Si–Si bond lengths are changed.

The variation of carrier density is shown to reflect the charge transfers among all orbitals (figures 6.4(e)–(h)). The electrons of pristine silicene are transferred from the blue region (near Si atoms) to the middle of two Si atoms, as indicated by the yellow region within a black rectangle in figure 6.4(e), namely covalent σ bonds. ρ also increases in the green region distant from the middle between two Si atoms (the purple rectangle), which can be treated as the creation of π bonding. In addition, all the hydrogenated structures exhibit a large amount of charge transfer from Si to H atoms (~ 0.9 e by Bader charge analysis), which is solid evidence of strong H–Si bonds, as indicated by the dashed rectangles in figures 6.4(f)–(h). This phenomenon also elucidates the weak bonding strength of the nearest Si–Si bond to the H atom.

The DOS and the partial DOS (PDOS) are used to understand the orbital-decomposed contributions of band structures. The DOS is zero at $E_F = 0$ for pristine silicene; it has a linear energy dependence near the Fermi level, and a symmetric logarithmic divergent peak exists at E ~ ±1 eV, as shown by the black curve in

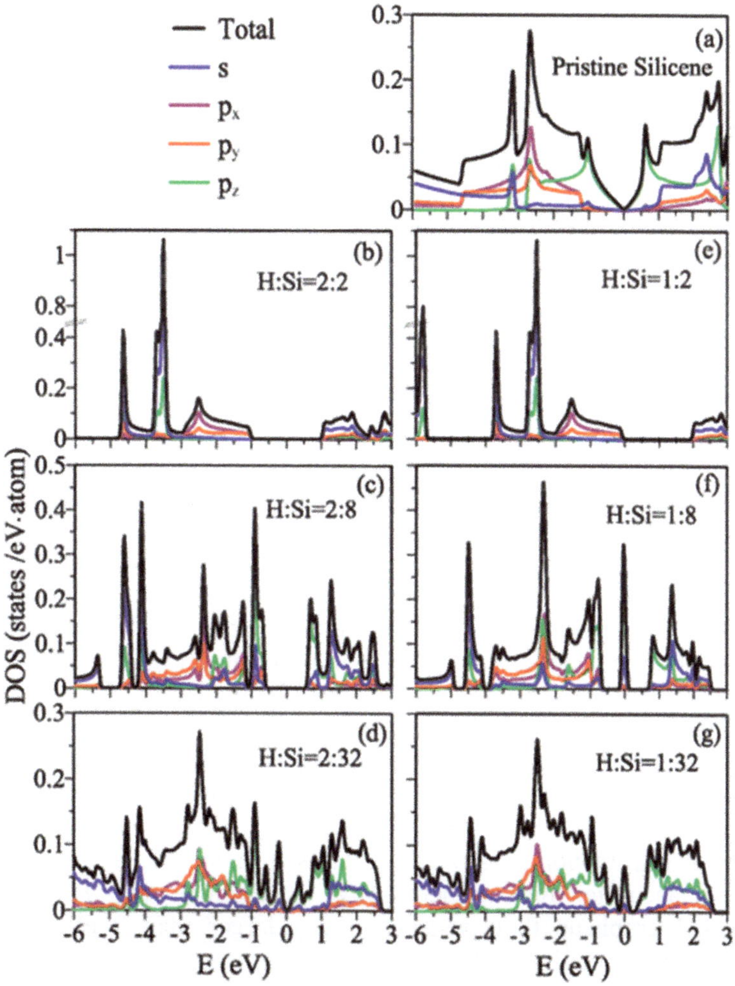

Figure 6.5. Densities of states of (a) pristine silicenes, chair configurations with (b) 2:2, (c) 2:8 and (d) 2:32 hydrogen coverage, and top configurations with (e) 1:2, (f) 1:8 and (g) 1:32 hydrogen coverage.

figure 6.5(a). These three features arise from the almost zero band gaps, the Dirac-cone structures and the saddle points of π bands, respectively. They are dominated purely by the $3p_z$ orbitals (the purple curve). A shoulder structure is presented in the middle-energy DOS (−1.2 eV), and a prominent peak is revealed at deeper energy (−2.7 eV) corresponding to, respectively, the saddle M point and the maximum point of the σ bands. Such features arise mainly through the $3p_x$ and $3p_y$ orbitals, as indicated by the green and blue curves in figure 6.5(a).

After hydrogenation, the DOS is changed drastically in the low-energy region. The DOS exhibits a stateless region at $E \leqslant \sim|1|$ eV and an extra prominent peak at about −3.5 and −4.5 eV, respectively, for fully hydrogenated silicene in the chair configuration arising from, respectively, the weak dispersive band and the partially flat band associated with H atoms, as indicated in figure 6.5(b). These features can be attributed to the strong bonding of the $3p_z$ and 1s orbitals (the purple and red curves). However, the prominent peak at −2.5 eV and the shoulder structure at −1.0 eV are very similar to those of pristine silicene because of the slight change in the σ band, which is composed of the $3p_x$ and $3p_y$ orbitals. Two prominent peaks are presented at ±0.6 eV in the 25% system as the hydrogen concentration decreases, as indicated in figure 6.5(c). The width of the peak is about 0.5 eV, and these states are related to π bands and thus the $3p_z$ orbitals. The $3p_x$ and $3p_y$ orbitals dominate the middle-energy states; however, there exist several subpeaks due to an increase of the number of subbands. The $3p_z$ orbitals contribute most of the states in the range of ±1 eV in the 6.3% structure, which suggests a similar band width of π electronic states to that in pristine silicene. Two strong peaks between −4.0 and 4.5 eV are revealed for the two lower concentrations, arising mainly from the $3p_z$ orbitals of Si atoms (the purple curve) and the 1s orbital of the H atoms (the red curve). The H concentration is roughly proportional to the intensities of these two peaks.

Common characteristics shared between the two systems in the DOS are found to be the strong peaks related to the planar σ bonds and the H-related bonds. Several analogies are observed between a top structure and its corresponding chair structure, including the widened π band width at low-energy (figures 6.5(d) and (g)), the prominent peaks dominated by two $3p_z$ orbitals at ±0.6 eV (figures 6.5(c) and (f)), the middle-energy peaks induced by $3p_x$ and $3p_y$ orbitals (figure 6.5(b)–(g)), and the strong peaks created by the H–Si bonds. These feature-rich peaks appear at almost the same energy. Interestingly, their peak intensities decrease gradually when hydrogen concentration decreases. The destruction of the mirror symmetry results in a delta-function-like peak at E_F (figures 6.5(e)–(g)), which is a critical distinction. The intensity of such peaks decreases gradually, which is responsible for the percentage of nearest non-passivated Si atoms. In addition, STS experiments can provide an accurate and efficient way to examine the H concentration. Hydrogenated graphene that possesses a gap larger than 2 eV has also been measured using this method [47–50].

6.2 Hydrogenated graphene

Graphene, a 2D allotrope of carbon structures, has attracted many theoretical and experimental studies [51–53] related mainly to its remarkable fundamental properties [54–56]. Graphene's zero-gap semiconductor properties can be tuned easily using adatom adsorption [57–59], different numbers of layers [60, 61], various stacking configurations [62, 63], external electronic or magnetic fields [64, 65] and significant mechanical strains [66, 67]. Atom adsorption is a relatively effective approach among these methods to induce metallic or semi-conducting properties.

Hydrogenated graphene was first reported by Elias *et al* [32]. Its electronic properties can be modulated by H atom adsorption, which has attracted a large number of experiments studies. Balog *et al* [34] discovered that H adsorption with a moiré pattern can open a band gap of 0.45 eV. A scanning tunneling microscopy (STM) image displaying the disordered adsorption pattern of a hydrogen-saturated surface was presented by Sessi *et al* [68]. The H concentration can be verified by the D and G band intensity in Raman spectroscopy [69]. Importantly, a H-atom-induced gap was discovered successfully using ARPES [34] and STS [70], which can be very useful for room temperature electronic applications.

Hydrogenated graphene is not as stable as a completely planar sheet. It is energetically favorable for a low buckled structure since the sp^3 bonding is more stable than the sp^2 bonding. The electronic structure of fully hydrogenated graphene has been proved to be an insulator using first principles calculations [71]. Other theoretical studies have explored the absence of an energy gap under special H atom distributions [72]. Varied H concentrations and distributions are also investigated [34, 73, 74]. These theoretical works indicate that the electronic structures of graphene could be altered easily.

DFT is used to investigate the optimal geometric configurations of hydrogenated graphene systematically. Three classical configurations with different hydrogen distributions are selected for 1D carbon nanotubes (figure 6.6(a)). Fully hydrogenated graphene (graphane) contains two carbon atoms and two hydrogen atoms in a unit cell, as shown in figure 6.6(b) for the chair-like structure. The geometric structures can be classified into three types based on the carbon atom arrangements between two hydrogen atoms. The first type, namely zigzag-hydrogenated graphene, includes 100%, 25% (figure 6.6(c)), 11.1%, 6.3%, 4% and 2% hydrogen concentrations. A primitive unit cell is enlarged in order to simulate the reduction of concentrations. The second type is the armchair-hydrogenated graphene, the concentrations of which vary as 33% (figure 6.6(d)), 8.3% and 3.7. The third type is chiral-hydrogenated graphene, where the concentration is 14.3% (figure 6.6(e)), 3.6% and 1.6%. These three types exhibit similar geometric variations under various concentrations, but they may have varied electronic structures.

The main features of optimized structures depend strongly on the distribution and concentration of hydrogen atoms, including the bond lengths of two carbon atoms and carbon–hydrogen atoms, the angles of the H–C–C bonds and the heights of the carbon atoms. The C–C bond lengths are non-uniform in hydrogenated graphene, but they are all 1.41 Å in pristine graphene. Measured from the hydrogen atom, the nearest- and next-nearest C–C bond lengths are about 1.47–1.53 Å and 1.36–1.40 Å, respectively. The H–C bond lengths remain almost unchanged around 1.12–1.15 Å under various hydrogen concentrations. The H–C–C angles change from 100.5° to 107.1° and decrease under a dilute H concentration in the chiral configuration, but for H–C–C angles in the armchair configuration the opposite is true. The nearest carbon atoms viewed from hydrogen atoms deviate from the graphene plane. When the H concentration decreases, such deviations increase from 0.22 to 0.57 Å.

STM measurements are able to provide the atomic arrangements of nanostructures. Studies have been conducted to observe the atomic structures of carbon-related materials, including 2D graphene [75], graphene compounds [34], 3D graphite [76], 1D graphene nanoribbons [77] and 1D carbon nanotubes [78]. The observations directly reveal the buckled and rippled structures of graphene [75], the distributions of doping atoms [79], the defects in graphite [76], the chiral structure of planar edges [80] and the cylindrical surface of nanotubes [78]. The predicted

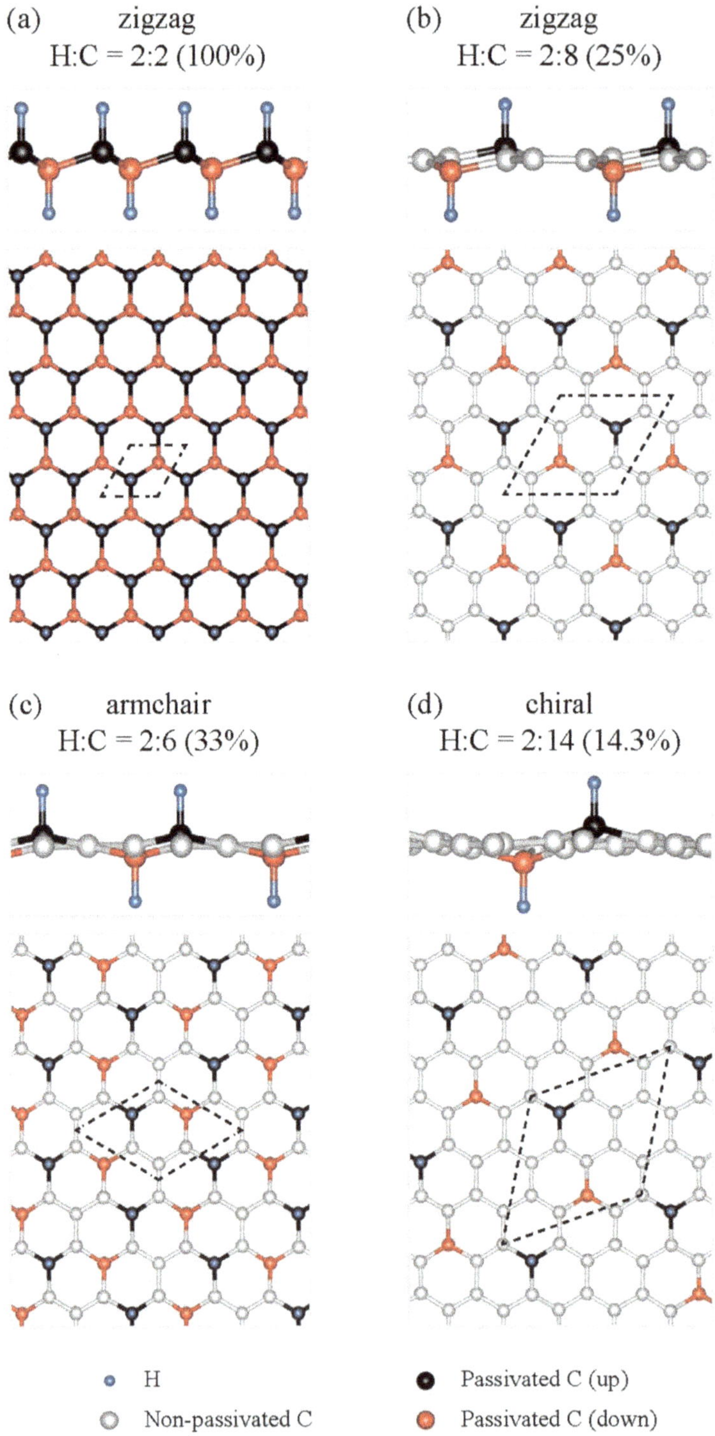

Figure 6.6. Geometric structures of hydrogenated graphene with top views and side views of (a) fully hydrogenated graphene, (b) 25% zigzag configuration, (c) 33% armchair configuration and (d) 14.3% chiral configuration.

structural properties of hydrogenated graphene deserve closer examination, including the stable H–C bonds and the large changes to the C–C bonds, C atom height and H–C–C angles.

The electronic structure is changed significantly due to the strong H–C bonding. The valence bands and the conduction bands are not symmetric around the Fermi level ($E_F = 0$), and the energy bands are mostly anisotropic along various high-symmetry directions, as shown in figures 6.7 and 6.8. Three kinds of energy dispersions are observed, including linear bands, parabolic bands and partially flat

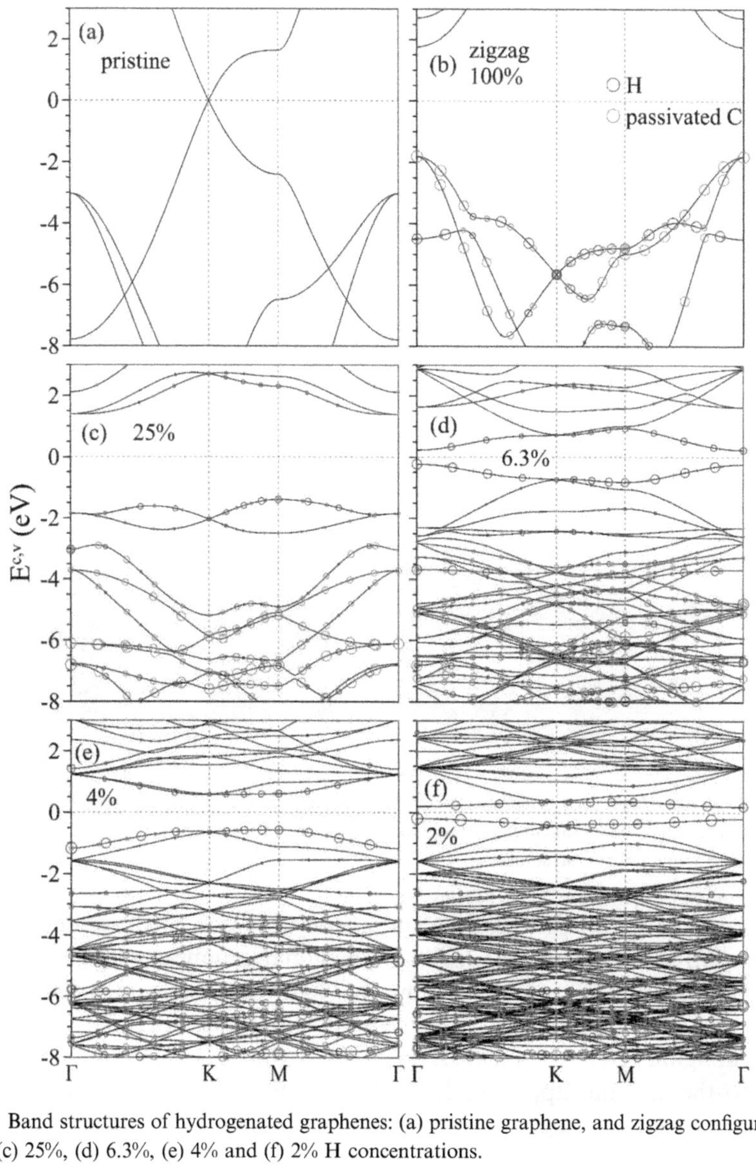

Figure 6.7. Band structures of hydrogenated graphenes: (a) pristine graphene, and zigzag configurations with (b) 100%, (c) 25%, (d) 6.3%, (e) 4% and (f) 2% H concentrations.

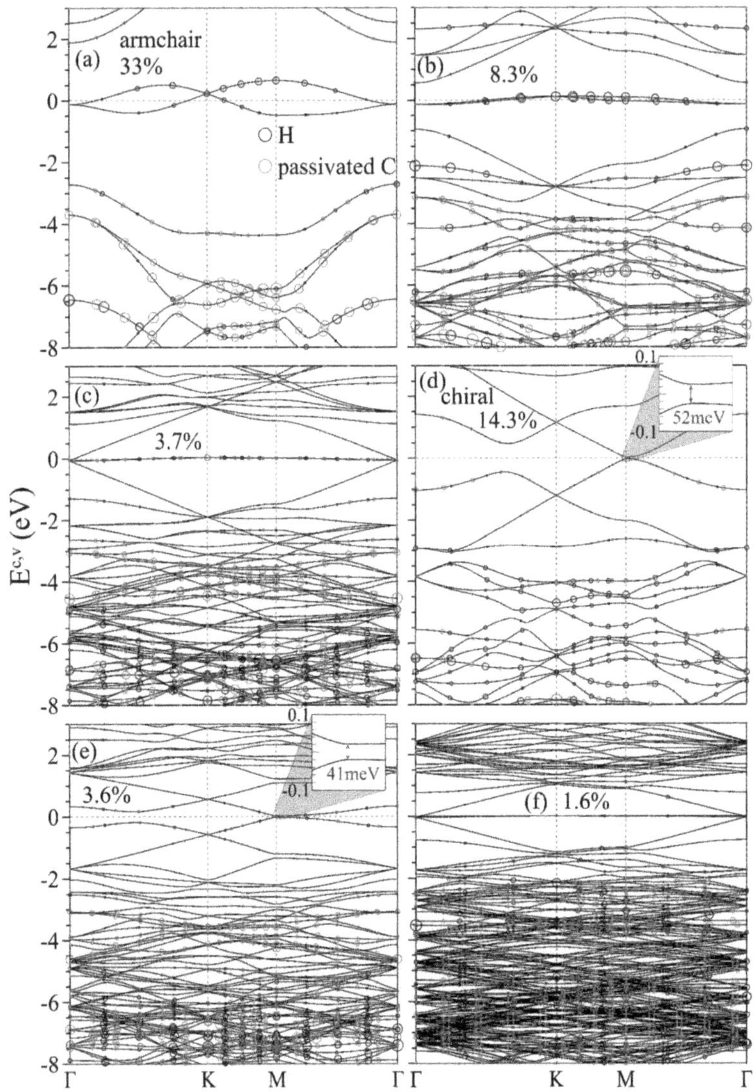

Figure 6.8. Band structures of hydrogenated graphenes: armchair configurations with (a) 33%, (b) 8.3% and (c) 3.7% H concentrations, and chiral configurations with (d) 14.3%, (e) 3.6% and (f) 1.6% H concentrations. The insets show the low-lying band structures.

bands. The linear isotropic bands of graphene (figure 6.7(a)) are destroyed in all other concentrations due to the lack of hexagonal symmetry. The energy gaps of zigzag systems can be modulated by the hydrogen concentration. The 100% hydrogenated structure exhibits the largest direct band gap of 3.5 eV at the Γ point, as shown in figure 6.7(b). Its HOS ($E^v = -1.7$ eV) and LUS ($E^c = 1.7$ eV) are related strongly to the $2p_x$ and $2p_y$ orbitals of the σ bands and the $2p_z$ orbitals of the π^* band, respectively. Hydrogen atoms contribute mostly to the deeper bands in the range $E^v \sim -3.5$ eV $\leqslant E \leqslant -8$ eV (the blue circles in figure 6.7(b)). The dramatic changes in low-energy bands can be ascribed to the strong H–C chemical bonding and the low bound state energy of hydrogen. Such features are similar to those of 100% hydrogenated silicene, as discussed in section 6.1.

The π bands approach the Fermi level gradually as the concentration decreases, while the σ bands do not. The π-electron-dominated HOS shifts to the M point for the 25% structure, and thus induces an indirect moderate gap of $E_g = 2.5$ eV. A direct band gap of 0.25 eV at the Γ point is observed for the 6.3% configuration with partially flat bands at ~−2.5 eV. The HOS and the LUS in the 4% system, respectively, occur at the M and K points. This results in an indirect band gap of 0.75 eV. As for the 2% structure, it possesses the smallest direct band gap (0.2 eV), which is contributed by non-passivated carbons.

The distribution of hydrogen affects the electronic properties. There are two weakly dispersive bands across E_F for the armchair-hydrogenated graphene, as shown in figures 6.8(a)–(c). These two bands originate mainly from the localized states of hydrogen and the second-, fourth- or sixth-nearest C atoms counted from the H atoms. These systems are gapless under various hydrogen concentrations. For the low concentration of 3.7%, a distorted Dirac cone is observed to be recovered at the Γ point due to the recovery of π electronic states, as shown in figure 6.8(c). On the other hand, chiral systems exhibit narrow-gap or gapless behavior, as indicated in figures 6.8(d)–(f). A pair of parabolic bands with the HOS and LUS at the M point determines a small direct gap. The E_g are 50 and 35 meV for, respectively, the 14% and 3.6% systems, as indicated by the insets in figures 6.8(d) and (f). Moreover, both the chiral and armchair systems present similar gapless band structures when the H concentration is sufficiently low. The linear energy dispersions of graphene have been identified using ARPES [81]. This experimental method has been used to verify the layer numbers [60, 82], stacking sequences [83], threshold voltages [84] and doping atom adsorption [85, 86]. In particular, the high-resolution method observes alkali-induced free electrons in the linear conduction band [87]. The above-mentioned energy spectra of hydrogenated graphenes are expected to be further tested using ARPES measurements.

The drastic changes in band structures can be understood fully by studying the charge density (ρ) and the charge density difference ($\Delta\rho$) [88, 89]. A strong covalent σ bond with a high charge density of the $2p_x$, $2p_y$ and 2s orbitals exists between two C atoms, as indicated by the rectangle in figure 6.9(a) for pristine graphene. After hydrogenization, the σ bonding is reduced slightly (figures 6.9(b)–(d)). Then a strong H–C bond is observed, as shown by the dashed rectangles in figures 6.9(b)–(d), which comes from the 1s orbitals of H atoms and the $2p_z$ orbitals of passivated C atoms. This H–C bond is insensitive to hydrogen concentration, reflected directly in the barely changed hydrogen induced bands (figures 6.7(b)–(f) and 6.8). On the other hand, $\Delta\rho$ reveals the weak π bonding of pristine graphene, as shown by the gray rectangle in figure 6.9(a). After 100% hydrogenation, all π bonds are terminated, as illustrated in figure 6.9(b). This is responsible for the absence of low-energy π band

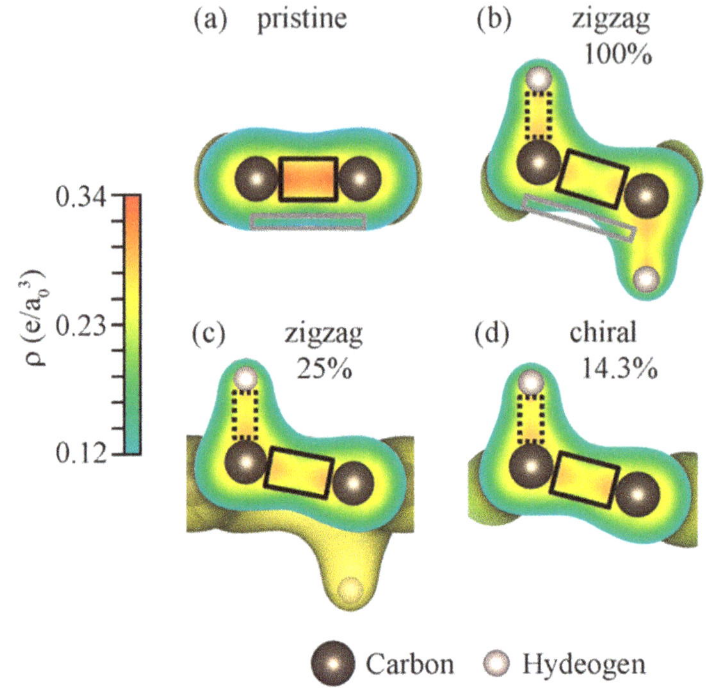

Figure 6.9. The charge density ρ of (a) pristine graphene, and (b) 100%, (c) 25% and (d) 14.3% hydrogenated graphene.

structures and the relatively large band gap (figure 6.7(b)). The first- and second-nearest neighbor carbons show terminations between each other, as indicated clearly by the arrows in figure 6.10(a). The π bonding is recovered steadily as the H concentration becomes low, but the nearest C–C bond is not (figures 6.10(b)–(d)), which accounts for the diminished energy gap in figures 6.7(b)–(f). For the chiral configurations, similar ρ and $\Delta\rho$ are also discovered in figures 6.10(e) and (f).

Hydrogenated graphenes present a lot of van Hove singularities in the DOS, as shown in figures 6.11(a)–(f), including sharp peaks, discrete shoulders and logarithmically divergent peaks. They arise from, respectively, the partially flat bands, the local minimum or local maximum points, and the saddle points of the energy bands. The DOS in pristine graphene is dominated fully by the π bonding in the range of -3 eV $\leqslant E \leqslant 3$ eV, as indicated by the red curve in figure 6.11(a). The DOS then presents a strong peak at -2.2 or 1.8 eV in the logarithmic form. The σ bonding starts to make more contributions to the DOS with increasing E, which includes a symmetric peak at -6.5 eV and a significant shoulder at -3 eV. The distinct structures of the π and σ bonding suggest that the interactions between the $2p_z$ and ($2p_x$, $2p_y$) orbitals are negligible.

The DOS is vanishing for fully hydrogenated graphene in the energy range ~ ±1.8 eV (figure 6.11(b)). The σ-electronic state induces a shoulder structure at -1.8 eV due to the $2p_x$ and $2p_y$ orbitals. The C–H bonding results in peak structures in the DOS in

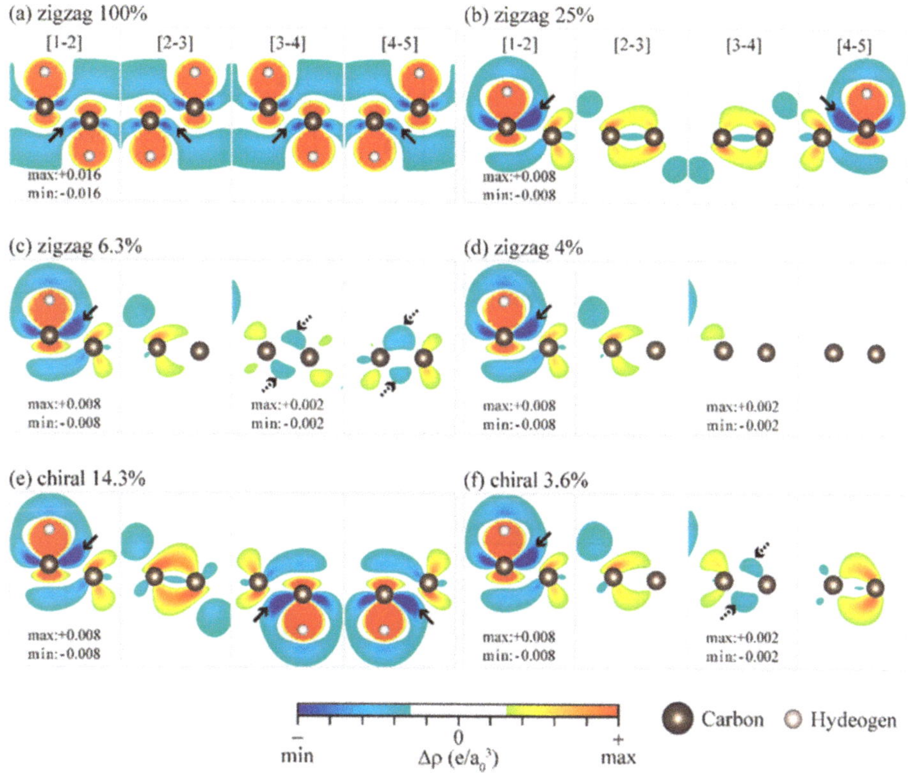

Figure 6.10. The charge density difference $\Delta\rho$ of hydrogenated graphene in (a) 100%, (b) 25%, (c) 6.3%, (d) 4%, (e) 14.3% and (f) 3.6% systems.

the range of -3.5 eV $\leqslant E \leqslant -8$ eV. The 25% structure exhibits two prominent peaks at about ± 1.3 eV, as shown in figure 6.11(c), which are related to the weak π electron bonds. The primary characteristics of the DOS are similar to those of pristine graphene for very low concentration, where the $2p_z$ orbitals dominate at low energy. For example, the strong symmetric peak at -6.5 eV and the shoulder-like structure at -3 eV in the 2% structure (figure 6.11(d)).

There are similar structures among the three different configurations in terms of the DOS due to strong chemical bonding, as shown in figures 6.11(b)–(h). These features include the bonding of 1s and $2p_z$ orbitals (figures 6.11(b), (e) and (g)) and prominent peaks from the planar σ bonding (figures 6.11(d), (f) and (h)). However, there also exist important differences between the three different configurations in the low-energy DOS. Two prominent peaks are revealed at $E \sim \pm 0.5$ for the 33% structure (the armchair system in figure 6.11(e)) and the 14% structure (the chiral system in figure 6.11(g)). They originate from two critical points of parabolic bands and two weakly dispersive bands (figures 6.8(a) and (d)). Such peaks become a delta-function-like peak at $E \sim 0$ for the 3.7% and 1.6% structures (figures 6.11(f) and (h)). Moreover, their DOSs show a linear E dependence at low energy, which is strong

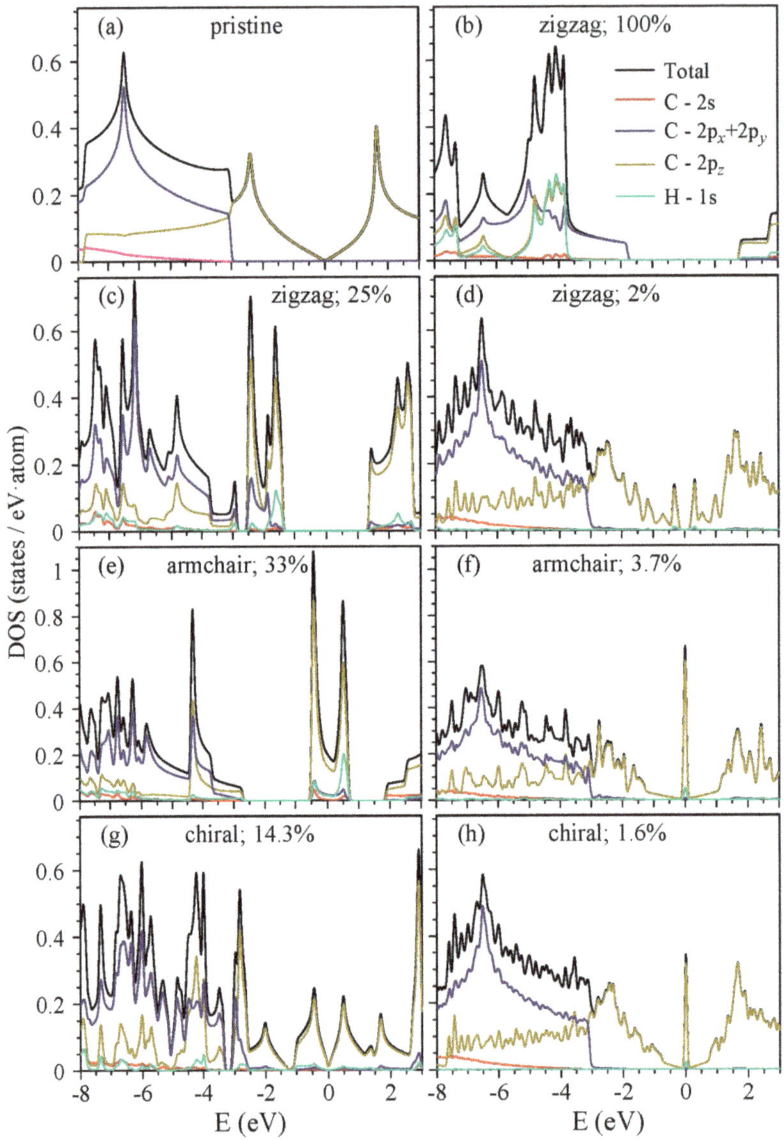

Figure 6.11. Orbital-projected DOS of (a) pristine graphene, zigzag configurations with (b) 100%, (c) 25% and (d) 2% H concentrations, armchair configurations with (e) 33% and (f) 3.7% H concentrations, and chiral configurations with (g) 14.3% and (h) 3.6% H concentrations.

evidence of a recovered Dirac-cone structure. STS measurements, where (*dI*/*dV*) is approximately proportional to the DOS, has been used to verify the DOS of adatom adsorption on graphene [90], 1D carbon nanotubes [78], 1D graphene nanoribbons [91], 2D few-layer graphene [92] and graphene irradiated with Ar^+ ions [93]. The above-mentioned main features can be investigated using STS.

References

[1] Geim A K and Novoselov K S 2010 The rise of graphene *Nanoscience and Technology: A Collection of Reviews from Nature Journals* (Singapore: World Scientific) pp 11–9

[2] Neto A C, Guinea F, Peres N M, Novoselov K S and Geim A K 2009 The electronic properties of graphene *Rev. Mod. Phys.* **81** 109

[3] Vogt P *et al* 2012 Silicene: compelling experimental evidence for graphene-like two-dimensional silicon *Phys. Rev. Lett.* **108** 155501

[4] Gao J and Zhao J 2012 Initial geometries, interaction mechanism and high stability of silicene on Ag (111) surface *Sci. Rep.* **2** 861

[5] Chen L *et al* 2012 Evidence for Dirac fermions in a honeycomb lattice based on silicon *Phys. Rev. Lett.* **109** 056804

[6] O'Hare A, Kusmartsev F V and Kugel K I 2012 A stable 'flat' form of two-dimensional crystals: could graphene, silicene, germanene be minigap semiconductors? *Nano Lett.* **12** 1045–52

[7] Lebegue S and Eriksson O 2009 Electronic structure of two-dimensional crystals from *ab initio* theory *Phys. Rev.* B **79** 115409

[8] Feng B *et al* 2013 Observation of Dirac cone warping and chirality effects in silicene *ACS Nano* **7** 9049–54

[9] Sprinkle M *et al* 2009 First direct observation of a nearly ideal graphene band structure *Phys. Rev. Lett.* **103** 226803

[10] Lu C L, Chang C P, Huang Y C, Chen R B and Lin M L 2006 Influence of an electric field on the optical properties of few-layer graphene with AB stacking *Phys. Rev.* B **73** 144427

[11] Huang H C, Lin S Y, Wu C L and Lin M F 2016 Configuration- and concentration-dependent electronic properties of hydrogenated graphene *Carbon* **103** 84–93

[12] Huang Y K, Chen S C, Ho Y H, Lin C Y and Lin M F 2014 Feature-rich magnetic quantization in sliding bilayer graphenes *Sci. Rep.* **4** 7509

[13] Tabert C J and Nicol E J 2013 Valley–spin polarization in the magneto-optical response of silicene and other similar 2D crystals *Phys. Rev. Lett.* **110** 197402

[14] Stille L, Tabert C J and Nicol E J 2012 Optical signatures of the tunable band gap and valley–spin coupling in silicene *Phys. Rev.* B **86** 195405

[15] Ezawa M 2012 Spin–valley optical selection rule and strong circular dichroism in silicene *Phys. Rev.* B **86** 161407

[16] Bechstedt F, Matthes L, Gori P and Pulci O 2012 Infrared absorbance of silicene and germanene *Appl. Phys. Lett.* **100** 261906

[17] Hwang E H, Adam S and Sarma S D 2007 Carrier transport in two-dimensional graphene layers *Phys. Rev. Lett.* **98** 186806

[18] Sarma S D, Adam S, Hwang E H and Rossi E 2011 Electronic transport in two-dimensional graphene *Rev. Mod. Phys.* **83** 407

[19] Kang J, Wu F and Li J 2012 Symmetry-dependent transport properties and magneto-resistance in zigzag silicene nanoribbons *Appl. Phys. Lett.* **100** 233122

[20] Li X, Mullen J T, Jin Z, Borysenko K M, Nardelli M B and Kim K W 2013 Intrinsic electrical transport properties of monolayer silicene and MoS_2 from first principles *Phys. Rev.* B **87** 115418

[21] Ni Z H, Yu T, Lu Y H, Wang Y Y, Feng Y P and Shen Z X 2008 Uniaxial strain on graphene: Raman spectroscopy study and band-gap opening *ACS Nano* **2** 2301–5

[22] Xia F, Farmer D B, Lin Y M and Avouris P 2010 Graphene field-effect transistors with high on/off current ratio and large transport band gap at room temperature *Nano Lett.* **10** 715–8
[23] Drummond N D, Zolyomi V and Fal'Ko V I 2012 Electrically tunable band gap in silicene. *Phys. Rev.* B **85** 075423
[24] Houssa M, van den Broek B, Scalise E, Pourtois G, Afanas' Ev V V and Stesmans A 2013 An electric field tunable energy band gap at silicene/(0001) ZnS interfaces *Phys. Chem. Chem. Phys.* **15** 3702–5
[25] Du Y *et al* 2014 Tuning the band gap in silicene by oxidation *ACS Nano* **8** 10019–25
[26] Georgakilas V *et al* 2012 Functionalization of graphene: covalent and non-covalent approaches, derivatives and applications *Chem. Rev.* **112** 6156–214
[27] Quhe R *et al* 2012 Tunable and sizable band gap in silicene by surface adsorption *Sci. Rep.* **2** 853
[28] Wang R, Pi X, Ni Z, Liu Y, Lin S, Xu M and Yang D 2013 Silicene oxides: formation, structures and electronic properties *Sci. Rep.* **3** 3507
[29] Zhang H *et al* 2011 Aryl functionalization as a route to band gap engineering in single layer graphene devices *Nano Lett.* **11** 4047–51
[30] Varykhalov A, Scholz M R, Kim T K and Rader O 2010 Effect of noble-metal contacts on doping and band gap of graphene *Phys. Rev.* B **82** 121101
[31] Ni Z *et al* 2014 Tunable band gap and doping type in silicene by surface adsorption: towards tunneling transistors *Nanoscale* **6** 7609–18
[32] Elias D C *et al* 2009 Control of graphene's properties by reversible hydrogenation: evidence for graphane *Science* **323** 610–3
[33] Haberer D V V D *et al* 2010 Tunable band gap in hydrogenated quasi-free-standing graphene *Nano Lett.* **10** 3360–6
[34] Balog R *et al* 2010 Bandgap opening in graphene induced by patterned hydrogen adsorption *Nat. Mater.* **9** 315
[35] Grassi R, Low T and Lundstrom M 2011 Scaling of the energy gap in pattern-hydrogenated graphene *Nano Lett.* **11** 4574–8
[36] Lalmi B, Oughaddou H, Enriquez H, Kara A, Vizzini S, Ealet B and Aufray B 2010 Epitaxial growth of a silicene sheet *Appl. Phys. Lett.* **97** 223109
[37] Aufray B, Kara A, Vizzini S, Oughaddou H, Léandri C, Ealet B and Le Lay G 2010 Graphene-like silicon nanoribbons on Ag (110): a possible formation of silicene *Appl. Phys. Lett.* **96** 183102
[38] Resta A *et al* 2013 Atomic structures of silicene layers grown on Ag (111): scanning tunneling microscopy and noncontact atomic force microscopy observations *Sci. Rep.* **3** 2399
[39] Meng L *et al* 2013 Buckled silicene formation on Ir (111) *Nano Lett.* **13** 685–90
[40] Nakano H *et al* 2006 Soft synthesis of single-crystal silicon monolayer sheets *Angew. Chem. Int. Ed.* **45** 6303–6
[41] Okamoto H *et al* 2010 Silicon nanosheets and their self-assembled regular stacking structure *J. Am. Chem. Soc.* **132** 2710–8
[42] Sugiyama Y, Okamoto H, Mitsuoka T, Morikawa T, Nakanishi K, Ohta T and Nakano H 2010 Synthesis and optical properties of monolayer organosilicon nanosheets *J. Am. Chem. Soc.* **132** 5946–7
[43] Cahangirov S, Topsakal M, Aktürk E, Şahin H and Ciraci S 2009 Two-and one-dimensional honeycomb structures of silicon and germanium *Phys. Rev. Lett.* **102** 236804

[44] Xu X *et al* 2014 Effects of oxygen adsorption on the surface state of epitaxial silicene on Ag (111) *Sci. Rep.* **4** 7543
[45] Houssa M, Scalise E, Sankaran K, Pourtois G, Afanas' Ev V V and Stesmans A 2011 Electronic properties of hydrogenated silicene and germanene *Appl. Phys. Lett.* **98** 223107
[46] Wei W, Dai Y, Huang B and Jacob T 2013 Many-body effects in silicene, silicane, germanene and germanane *Phys. Chem. Chem. Phys.* **15** 8789–94
[47] Li G, Luican A and Andrei E Y 2009 Scanning tunneling spectroscopy of graphene on graphite *Phys. Rev. Lett.* **102** 176804
[48] Brar V W *et al* 2007 Scanning tunneling spectroscopy of inhomogeneous electronic structure in monolayer and bilayer graphene on SiC *Appl. Phys. Lett.* **91** 122102
[49] Chae J *et al* 2012 Renormalization of the graphene dispersion velocity determined from scanning tunneling spectroscopy *Phys. Rev. Lett.* **109** 116802
[50] Premlal B, Cranney M, Vonau F, Aubel D, Casterman D, De Souza M M and Simon L 2009 Surface intercalation of gold underneath a graphene monolayer on SiC (0001) studied by scanning tunneling microscopy and spectroscopy *Appl. Phys. Lett.* **94** 263115
[51] Novoselov K S *et al* 2005 Two-dimensional gas of massless Dirac fermions in graphene *Nature* **438** 197
[52] Cao Y, Fatemi V, Fang S, Watanabe K, Taniguchi T, Kaxiras E and Jarillo-Herrero P 2018 Unconventional superconductivity in magic-angle graphene superlattices *Nature* **556** 43
[53] Soavi G *et al* 2018 Broadband, electrically tunable third-harmonic generation in graphene *Nat. Nanotechnol.* **13** 583
[54] Berger C *et al* 2006 Electronic confinement and coherence in patterned epitaxial graphene *Science* **312** 1191–6
[55] Balandin A A, Ghosh S, Bao W, Calizo I, Teweldebrhan D, Miao F and Lau C N 2008 Superior thermal conductivity of single-layer graphene *Nano Lett.* **8** 902–7
[56] Lee C, Wei X, Kysar J W and Hone J 2008 Measurement of the elastic properties and intrinsic strength of monolayer graphene *Science* **321** 385–8
[57] Liu H, Liu Y and Zhu D 2011 Chemical doping of graphene *J. Mater. Chem.* **21** 3335–45
[58] Wei D, Liu Y, Wang Y, Zhang H, Huang L and Yu G 2009 Synthesis of N-doped graphene by chemical vapor deposition and its electrical properties *Nano Lett.* **9** 1752–8
[59] Joucken F *et al* 2015 Charge transfer and electronic doping in nitrogen-doped graphene *Sci. Rep.* **5** 14564
[60] Sutter P, Hybertsen M S, Sadowski J T and Sutter E 2009 Electronic structure of few-layer epitaxial graphene on Ru (0001) *Nano Lett.* **9** 2654–60
[61] Lauffer P, Emtsev K V, Graupner R, Seyller T, Ley L, Reshanov S A and Weber H B 2008 Atomic and electronic structure of few-layer graphene on SiC (0001) studied with scanning tunneling microscopy and spectroscopy *Phys. Rev.* B **77** 155426
[62] Zhong X, Pandey R and Karna S P 2012 Stacking dependent electronic structure and transport in bilayer graphene nanoribbons *Carbon* **50** 784–90
[63] Tran N T T, Lin S Y, Glukhova O E and Lin M F 2015 Configuration-induced rich electronic properties of bilayer graphene *J. Phys. Chem.* C **119** 10623–30
[64] Lai Y H, Ho J H, Chang C P and Lin M F 2008 Magnetoelectronic properties of bilayer Bernal graphene *Phys. Rev.* B **77** 085426
[65] Lin C Y, Wu J Y, Ou Y J, Chiu Y H and Lin M F 2015 Magneto-electronic properties of multilayer graphenes *Phys. Chem. Chem. Phys.* **17** 26008–35

[66] Lin S Y, Lin Y T, Tran N T T, Su W P and Lin M F 2017 Feature-rich electronic properties of aluminum-adsorbed graphenes *Carbon* **120** 209–18
[67] Pereira V M and Neto A C 2009 Strain engineering of graphene's electronic structure *Phys. Rev. Lett.* **103** 046801
[68] Sessi P, Guest J R, Bode M and Guisinger N P 2009 Patterning graphene at the nanometer scale via hydrogen desorption *Nano Lett.* **9** 4343–7
[69] Jaiswal M, Lim C H Y X, Bao Q, Toh C T, Loh K P and Ozyilmaz B 2011 Controlled hydrogenation of graphene sheets and nanoribbons *ACS Nano* **5** 888–96
[70] Castellanos-Gomez A, Wojtaszek M, Tombros N and van Wees B J 2012 Reversible hydrogenation and bandgap opening of graphene and graphite surfaces probed by scanning tunneling spectroscopy *Small* **8** 1607–13
[71] Sofo J O, Chaudhari A S and Barber G D 2007 Graphane: a two-dimensional hydrocarbon *Phys. Rev.* B **75** 153401
[72] Chandrachud P, Pujari B S, Haldar S, Sanyal B and Kanhere D G 2010 A systematic study of electronic structure from graphene to graphane *J. Phys. Condens. Matter* **22** 465502
[73] Yang M, Nurbawono A, Zhang C, Feng Y P and Ariando 2010 Two-dimensional graphene superlattice made with partial hydrogenation *Appl. Phys. Lett.* **96** 193115
[74] Gao H, Wang L, Zhao J, Ding F and Lu J 2011 Band gap tuning of hydrogenated graphene: H coverage and configuration dependence *J. Phys. Chem.* C **115** 3236–42
[75] De Parga A V, Calleja F, Borca B M C G, Passeggi M C G Jr, Hinarejos J J, Guinea F and Miranda R 2008 Periodically rippled graphene: growth and spatially resolved electronic structure *Phys. Rev. Lett.* **100** 056807
[76] Červenka J, Katsnelson M I and Flipse C F J 2009 Room-temperature ferromagnetism in graphite driven by two-dimensional networks of point defects *Nat. Phys.* **5** 840
[77] Ruffieux P *et al* 2012 Electronic structure of atomically precise graphene nanoribbons *ACS Nano* **6** 6930–5
[78] Wilder J W, Venema L C, Rinzler A G, Smalley R E and Dekker C 1998 Electronic structure of atomically resolved carbon nanotubes *Nature* **391** 59
[79] Kondo T *et al* 2012 Atomic-scale characterization of nitrogen-doped graphite: effects of dopant nitrogen on the local electronic structure of the surrounding carbon atoms *Phys. Rev.* B **86** 035436
[80] Tao C *et al* 2011 Spatially resolving edge states of chiral graphene nanoribbons *Nat. Phys.* **7** 616
[81] Ohta T *et al* 2008 Morphology of graphene thin film growth on SiC (0001) *New J. Phys.* **10** 023034
[82] Ohta T, Bostwick A, McChesney J L, Seyller T, Horn K and Rotenberg E 2007 Interlayer interaction and electronic screening in multilayer graphene investigated with angle-resolved photoemission spectroscopy *Phys. Rev. Lett.* **98** 206802
[83] Ohta T, Bostwick A, Seyller T, Horn K and Rotenberg E 2006 Controlling the electronic structure of bilayer graphene *Science* **313** 951–4
[84] Zhou S Y *et al* 2007 Substrate-induced bandgap opening in epitaxial graphene *Nat. Mater.* **6** 770
[85] Zhou S Y, Siegel D A, Fedorov A V and Lanzara A 2008 Metal to insulator transition in epitaxial graphene induced by molecular doping *Phys. Rev. Lett.* **101** 086402
[86] Schulte K, Vinogradov N A, Ng M L, Mårtensson N and Preobrajenski A B 2013 Bandgap formation in graphene on Ir (111) through oxidation *Appl. Surf. Sci.* **267** 74–6

[87] Papagno M *et al* 2011 Large band gap opening between graphene Dirac cones induced by Na adsorption onto an Ir superlattice *ACS Nano* **6** 199–204

[88] Lin S Y, Chang S L, Shyu F L, Lu J M and Lin M F 2015 Feature-rich electronic properties in graphene ripples *Carbon* **86** 207–16

[89] Lin S Y, Chang S L, Tran N T T, Yang P H and Lin M F 2015 H–Si bonding-induced unusual electronic properties of silicene: a method to identify hydrogen concentration *Phys. Chem. Chem. Phys.* **17** 26443–50

[90] Gyamfi M, Eelbo T, Waśniowska M and Wiesendanger R 2011 Fe adatoms on graphene/Ru (0001): adsorption site and local electronic properties *Phys. Rev.* B **84** 113403

[91] Huang H, Wei D, Sun J, Wong S L, Feng Y P, Neto A C and Wee A T S 2012 Spatially resolved electronic structures of atomically precise armchair graphene nanoribbons *Sci. Rep.* **2** 983

[92] Choi J, Lee H and Kim S 2010 Atomic-scale investigation of epitaxial graphene grown on 6H–SiC (0001) using scanning tunneling microscopy and spectroscopy *J. Phys. Chem.* C **114** 13344–8

[93] Tapaszto L, Dobrik G, Nemes-Incze P, Vertesy G, Lambin P and Biro L P 2008 Tuning the electronic structure of graphene by ion irradiation *Phys. Rev.* B **78** 233407

Chapter 7

Halogenated silicene systems

7.1 Introduction

In the periodic table, halogen atoms possess very strong electron affinities, illustrating the remarkable charge redistributions in any chemical bonding, e.g. the unique charge densities in X–C and X–Si bonds (X = F, Cl, Br, I, At) [1]. They could serve as outstanding guest adatoms in the chemical modification of 2D group-IV materials. To date, many experimental [2] and theoretical [3] studies have been conducted on halogenated graphene systems. Generally, halogen-absorbed materials present fluorination [4] and other halogenation effects [5], according to the varied fundamental properties. That is, halogenated graphenes can be classified into fluorinated and other graphenes through their completely different electronic properties. Interestingly, fluorinated graphenes have been synthesized successfully using the fluorination methods of direct gas [6], plasma [7], photochemistry [8] and hydrothermal reactions [9]. Careful analyses and studies show that such systems exhibit unusual semiconducting or p-type behaviors, which are sensitive to the fluorination concentrations and configurations. Large energy gaps, which frequently survive high-level fluorination, are clearly present in transport measurements [10], optical transmission coefficients [11] and photoluminescence spectra [12]. Furthermore, p-type doping is verified from the redshift of the Fermi level in angle-resolved photoemission spectroscopy (ARPES) measurements [13]. The tunable band gaps suggest many potential applications in various areas, e.g. Li^+-based batteries [14], supercapacitors [15], biological scaffold materials [16], printing technologies [17], molecule detectors [18] and biosensors [19]. For magnetic spin configurations, first-principles calculations predict ferromagnetic phenomena [20], while measured magnetic moments present temperature-dependent paramagnetism [21]. This significant difference might be associated with non-uniform adatom nanoclusters in rippled graphene. Also, experimental synthesis has been achieved

doi:10.1088/978-0-7503-3299-6ch7

for chlorinated [22], brominated [23] and iodinated graphenes [24]. The theoretical predictions show sharp contrasts between the Cl-, Br-, I- and At-absorbed graphenes and the F-absorbed graphenes in terms of their geometric, electronic and magnetic properties [25]. The significant differences lie in the critical multi-orbital hybridization of the X–C bonds. Most importantly, non-fluorinated graphene systems possess many free holes under arbitrary adatom distributions and concentrations, as supported by the upshift of the graphitic G-band phonon [1].

Compared to the halogenation effects in graphene systems [1], halogenated silicene materials are expected to present more complicated physical, chemical and material phenomena, owing mainly to the buckled honeycomb lattices and the weak but significant sp^3 bonding, the different multi-orbital hybridization cases, and the host- and guest-atom-induced spin configurations [26]. Current experimental synthesis methods have achieved the successful chlorination of monolayer silicene through molecular beam epitaxy and scanning tunneling microscope (STM) surface observations [27]. Single-layer silicene on Ag(111) with different phases is prepared by depositing silicon atoms under the different substrate temperatures. STM measurements display three typical moiré superlattices [28–30]. Then, chlorinated silicenes are produced by exposing the samples to a Cl_2–Ar mixture gas ($Cl_2/Ar \times 100\% = 2\%$) at room temperature. The Cl-adsorption concentration is proportional to the exposure time. According to STM results after chlorination, the buckled geometry in silicene may be modified strongly and become a well-behaved structure at a sufficiently high coverage. Scanning tunneling spectroscopy (STS) and ARPES experiments would be very useful in determining semiconducting or p-type metallic behaviors. Furthermore, the fluorinated silicene materials could provide more information on the various halogenation phenomena [31].

On the theoretical side, first-principles calculations have been utilized to investigate the geometric, electronic and mechanical properties. The previous theoretical studies were confined to a few halogen concentrations, in particular to double-sided saturated chemical adsorption [32]. The calculated results, which were conducted on F-, Cl-, Br- and I-adsorbed monolayer silicene, include the lattice constants, silicon-halogen bond lengths, height differences between A and B sublattices in a honeycomb structure, adatom-enriched binding energies, low-lying valence and conduction bands, effective masses of energy dispersions near the Fermi level and mechanical constants. The surface functionalization and mechanical strain appear to create the middle-energy gaps of $E_g \sim 1.5$–2.0 eV for the full suppression of $3p_z$–π bonding under all X–Si chemical bonds. Their magnitudes are smaller than that of graphane, $E_g \sim 3.5$ eV [33]. Such materials are expected to be potential candidates for field-effect transistors [34]. However, the above-mentioned predictions do not include all the critical factors (the various concentrations and configurations of halogen adatoms), the multi-orbital hybridization in the Si–Si, halogen–Si and halogen–halogen bonds, and the spin-dependent magnetism (the effective magnetic moments). These would induce greatly varied fundamental properties.

In this chapter, the geometric, magnetic and electronic properties of halogen-adsorbed silicene systems are investigated thoroughly through first-principles calculations, including concise multi-orbital hybridization in the Si–Si, halogen–Si and halogen–halogen bonds, and the silicon- and halogen-induced spin configurations. The double-sided and single-sided adsorption configurations in figures 7.1(a) and (b) are included in the numerical calculations in which they can lead to, respectively, different electronic and magnetic properties. The main features are determined by the adatom-varied geometric structures, atom-dominated energy bands, spin density distributions, spatial charge densities and their variation after adsorption, and spin- and orbital-projected densities of states. The similar and different characteristics of the double- and single-sided adsorption cases are identified fully through halogen-related weakly dispersed bands at deep energies, adatom-modified middle-energy σ bands and the recovery of low-energy π bands during the decrease of the halogen concentrations. However, the breaking of the mirror symmetry in the single-sided configurations results in a critical discrepancy, i.e. there appear valence bands with or without spin-splitting intersecting with the Fermi level. Also, whether the fluorinated silicene systems and other halogenated materials are completely different to each other will be clarified using the above-mentioned physical quantities. The significant differences between the halogenated silicenes and graphenes are illustrated clearly by the buckled or planar honeycomb lattices, different strengths of chemical bonding, and important or negligible spin-dependent interactions [1, 32]. All the theoretical predictions can be verified

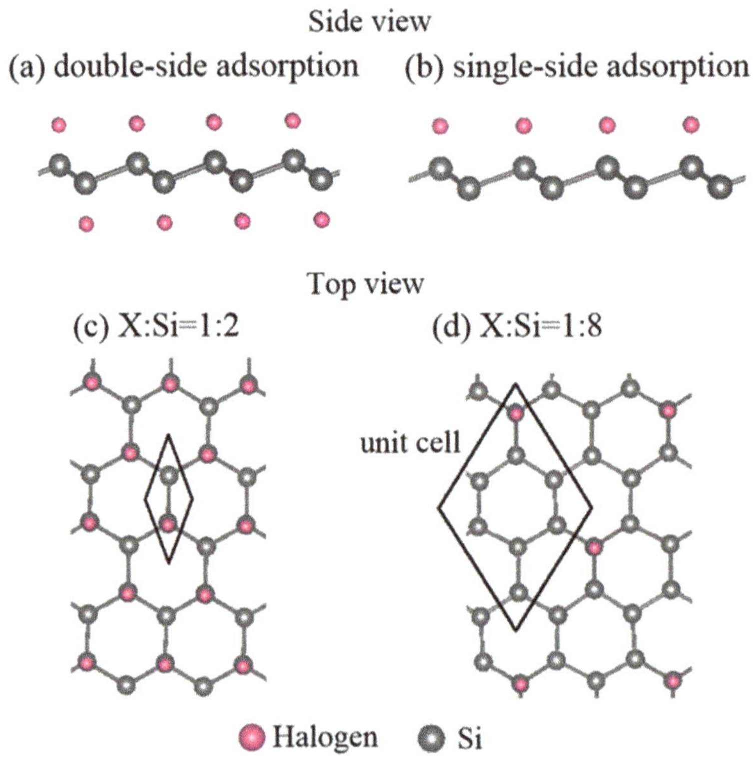

Figure 7.1. Side-view structures of halogenated silicene for (a) double-sided adsorption and (b) single-sided adsorption. Top-view structures for various unit cells of (c) X:Si = 1:2 and (d) X:Si = 1:8.

by high-resolution measurements obtained using STM, transmission electron microscopy (TEM), low-energy electron diffraction (LEED), ARPES and scanning tunneling spectroscopy (STS) experiments (details in chapter 3).

7.2 Halogenated silicenes

7.2.1 Geometric structure

Two different adsorption configurations are used in this systematic investigation, double-sided and single-sided adsorption, as shown in figures 7.1(a) and (b), respectively. The various adatom concentrations include 50% and 12.5% adsorption, as illustrated in figures 7.1(c) and (d), respectively. The structural stability is evaluated from the binding energy of $E_b = E_T - E_P - nE_A/n$, in which E_T, E_P and E_A represent the ground state energies of the total system, the pristine system and isolated halogen adatoms, respectively. The symbol n is used to signify the number of halogen adatoms. As a result, a greater geometric stability corresponds to a smaller binding energy. Through specific calculations, the top site is determined to be the most optimal position for halogen adatoms adsorbed on silicene surfaces, regardless of the doping case [35]. E_b decreases with the enhancement of the atomic number for the various halogen adatoms shown in tables 7.1 and 7.2. This indicates clearly that stronger chemical bonds achieve higher structure stability. The F-adsorbed silicene systems retain the highest structural stability among the halogen adatoms. Also, it should be mentioned that halogenated silicene systems exhibit greater geometric stability than halogenated graphene, owing mainly to the highly reactive buckled surface of silicene systems. The transfer of electrons from Si atoms to halogen adatoms determines the free hole density in the system and the significant halogen–Si bonds, which are insensitive to the adatom concentration. Moreover, the

Table 7.1. Binding energy E_b (eV), magnetic moment/magnetism, energy gap $E_g^{d(i)}$/metal (M)/semi-metal (SM), halogen–Si bond length (Å), Si–Si bond length (Å), adatom–Si–Si bond angle (°) and buckling Δ (Å) for the various double-sided adsorption configurations. NM and FM correspond to non-magnetism and ferromagnetism, respectively. ([27] Copyright 2019, Springer Nature. With permission of Springer.)

Ratio of adatoms and Si atoms	E_b (eV)	Magnetic moment/ magnetism	$E_g^{d(i)}$ (eV)/ M/SM	Adatom–Si bond length (Å)	Nearest Si–Si bond length (Å)	Adatom–Si–Si bond angle (°)	Δ (Å)
Pristine	—	0/NM	$E_g^d = 0.01$	—	2.255	—	0.49
F:Si = 2:2	−5.294 90	0/NM	$E_g^d = 0.47$	1.631	2.322	109.12	0.76
F:Si = 2:8	−5.314 50	0/NM	$E_g^i = 0.75$	1.639	2.319	110.42	0.77
F:Si = 2:18	−5.391 29	0/NM	M	1.637	2.318	111.92	0.89
F:Si = 2:32	−5.458 32	0/NM	M	1.641	2.318	113.96	0.91
Cl:Si = 2:2	−3.226 31	0/NM	$E_g^d = 1.16$	2.071	2.328	109.52	0.78
Br:Si = 2:2	−2.412 61	0/NM	$E_g^d = 1.09$	2.240	2.326	109.42	0.77
I:Si = 2:2	−1.293 62	0/NM	$E_g^d = 0.51$	2.495	2.326	109.41	0.77
At:Si = 2:2	−0.710 32	0/NM	SM	2.584	2.326	109.43	0.77

Table 7.2. Similar presentation as in table 7.1, but shown for the single-sided adsorption configurations. ([27] Copyright 2019, Springer Nature. With permission of Springer.)

Ratio of adatom and Si atom	E_b (eV)	Magnetic moment/ magnetism	$E_g^{d(i)}$ (eV)/ M/SM	Adatom–Si bond length (Å)	Nearest Si–Si bond length (Å)	Adatom–Si–Si bond angle (°)	Δ (Å)
F:Si = 1:2	−4.768 99	0/NM	M	1.633	2.321	109.04	0.76
F:Si = 1:8	−5.128 32	0.45/FM	M	1.636	2.315	114.24	0.95
F:Si = 1:18	−5.235 63	0/NM	M	1.636	2.313	116.96	1.05
F:Si = 1:32	−5.310 26	0/NM	M	1.637	2.313	117.40	1.06
Cl:Si = 1:2	−2.587 19	0.72/FM	M	2.091	2.328	109.56	0.78
Cl:Si = 1:8	−3.188 66	0.64/FM	M	2.096	2.320	113.96	0.94
Br:Si = 1:2	−1.783 24	0.89/FM	M	2.296	2.326	109.43	0.77
I:Si = 1:2	−0.832 02	0.79/FM	M	2.566	2.336	110.08	0.80
At:Si = 1:2	−0.406 16	0.82/FM	M	2.830	2.319	108.87	0.75

transfer of electrons from Si atoms to halogen adatoms leads to weakness of the nearest Si–Si bonds, i.e. the nearest Si–Si bond lengths are extended. The buckled structures are enhanced when the adatom concentrations decrease gradually, as shown in tables 7.1 and 7.2. Note that the buckled silicene systems are stabilized by a mix of sp^2 and sp^3 hybridization instead of the sp^2 hybridization in planar graphene, such that the chemisorption of silicene systems can acquire a higher structural stability than the chemisorption of graphene. In addition, the combination of sp^2 and sp^3 hybridization is changed to sp^3 hybridization when the halogen adatoms are bonded with Si atoms. The orbital hybridization in Si-3p_z and halogen-p_x, -p_y and -p_z and the weak sp^3 orbital hybridization of Si-3s, -3p_x, -3p_y and -3p_z determine diverse phenomena directly. Similar geometric deformations occur in the double-sided and single-sided configurations, in which they can exhibit different fundamental physical properties due mainly to the breaking of mirror symmetry in the single-sided adsorption configuration.

7.2.2 Electronic band structure

The electronic band structure of a pristine silicene system presents a Dirac cone structure at the K point, which is made of the low-lying π bands of perpendicular orbitals (Si-3p_z), as shown in figure 7.2(a). Also, a band gap of 1 meV arises at the K point. The low-lying linear bands become parabolic dispersions with a saddle point at the M point, in which they are dominated mainly by Si-3p_z orbitals, namely π bands. In addition, there is an obvious separation of the π bands and σ bands within ±2 eV owing to the weak sp^3 hybridization. The occupied and unoccupied σ parabolic bands are initiated at, respectively, −1 and 1.2 eV, as illustrated by the black and red rectangles in figure 7.2(a). The highest occupied bands exhibit a double degeneration at the Γ point, while the lowest unoccupied bands are non-degenerate at the M point. The former bands are contributed mainly by the Si-3p_x and Si-3p_y orbitals, and they are hybridized with the Si-3p_z orbitals at the lower

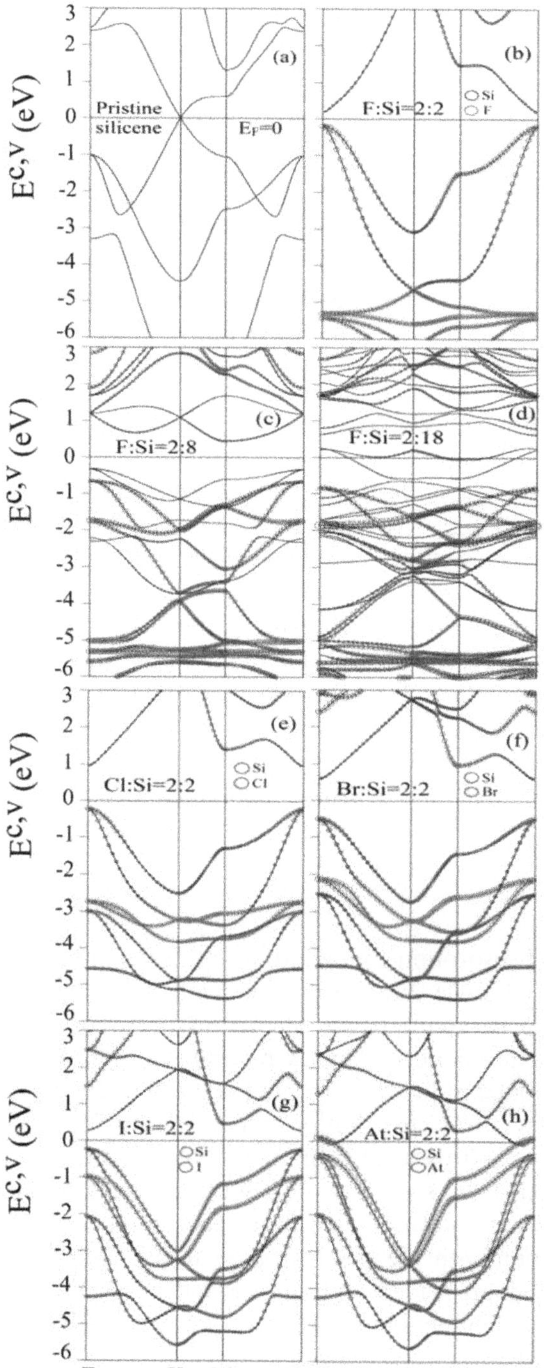

Figure 7.2. Electronic band structures of (a) pristine silicene, and halogenated silicene under double-sided adsorption for (b) F:Si = 2:2, (c) F:Si = 2:8, (d) F:Si = 2:18, (e) Cl:Si = 2:2, (f) Br:Si = 2:2, (g) I:Si = 2:2 and (h) At:Si = 2:2. The violet and green circles represent the contributions of passivated Si atoms and halogen adatoms, respectively. ([27] Copyright 2019, Springer Nature. With permission of Springer.)

energy of −2.5 eV. The third occupied σ band initiates at −1 eV, dominated by the Si-3s orbitals, and becomes gradually parabolically dispersed. This band becomes partially flat at −3.2 eV near the Γ point, in which there exists an orbital hybridization between the Si-3s orbitals and Si-3p_z orbitals. The unique band structure of silicene is reshaped fully by the various halogen-adsorption configurations.

The various double-sided configurations can belong to the semiconductors or to p-type metals, and depends strongly on the specific adatom concentrations. Direct band gaps of 0.47, 1.16, 1.09 and 0.51 eV are induced in the 100% fluorinated, chlorinated, brominated and iodinated silicene systems, as illustrated in figures 7.2(b), (e)–(g), respectively. The energy gaps are determined by the highest occupied states (HOSs) and lowest unoccupied states (LUSs) at the Γ point, and their magnitudes do not have a linear relationship with their atomic numbers. On the other hand, the 100% astatinated silicene system exhibits semi-metallic behavior due to the weak At–Si bonds, as illustrated in figure 7.2(h). The F-related weakly dispersed bands dominate at the deepest energies due to its shortest F–Si bond length among the halogen adatoms, as shown by the green circles in figure 7.2(b). This band dispersion becomes more obvious and dominates gradually at higher energies with an increase of its atomic number, as illustrated by the green circles in figures 7.2(e)–(h). As the adatom concentrations decrease, an indirect band gap of 0.75 eV is created in the 25% double-sided system shown in figure 7.2(c). At the critical concentration of 11%, it displays p-type metallic behavior, as illustrated in figure 7.2(d). Also, the weakly dispersed bands near the Fermi level are dominated by the 3p_z orbitals of non-passivated Si atoms, as shown in figures 7.2(c) and (d). The middle-energy σ bands of pristine silicene system in figure 7.2(a) are situated at −0.3 eV in figure 7.2(c) and −0.7 eV in figure 7.2(d). The middle-energy σ bands are dominated by both passivated Si atoms and non-passivated Si atoms, in which the non-passivated Si atoms present a higher dominance in the 11% double-sided configuration.

The single-sided adsorption configurations all belong to p-type metals with a free hole density. The full single-sided adsorption configurations in figures 7.3(a), (d), (f)–(h) present valence parabolic bands intersecting with the Fermi level. When the concentrations are further decreased, the parabolic dispersion is changed to partially flat bands, as indicated in figure 7.3(b) and (e), and figure 7.3(c) for the 12.5% and 5.6% systems, respectively. In addition, the π bonds of 3p_z orbitals between non-passivated Si atoms and passivated Si atoms are deformed significantly due to the transfer of electrons from the passivated Si atoms and the halogen adatoms. This demonstrates that the non-passivated Si atoms cannot form extended π electronic states via interaction with the passivated Si atoms. The most distinct difference between the fully double-sided and single-sided adsorption configurations is the valence energy bands intersecting with the Fermi level. In addition, the halogen-related weakly dispersed bands in the double-sided adsorption configurations in figure 7.2 are reduced in number compared to that in the single-sided adsorption configurations in figure 7.3. Also, there exist similar features in the double-sided and single-sided adsorption configurations, such as the recovery of low-energy π bands in the low concentration configurations, the adatom-modified middle-energy σ bands

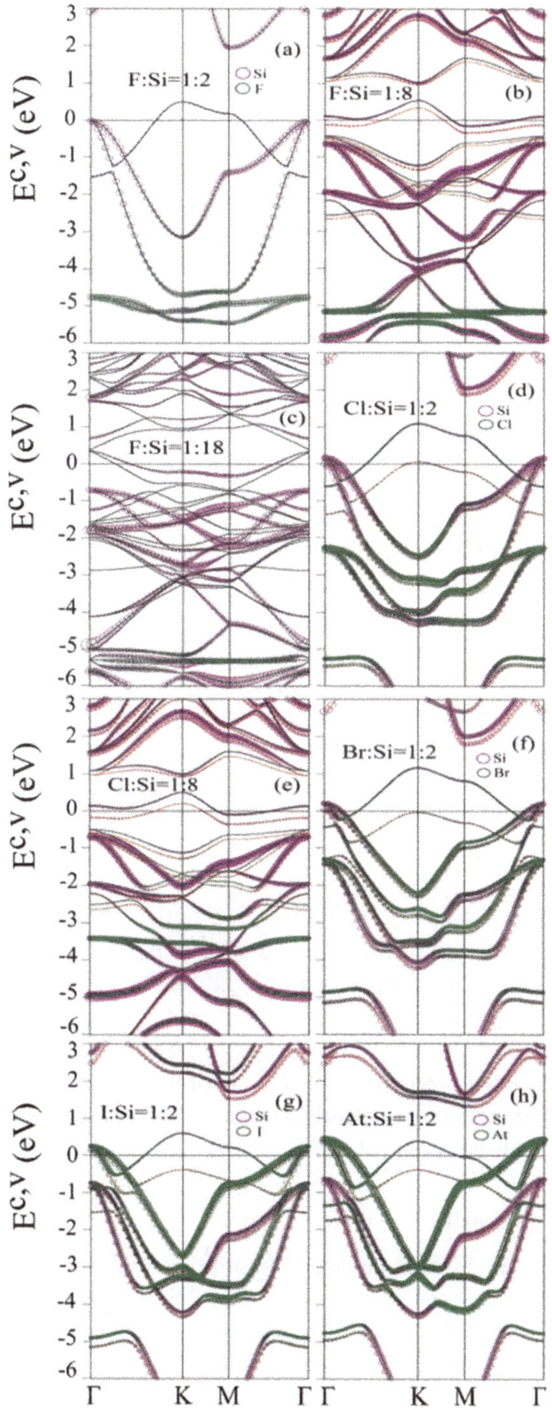

Figure 7.3. A similar plot as in figure 7.2, but shown for single-sided adsorption of (a) F:Si = 1:2, (b) F:Si = 1:8, (c) F:Si = 1:18, (d) Cl:Si = 1:2, (e) Cl:Si = 1:8, (f) Br:Si = 1:2, (g) I:Si = 1:2 and (h) At:Si = 1:2. The red and black curves correspond, respectively, to the spin-up and spin-down energy bands. ([27] Copyright 2019, Springer Nature. With permission of Springer.)

and the halogen-induced weakly dispersed bands at deep energies. When the adatom concentration decreases, both the 5.6% single-sided (figure 7.3(c)) and the 11% double-sided (figure 7.2(d)) systems present low-lying π bands with large energy widths. These low-energy π bands are contributed mainly by the non-passivated Si-3p_z orbitals, indicating that the π bonds are recovered among the Si atoms. The middle-energy σ bands in figures 7.3(a)–(c) and 7.2(b)–(d) are very similar to each other in terms of their band width, initiated energy and orbital contribution, owing mainly to the barely affected σ bonds. Furthermore, the halogen-related weakly dispersed bands are confined at similar energies in both the single-sided and double-sided configurations since the halogen–Si bonds are insensitive to the adatom concentrations.

7.2.3 Spatial spin density

When the mirror symmetry no longer exists in figure 7.1(b), the various spin arrangements can be presented in the single-sided adsorption configurations, which can be confirmed from the spin-splitting energy bands in figure 7.3. Only the 50% fluorinated single-sided configuration in figure 7.3(a) displays parabolic dispersed bands without spin-splitting. The other 50% single-sided configurations exhibit parabolic dispersed bands with spin-splitting, which can be verified by the red and black curves in figures 7.3(d), (f)–(h). The parabolic dispersed bands with spin-splitting are contributed mainly by the non-passivated Si atoms, which is confirmed by the spin arrangements shown in figures 7.4(a)–(c). However, the spin-splitting bands become weaker in the 12.5% single-sided configuration illustrated in figures 7.3(b) and (e). This indicates that the spin-up and spin-down states can co-exist in its spin density distributions under a decrease of the adatom concentration, as illustrated by the red sphere (the spin-up state) and the green sphere (the spin-down state) in figures 7.4(d)–(f). The difference between the spin-up and spin-down states determines the strength of the net magnetic moments directly, i.e. 0.72 μ_B, 0.89 μ_B and 0.79 μ_B in the 50% configurations and 0.64 μ_B, 0.67 μ_B and 0.45 μ_B in the 12.5% configurations, as shown in table 7.2. It should be noted that the spin orientations vanish fully at the critical concentration of 5.6%, leading to the vanishing of the net magnetic moment, presented in table 7.2. Moreover, any adatom concentration beyond the critical concentration leads to the absence of the spin orientations. This result is similar for all halogen adatoms.

7.2.4 Spatial charge density

The strength of the Si–Si and halogen–Si bonds can be illustrated clearly by the spatial charge density (ρ), as shown in figures 7.5(a)–(g). ρ presents very strong σ bonds between two Si atoms in the pristine silicene system shown in figure 7.5(a) or two non-passivated Si atoms in the low concentration configurations, as presented by the black rectangle in figures 7.5(f) and (g). When Si atoms are bonded to halogen adatoms, the strength of such σ bonds becomes weaker, as shown by the gray rectangle in figures 7.5(b)–(g), leading to extended Si–Si bond lengths. ρ is reduced greatly between two Si atoms away from the horizontal line. However, its charge

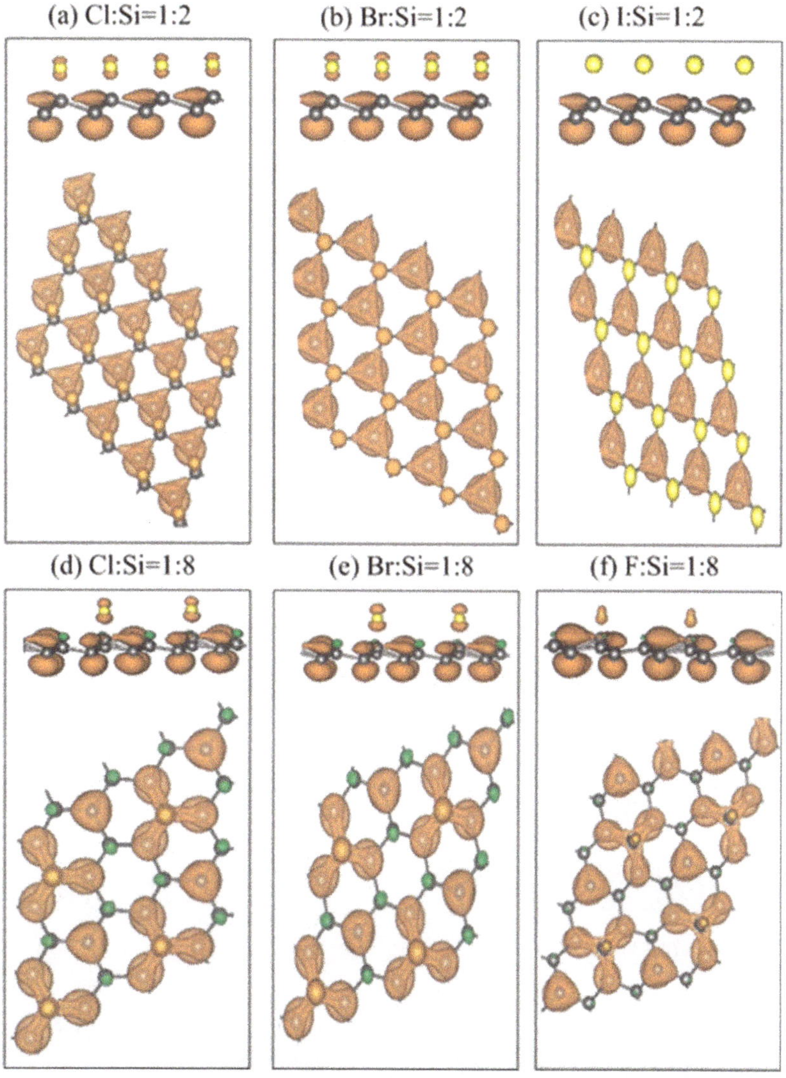

Figure 7.4. Spatial spin arrangement with side-views and top-views for (a) Cl:Si = 1:2, (b) Br:Si = 1:2, (c) I: Si = 1:2, (d) Cl:Si = 1:8, (e) Br:Si = 1:8 and (f) F:Si = 1:8. The red and green spheres, respectively, correspond to the spin-up and spin-down orientations. ([27] Copyright 2019, Springer Nature. With permission of Springer.)

distribution is extended. This accounts for the π-electronic states, as displayed by the red rectangle in figures 7.5(a), (f) and (g). In addition, the barely extended states, corresponding to the π-charge-depleted region, are recognized between the passivated and non-passivated Si atoms shown by the purple rectangle in figures 7.5(d)–(g). Moreover, the strength of the halogen–Si bonds is much higher than that of the Si–Si bonds, as illustrated by the dashed black rectangle in figures 7.5(b)–(g). As a result, the significant halogen–Si bonds are insensitive to the halogen concentrations. Furthermore, the charge transfers among all the orbitals can be evaluated by the

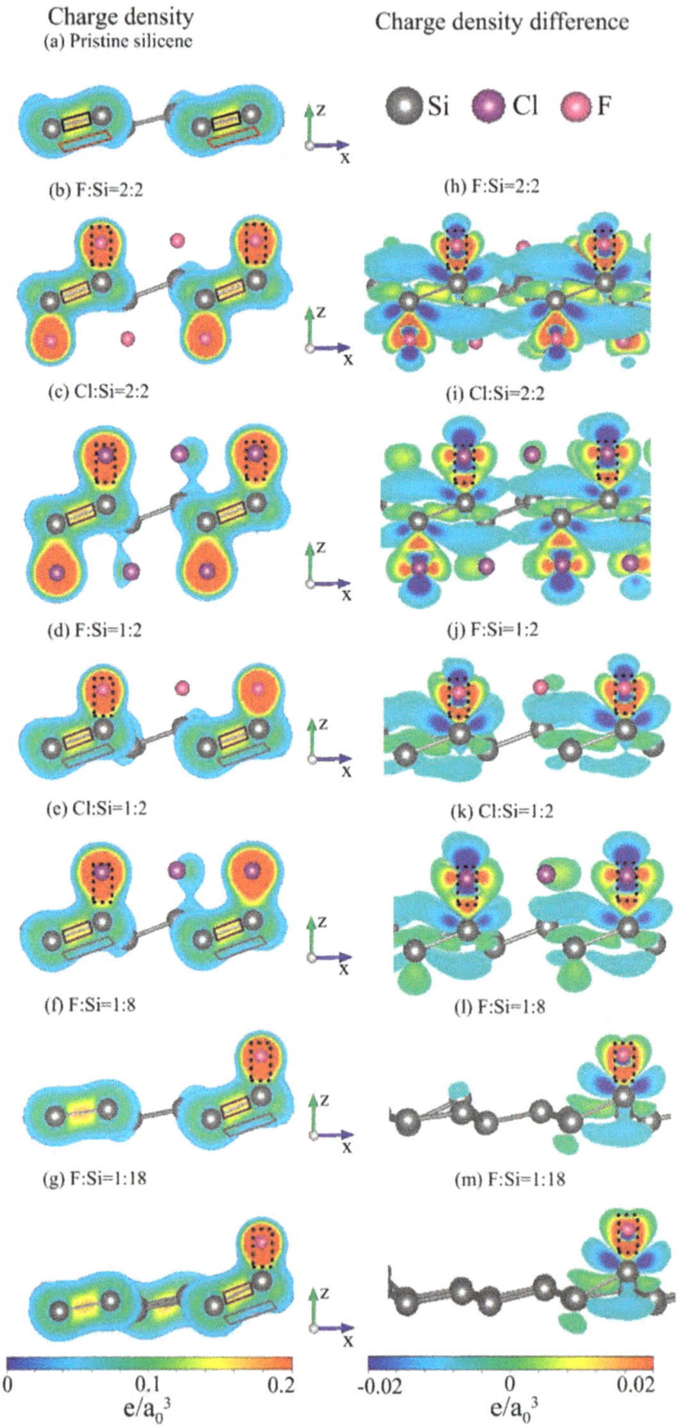

Figure 7.5. Spatial charge density (left column) for (a) pristine silicene, and halogenated silicene with (b) F:Si = 2:2, (c) Cl:Si = 2:2, (d) F:Si = 1:2, (e) Cl:Si = 1:2, (f) F:Si = 1:8 and (g) F:Si = 1:18; charge density difference (right column) for (h) F:Si = 2:2, (i) Cl:Si = 2:2, (j) F:Si = 1:2, (k) Cl:Si = 1:2, (l) F:Si = 1:8 and (m) F:Si = 1:18. a_0 is the Bohr radius. ([27] Copyright 2019, Springer Nature. With permission of Springer.)

charge density difference ($\Delta\rho$), as shown in figures 7.5(h)–(m). $\Delta\rho$ is found by subtracting the total charge density of the pristine silicene and isolated halogen adatoms from that of the halogen-adsorbed silicene system. The significant transfer of electrons from Si atoms to halogen adatoms in the halogen-adsorbed silicene systems forms strong halogen–Si bonds, as shown by the dashed black rectangle in figures 7.5(h)–(m). In addition, the weaker strength of the nearest Si–Si bonds also originates from the transfer of electrons from Si atoms to halogen adatoms.

7.2.5 Density of states

The main features in the electronic band structures and the orbital hybridization in the chemical bonds can be verified by the orbital-projected density of states (DOS). For pristine silicene systems, the vanished electronic states at $E_F = 0$, the V-shaped structure and two symmetric logarithmic divergent peaks at $E^{c,v} = \pm 1$ eV are formed by the Si-3p$_z$ orbitals, as shown by the red dashed curve in figure 7.6(a). The unique features of the low-lying DOSs come mainly from the zero band gap, the Dirac cone structure and the saddle points of π bands at the M point in the corresponding electronic band structure in figure 7.2(a). The middle-energy DOS presents a shoulder structure at −1 eV and a prominent peak at −2.6 eV that originates from the maximum point and the saddle point of the σ bands, respectively. They are contributed by mainly the Si-3p$_x$ and Si-3p$_y$ orbitals, as shown by the green and blue dashed curves in figure 7.6(a). Furthermore, the hybridization between Si-3s orbitals, illustrated by the purple dashed curve in figure 7.6(a), and Si-3p$_z$ orbitals induces a prominent peak at −3.2 eV, resulting from a partially flat band with an initiated energy at the Γ point. Also, it should be noted that the interactions between the π and σ bands at low energy are relatively weak.

Under the various adsorption configurations, the low-energy DOS is reshaped fully. For the 100% double-sided adsorption configurations, vacant states in a specific region centered at E_F arise in the fluorinated, chlorinated, brominated and iodinated systems, as presented in figures 7.6(b), (e)–(g), respectively. In contrast, a finite DOS at E_F appears in the astatinated system shown in figure 7.6(h) as result of the semi-metallic behavior presented in figure 7.2(h). The middle-energy DOS exhibits a shoulder structure at −0.2 eV and a prominent peak at −1.5 eV, which are dominated by the Si-3p$_x$ and Si-3p$_y$ orbitals shown by the green and blue dashed curves in figure 7.6(b), and are very similar to those of pristine silicene in figure 7.6(a). In addition, a modified middle-energy DOS is present in figures 7.6(e)–(h). Moreover, the extra strong peaks at deep energies are dominated by halogen-2p$_x$ and halogen-2p$_y$ orbitals (figure 7.6(b)), accounting for the halogen-related weakly dispersed bands at deep energies (figure 7.2(b)). The strong peaks become weaker and dominate at higher energies as their atomic numbers increase, as illustrated in figures 7.6(e)–(h). When the adatom concentration decreases, three low-energy prominent peaks are present at −0.6, 0.8 and 1.4 eV in the 25% double-sided configuration, as shown in figure 7.6(c). The band width of these prominent peaks around 0.6 eV and their states are related to the Si-3p$_z$ orbitals shown by the red dashed curve. Furthermore, the middle-energy DOSs are dominated mainly by

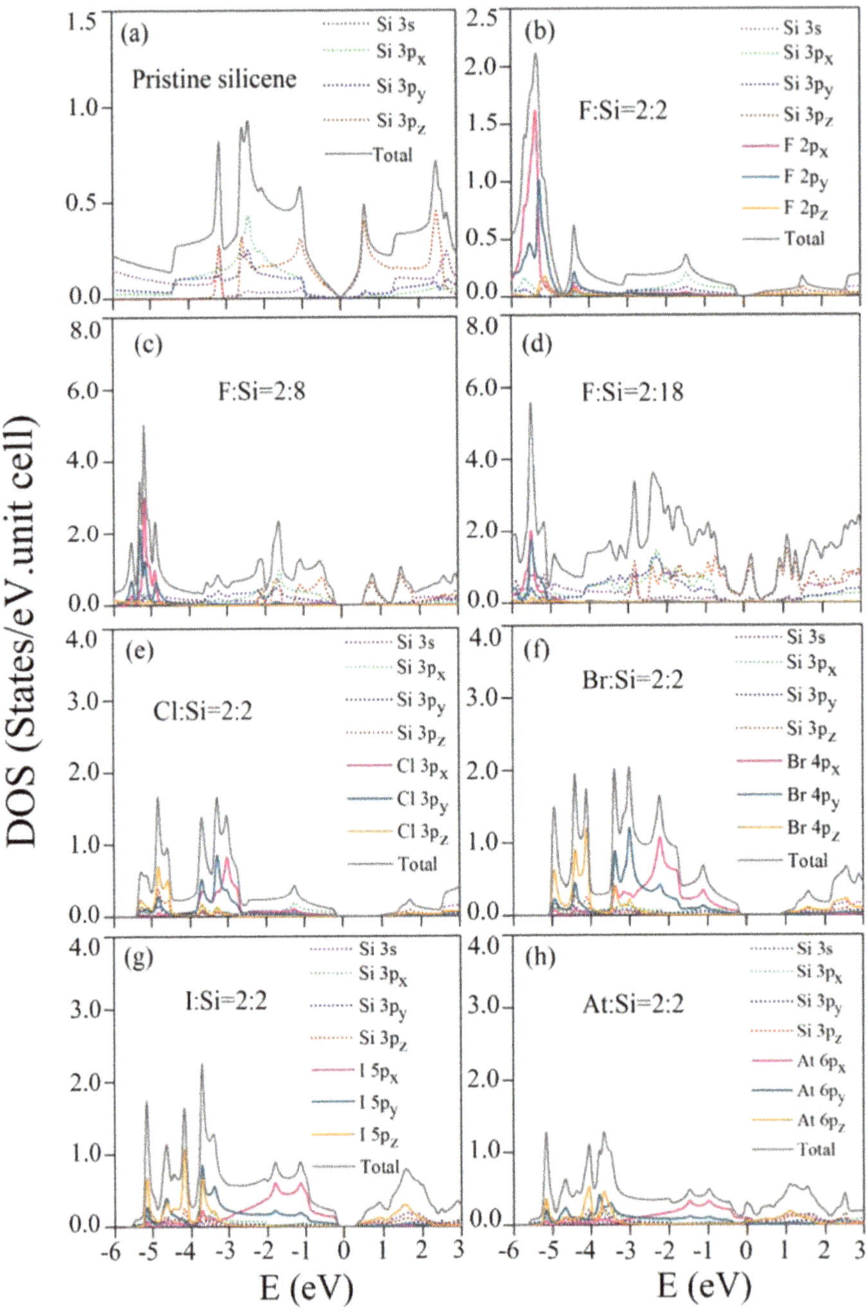

Figure 7.6. Orbital-projected DOSs for (a) pristine silicene, and double-sided adsorption of (b) F:Si = 2:2, (c) F:Si = 2:8, (d) F:Si = 2:18, (e) Cl:Si = 2:2, (f) Br:Si = 2:2; (g) I:Si = 2:2 and (h) At:Si = 2:2. ([27] Copyright 2019, Springer Nature. With permission of Springer.)

the Si-3p_x and Si-3p_y orbitals. However, the single-peak structure in the fully double-sided adsorption configuration separates into several sub-peaks due to the increase of the number of sub-bands. At the critical concentration of 11%, most of the low-energy DOSs within ±1 eV are dominated by the non-passivated Si-3p_z orbitals, as illustrated in figure 7.6(d). This shows the recovery of low-energy π bands in the low concentration configurations, in which they also reveal halogen-related strong peaks at deep energies similar to the 100% configuration.

The halogen adatom- and silicon atom-induced strong peaks that dominate at the deep energies are the fundamental features of the DOSs in the single-sided and double-sided adsorption configurations. The extended low-energy π band widths (figures 7.7(c) and 7.6(d)), the three π-orbital-related prominent peaks within −1 to 2 eV (figures 7.7(b) and 7.6(c)), the σ orbital-related single-peak structures (figures 7.7(a) and 7.6(b)) and the halogen adatom-created extra strong peaks at deep energies (figures 7.7(a) and 7.6(b); figures 7.7(b) and 7.6(c); figures 7.7(c) and 7.6(d)) are similar features in the single-sided and double-sided adsorption configurations. Such extra strong peaks are confined to the same energy range. However, the intensities of these peaks decrease gradually corresponding to the reduction of the adatom concentration. Also, the above similar DOS features are observed in other halogen-adsorption configurations (figures 7.7(d)–(h) and 7.6(e)–(h)). Otherwise, the broken mirror symmetry in the single-sided adsorption configurations results in a finite value of the DOS at the Fermi level, as shown in figures 7.7(a)–(h). Furthermore, the asymmetric prominent peaks near the Fermi level (figures 7.7(b), (d)–(h)) originate from the ferromagnetic energy bands.

STS is a very useful experimental tool to verify the theoretical predictions for DOSs. Semiconducting, semi-metallic and metallic behaviors have been confirmed fully by high-resolution STS measurements, which can also be used to identify the close relationships between the electronic energy spectra and the orbital hybridization of the chemical bonds. To date, the electronic band structures and dimension-varied van Hove singularities in monolayer graphene [36] and few-layer graphene systems [37–39], 1D graphene nanoribbons [39, 40], adatom-adsorbed graphene [41], monolayer silicene [42] and hydrogenated silicene [43] have been verified successfully by such STS measurements. The predicted DOSs of the halogen adatom-adsorbed silicene systems can be examined fully using STS experiments, such as the vanishing or retention of the finite value of the DOSs at the Fermi level, the Si- or halogen-induced strong peaks and the halogen-modified σ shoulder structures. Furthermore, the asymmetric peaks near the Fermi level in the DOSs can be also confirmed using spin-polarized STS.

7.3 Conclusion

The geometric structures and the magnetic and electronic properties of halogen-adsorbed silicene are explored using first-principles calculations. The fundamental geometric, magnetic and electronic properties are discovered through detailed analyses of the adatom-varied geometric structure, spatial spin density distribution, atom-dominated band structure, spatial charge density distribution and orbital-

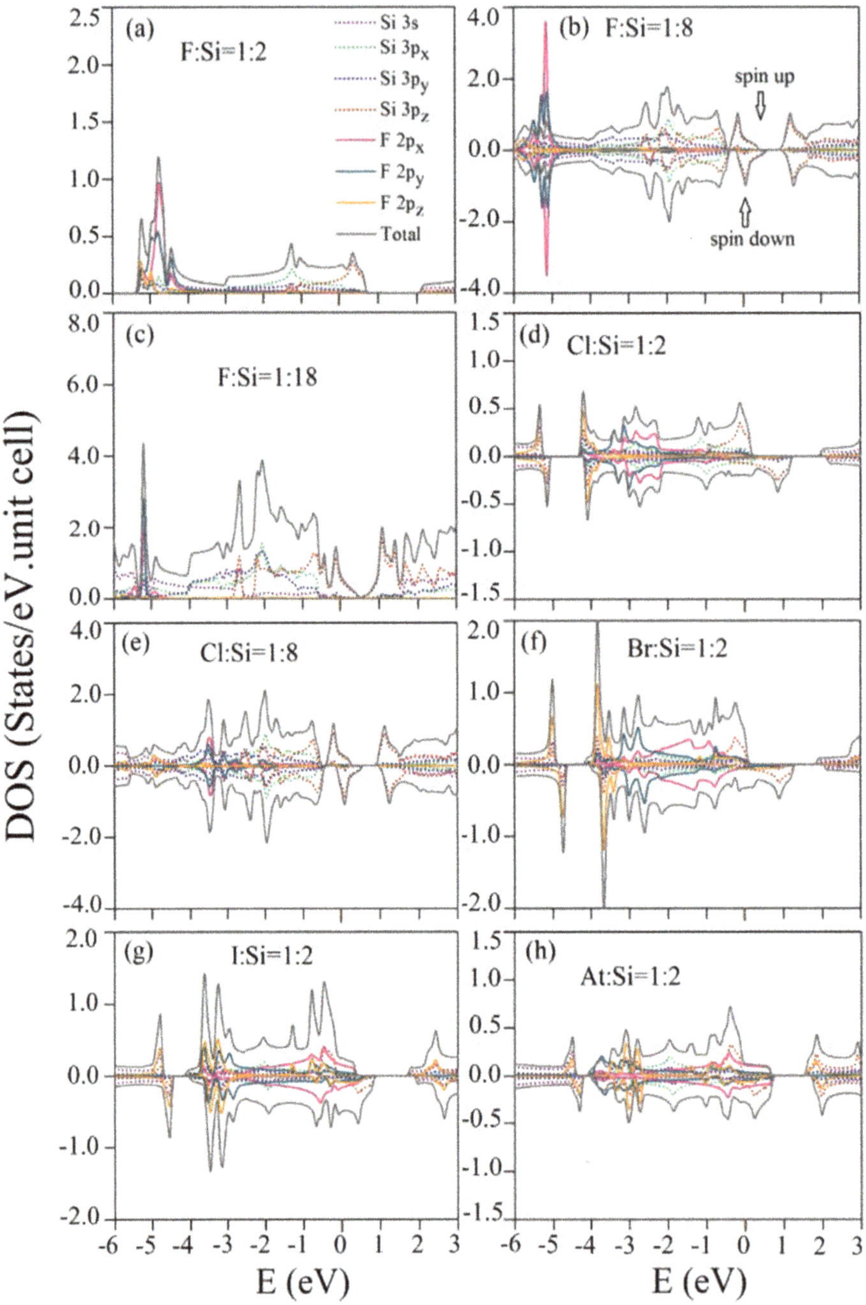

Figure 7.7. A similar plot to figure 7.6, but shown for single-sided adsorption of (a) F:Si = 1:2, (b) F:Si = 1:8, (c) F:Si = 1:18, (d) Cl:Si = 1:2, (e) Cl:Si = 1:8; (f) Br:Si = 1:2, (g) I:Si = 1:2 and (h) At:Si = 1:2. ([27] Copyright 2019, Springer Nature. With permission of Springer.)

projected density of states, respectively, which are varied by various adsorption configurations. The main factor, halogen–Si bonds, induces the diverse phenomena directly. The free hole density is created by the transfer of electrons from Si atoms to halogen adatoms. Adatom-enriched electronic band structures are present in the destruction or recovery of low-lying π bands, the halogen-induced weakly dispersed bands at deep energies, and the valence energy bands with or without spin-splitting intersecting with the Fermi level. The significant competition between the halogen–Si bonds and the weak sp^3 hybridization determines the first feature. In the 100% double-sided systems, the band gaps originate from the σ bands. When the adatom concentration decreases, the low-lying π bands are restored. Smaller energy gaps arise in the 25% system. At the critical concentration of 11% p-type metallic behavior is presented. The significant halogen–Si bonds, the second feature, which are insensitive to the adatom concentrations, are a similar feature of the double-sided and single-sided configurations. The broken mirror symmetry, the third feature, is a feature unique to the single-sided configuration. Furthermore, ferromagnetic behavior is exhibited only in the various single-sided configurations, which can be recognized in the spin-splitting electronic band structures and verified from the asymmetric prominent peaks in the spin- and orbital-projected densities of states (DOSs). The main features in the diverse electronic band structures are verified completely by their DOSs. On the experimental side, the ARPES and STS measurements are available to confirm directly the feature-rich band structures and the diverse DOSs, respectively. The rich fundamental properties of halogen-doped silicene systems are very promising for practical applications.

References

[1] Karlicky F, Kumara Ramanatha Datta K, Otyepka M and Zboril R 2013 Halogenated graphenes: rapidly growing family of graphene derivatives *ACS Nano* **7** 6434–64

[2] Sahin H, Topsakal M and Ciraci S 2011 Structures of fluorinated graphene and their signatures *Phys. Rev.* B **83** 115432

[3] Peelaers H, Hernandez-Nieves A D, Leenaerts O, Partoens B and Peeters F M 2011 Vibrational properties of graphene fluoride and graphane *Appl. Phys. Lett.* **98** 051914

[4] Nguyen D K, Lin Y T, Lin S Y, Chiu Y H, Tran N T T and Lin M F 2017 Fluorination-enriched electronic and magnetic properties in graphene nanoribbons *Phys. Chem. Chem. Phys.* **19** 20667–76

[5] Nguyen D K, Tran N T T, Nguyen T T and Lin M F 2018 Diverse electronic and magnetic properties of chlorination-related graphene nanoribbons *Sci. Rep.* **8** 17859

[6] Wang X *et al* 2013 High-yield production of highly fluorinated graphene by direct heating fluorination of graphene-oxide *ACS Appl. Mater. Interfaces* **5** 8294–9

[7] Zhang H *et al* 2016 Spectroscopic investigation of plasma-fluorinated monolayer graphene and application for gas sensing *ACS Appl. Mater. Interfaces* **8** 8652–61

[8] Gong P *et al* 2013 Photochemical synthesis of fluorinated graphene via a simultaneous fluorination and reduction route *RSC Adv.* **38** 6327–30

[9] Feng W, Long P, Feng Y and Li Y 2016 Two-dimensional fluorinated graphene: synthesis, structures, properties and applications *Adv. Sci.* **3** 1500413

[10] Cheng S H *et al* 2010 Reversible fluorination of graphene: evidence of a two-dimensional wide bandgap semiconductor *Phys. Rev.* B **81** 205435
[11] Chantharasupawong P *et al* 2012 Optical power limiting in fluorinated graphene oxide: an insight into the nonlinear optical properties *J. Phys. Chem.* C **116** 25955–61
[12] Gong P *et al* 2014 Tunable photoluminescence and spectrum split from fluorinated to hydroxylated graphene *Nanoscale* **6** 3316–24
[13] Haberer D *et al* 2011 Direct observation of a dispersionless impurity band in hydrogenated graphene *Phys. Rev.* B **83** 165433
[14] Chen W, Zhu Z, Li S, Chen C and Yan L 2012 Efficient preparation of highly hydrogenated graphene and its application as a high-performance anode material for lithium ion batteries *Nanoscale* **4** 2124–9
[15] Qiu Y, Zhang X and Yang S 2011 High performance supercapacitors based on highly conductive nitrogen-doped graphene sheets *Phys. Chem. Chem. Phys.* **13** 12554–8
[16] Feng L and Liu Z 2011 Graphene in biomedicine: opportunities and challenges *Nanomedicine* **6** 317–24
[17] Secor E B *et al* 2013 Inkjet printing of high conductivity, flexible graphene patterns *J. Phys. Chem. Lett.* **4** 1347–51
[18] Schedin F *et al* 2007 Detection of individual gas molecules adsorbed on graphene *Nat. Mater.* **6** 652
[19] Kuila T, Bose S, Khanra P, Mishra A K, Kim N H and Lee J H 2011 Recent advances in graphene-based biosensors *Biosens. Bioelectron.* **26** 4637–48
[20] Dai J and Yuan J 2010 Adsorption of molecular oxygen on doped graphene: atomic, electronic, and magnetic properties *Phys. Rev.* B **81** 165414
[21] Zhu J *et al* 2016 Magnetic properties of sulfur-doped graphene *J. Magn. Magn. Mater.* **401** 70–6
[22] Sahin H and Ciraci S 2012 Chlorine adsorption on graphene: chlorographene *J. Phys. Chem.* C **116** 24075–83
[23] Li Y *et al* 2012 Synthesis of partially hydrogenated graphene and brominated graphene *J. Mater. Chem.* **22** 15021–4
[24] Šimek P, Klímová K, Sedmidubský D, Jankovský O, Pumera M and Sofer Z 2015 Towards graphene iodide: iodination of graphite oxide *Nanoscale* **7** 261–70
[25] Tran N T T, Lin S Y, Lin C Y and Lin M F 2017 *Geometric and Electronic Properties of Graphene-related Systems: Chemical Bonding Schemes* (Boca Raton, FL: CRC Press)
[26] Nguyen D K, Tran N T T, Chiu Y H and Lin M F 2019 Concentration-diversified magnetic and electronic properties of halogen-adsorbed silicene *Sci. Rep.* **9** 13746
[27] Li W, Sheng S, Chen J, Cheng P, Chen L and Wu K 2016 Ordered chlorinated monolayer silicene structures *Phys. Rev.* B **93** 155410
[28] Feng B *et al* 2012 Evidence of silicene in honeycomb structures of silicon on Ag (111) *Nano Lett.* **12** 3507–11
[29] Fleurence A *et al* 2012 Experimental evidence for epitaxial silicene on diboride thin films *Phys. Rev. Lett.* **108** 245501
[30] Lin C L *et al* 2012 Structure of silicene grown on Ag (111) *Appl. Phys. Express* **5** 045802
[31] Wang X, Liu H and Tu S T 2015 First-principles study of half-fluorinated silicene sheets *RSC Adv.* **5** 6238–45
[32] Zhang W B, Song Z B and Dou L M 2015 The tunable electronic structure and mechanical properties of halogenated silicene: a first-principles study *J. Mater. Chem.* C **3** 3087–94

[33] Osborn T H, Farajian A A, Pupysheva O V, Aga R S and Voon L L Y 2011 *Ab initio* simulations of silicene hydrogenation *Chem. Phys. Lett.* **511** 101–5
[34] Tao L *et al* 2015 Silicene field-effect transistors operating at room temperature *Nat. Nanotechnol.* **10** 227
[35] Sivek J, Sahin H, Partoens B and Peeters F M 2013 Adsorption and absorption of boron, nitrogen, aluminum, and phosphorus on silicene: stability and electronic and phonon properties *Phys. Rev.* B **87** 085444
[36] Gao L 2014 Probing electronic properties of graphene on the atomic scale by scanning tunneling microscopy and spectroscopy *Graphene 2D Mater.* **1** 23–46
[37] Lauffer P *et al* 2008 Atomic and electronic structure of few-layer graphene on SiC(0001) studied with scanning tunneling microscopy and spectroscopy *Phys. Rev.* B **77** 155426
[38] Yankowitz M, Wang F, Lau C N and LeRoy B J 2013 Local spectroscopy of the electrically tunable band gap in trilayer graphene *Phys. Rev.* B **87** 165102
[39] Pierucci D *et al* 2015 Evidence for flat bands near the Fermi level in epitaxial rhombohedral multilayer graphene *ACS Nano* **9** 5432–9
[40] Huang H *et al* 2012 Spatially resolved electronic structures of atomically precise armchair graphene nanoribbons *Sci. Rep.* **2** 983
[41] Gyamfi M, Eelbo T, Wasniowska M and Wiesendanger R 2011 Fe adatoms on graphene/Ru (0001): adsorption site and local electronic properties *Phys. Rev.* B **84** 113403
[42] Chiappe D, Grazianetti C, Tallarida G, Fanciulli M and Molle A 2012 Local electronic properties of corrugated silicene phases *Adv. Mater.* **24** 5088–93
[43] Lin C *et al* 2015 Direct observation of ordered configurations of hydrogen adatoms on graphene *Nano Lett.* **15** 903–8

Chapter 8

Sodium-, magnesium- and aluminum-doped silicene systems

Metal adatom adsorption and intercalation in few-layer graphene systems have attracted a lot of experimental [1–3] and theoretical attention [4, 5, 6], and can have greatly varied physical, chemical and material phenomena under n-type doping, e.g. the blue shift of the Fermi level in potassium-doped graphene systems based on angle-resolved photoemission spectroscopy (ARPES) measurements [7]. The significant orbital hybridizations in metal–carbon bonds lead to dramatic transformations from high numbers of valence charges into conduction carriers. The critical mechanisms can be identified thoroughly using sensitive VASP calculations and analyses, as has been done for alkali-absorbed 2D graphene systems [8, 9, 10] and graphene nanoribbons [11, 12]. The calculated results show that single s-2p_z orbital hybridizations of alkali–carbon bonds are responsible for the rich and unique fundamental properties, including the optimal hollow-site adsorption position, the planar honeycomb lattice, the full charge transfer from alkali to carbon atoms, the partially occupied conduction bands after chemisorption, the presence of only a few alkali-dominated conduction bands, and the roughly rigid-band structures for the π and σ bands [8]. Most importantly, such materials are predicted to provide the highest amount of free carriers among all the layered materials and compounds, which has been examined thoroughly for 1D alkali-adsorbed graphene nanoribbons [11].

Aluminum-doped graphene systems are available for creating high free carrier density [13, 14], as has been done for alkali atoms. Previous first-principles calculations [13] show clearly that the π- and σ-bonding of carbon atoms present only a slight modification after Al adatom adsorption, while electrons are transferred largely from Al to C. Most importantly, there exist significant hybridizations between the (3s, 3p_x, 3p_y) and 2p_z orbitals in the Al–C bonds. Such chemical pictures agree with the quasi-rigid shifts of the C-created energy bands, the free electrons in the conduction Dirac cone, and the Al-dominated valence and conduction bands. In particular, the Al-induced high free carrier density is almost the same as that of the alkali doping, owing mainly to the almost identical red-shift Dirac-cone structures. However, the adatom-created valence bands are absent in alkali-doped systems because of the single electron in the outermost s orbital. The above-mentioned systems might have great potential for future technological applications, such as high-capacity batteries [15] and energy storage devices [16, 17].

Previous first-principles calculations [18, 19, 22–24] on metal adatom absorption include alkalis (Li, Na and K), alkaline-earths (Be, Mg and Ca), group-III (B and Al) and group-IV adatoms (N and P), and transition metals (Ti, V, Cr, Mn, Fe, Co, Mo and W), in which only specific configurations and concentrations were explored for their geometric, electronic and magnetic properties. The optimal positions—the hollow, the top and bridge, and the hollow and bridge sites—depend on complicated competition and/or cooperation between the Si–Si, Si–M and M–M bonds. Specifically, alkali-absorbed silicene systems present well-defined Dirac-cone structures, metallic behaviors with a blue shift of the Fermi level in the conduction band, slightly separated Dirac points and almost identical charge transfers for Li, Na and K. Similar metallization phenomena have been revealed in the graphene chemisorption of alkali adatoms [8, 20], while the latter have higher conduction electron densities. Also, such behaviors appear in Al-, B-, N- and P-adsorbed silicene systems, where Al is an n-type metal and the others are p-type metals. However, the Be-, Mg- and Ca-doped silicene materials may belong to the narrow-gap semiconductors. The rich band properties for II-, III- and V-group chemisorption on silicene surfaces, which are expected to be sensitive to changes in adatom concentrations and configurations, will be examined thoroughly. Moreover, transition-metal chemisorption creates metallic, half-metallic and semiconducting behaviors, accompanied by the ferromagnetic or non-magnetic spin distributions. In addition, the rich multi-orbital hybridizations in the critical Si–M bonds should play an important role in the calculated physical quantities.

In this chapter, sodium, magnesium and aluminum adatom absorption on silicene surfaces is demonstrated to be highly suitable for exploring n-type doping behaviors, since they have, respectively, atomic configurations of of 3s, $3s^2$, and ($3s^2$, $3p^2$), and the atomic numbers 12, 13 and 14 (distinct dangling bonds). The single- and double-sided chemisorption cases, various adatom concentrations (including the highest one), the hollow, top, valley and bridge sites, and the buckled structures are included in the VASP calculations. The focuses of study are the metallization phenomena (n-type doping), the specific conduction bands across the Fermi level with direction-dependent Fermi momenta, the 2D free carrier density, the low-lying extended states (the π-bonding states) on a honeycomb lattice, the well-defined, modified and thoroughly suppressed Dirac cones, the rigid shift of σ bands and undefined bands, and the spin-degenerate or spin-split energy bands (the vanishing or finite magnetic moments). They are expected to vary, depending greatly on the different single- and multi-orbital hybridizations in the Na–, Mg– and Al–Si bonds. According to detailed calculations and analyses of the optimal metal adatom positions, the Si- and adatom-dominated energy bands, the spatial charge distributions and their variation after chemisorption, the spin densities, and the atom-, orbital- and spin-decomposed van Hove singularities, concise chemical pictures are proposed to account for the varied electronic and magnetic properties. How many orbitals of the metal adatoms and silicon host atoms take part in significant chemical bonding will be resolved in the current work. Interestingly, whether the ferromagnetic spin configuration can survive under specific metal adatom additions, adsorption configurations and concentrations is examined thoroughly using the spin-up and/or spin-down density distributions, such as their localized regions, enclosed volumes, and symmetric or highly asymmetric arrangements. The similarities and important differences between silicene- and graphene-related systems are investigated in detail by comparing the strengths of chemical bonding and the buckling degrees [8, 21].

8.1 Na adsorption

Sodium, magnesium and aluminum adatom adsorption on monolayer silicene surfaces presents diverse geometric structures, owing mainly to the different types of interlayer chemical bonding. The Na-adsorbed silicenes present a hollow-site optimal position, as shown clearly in figures 8.1(a)–(f) and table 8.1. These systems are stable under double- and single-sided absorption with specific concentrations and distributions. The highest saturation corresponds to the $6{:}6_d$ case with upper and lower adatoms (similar to the arrangement in figure 8.1(a) but both-side adsorption). The buckling degree is comparable to that (0.48 Å) of a pristine system at a sufficiently low concentration, <12.5%, while it might be enhanced greatly for other conditions, e.g. 0.86, 0.62 and 0.58 Å, respectively, in the 6:6, 3:6 and 2:6 cases. That the π and σ bonding are non-orthogonal to each other becomes more obvious with the increase of Na concentration.

The Na–Si bonds lengths lie in the range of ~3.10–3.26 Å, depending weakly on the adsorption. The nearest Si–Si bonds are stretched slightly by the observable change from 2.28 Å to ~2.30–2.33 Å. This indicates a partial charge transfer through the interlayer chemical bonding. According to the above-mentioned geometric features, the σ-electronic chemical bonding, which arises from the 3s, $3p_x$ and $3p_y$ orbitals of silicon atoms, is deduced to take little part in the alkali adatom chemisorption. As a result, the single-orbital hybridization of Na- and Si-3s will dominate the adsorption-induced chemical bonding and thus the other fundamental properties. The theoretical predictions for the optimal top views could be examined using STM and the side views using TEM and LEED measurements (details in sections 3.1 and 3.2). In contrast, alkali-adsorbed graphenes have planar structures, shorter Na–C bond lengths of ~2.35–2.37 Å and lower ground state energies (higher adsorption energies). It is relatively easy to synthesize such materials.

Compared to pristine silicene (figure 8.2(a)), sodium-adsorbed systems exhibit unusual valence and conduction bands (figures 8.2(b)–(h)). The main features of the band structures include highly asymmetric energy spectra around the Fermi level, a modified Dirac-cone structure with or without energy spacing, a blue shift of the Fermi level relative to the Dirac point (metallic behavior), roughly rigid σ valence bands, splitting and anti-crossing states, Na-dominated or Na- and Si-dominated

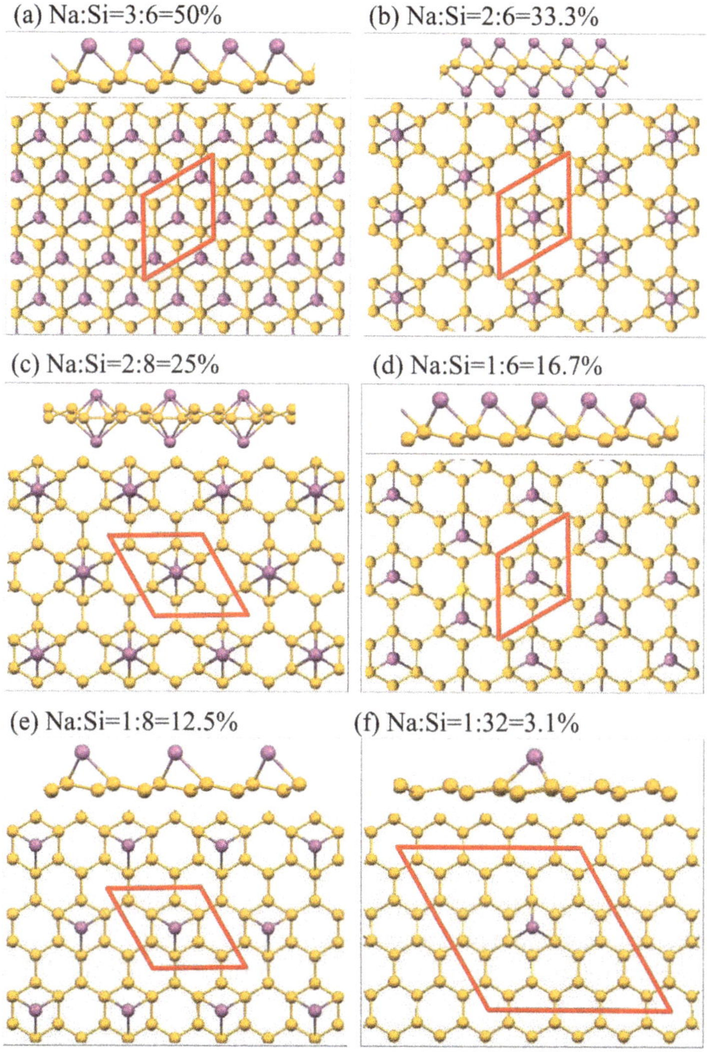

Figure 8.1. Side and top views of sodium-adsorbed silicene systems for different cases: (a) 3:6, (b) 2:6, (c) 2:8, (d) 1:6, (e) 1:8 and (f) 1:32.

conduction and/or valence states, and extra critical points in the energy–wave-vector space related to band hybridizations, in which the electronic-state dominance of Na and Si atoms is indicated by, respectively, the red and blue circles. The anisotropic and distorted Dirac cones, which arises mainly from the $3p_z$ orbitals (the orbital-decomposed density of states in figure 8.4), are initiated from the stable Γ or K valley. The initial or bottom valence band is very close to the first σ energy subbands with concave-downward dispersion relations, where an energy spacing of ~0.20 eV corresponds to the deeper-state σ band. Interestingly, Na-created states serve as conduction bands above the Fermi level, except that at very high concentrations, >50% in figures 8.2(b) and (c), they display the opposite behavior. Also, the Na

Table 8.1. The calculated nearest passivated Si–Si and Na–Si bond lengths, Na heights, total magnetic moments, buckle height and binding energies of Na-absorbed silicene systems.

Na:Si	Si–Si bond length (passivation) (Å)	Na–Si bond length (Å)	Na height (Å)	Magnetic moment (μ_B)	Buckle height (Å)	E_b (eV)
Pristine Si	2.28	—	—	0	0.48	—
6:6 = 100%	2.41	2.94	2.08	0	0.86	−1.74
3:6 = 50%	2.33	3.15	2.28	0	0.62	−3.03
2:6 = 33.3%	2.31	3.10	2.18	0	0.58	−1.48
2:8 = 25%	2.30	3.10	2.12	0	0.57	−2.14
1:6 = 16.7%	2.31	3.12	2.17	0	0.52	−1.44
1:8 = 12.5%	2.30	3.17	2.22	0	0.48	−1.60
1:32 = 3.1%	2.30	3.26	2.35	0	0.48	−1.86

adatoms make minor contributions to the deeper valence bands. It should be noted that similar results are also revealed in any alkali adatom adsorption because of the comparable electron affinities. Generally, the π, $\pi*$ and σ bands of the host atoms and the adatom-dominated conduction bands can be characterized from their significant features, which are further supported by the charge density distributions in figure 8.3 and the density of states in figure 8.4. This suggests a very slight distortion of the extended π bonding on the silicene surface and almost unchanged σ bonding. Furthermore, it is very useful in simulating electronic states through the phenomenological model (the tight-binding model) and thus understanding the other fundamental properties, such as the magnetic rich quantization in alkali-adsorbed silicene systems [22, 24, 25]. Specifically, the spin-split electronic structures are absent for various Na adsorption cases, i.e. there are no spin-dependent ferromagnetic or anti-ferromagnetic configurations.

Sodium adsorption on a silicene surface is able to create free conduction electrons with sufficiently high carrier densities (the so-called n-type doping phenomenon). The Fermi level appears to be changed dramatically from the middle of the well-behaved Dirac-cone structure (figure 8.2(a)) into the conduction cone of the modified structure. The blue shift of E_F is ~0.30 eV, and its magnitude is about one third larger than those in graphene-related systems [24]. That is to say, only the partial 3s orbitals in sodium adatoms serve as conduction electrons. The Na-induced free electrons correspond to the occupied states between E_F and the conduction Dirac point. However, it is impossible to evaluate the 2D carrier density from the k-space area covered by the first-principles calculations [11], owing mainly to the anisotropic energy spectra and the significant contributions of Si- and Na-created conduction bands. In addition, the specific charge transfer of each sodium adatom could be estimated using silicene nanoribbons, as has been done for 1D graphene nanoribbons [11]. On the experimental side, optical [26] and transport [27] measurements might be efficient and reliable in examining the Na-created free electron density. The modified Dirac-cone structure arises only at stable K or Γ valleys as a

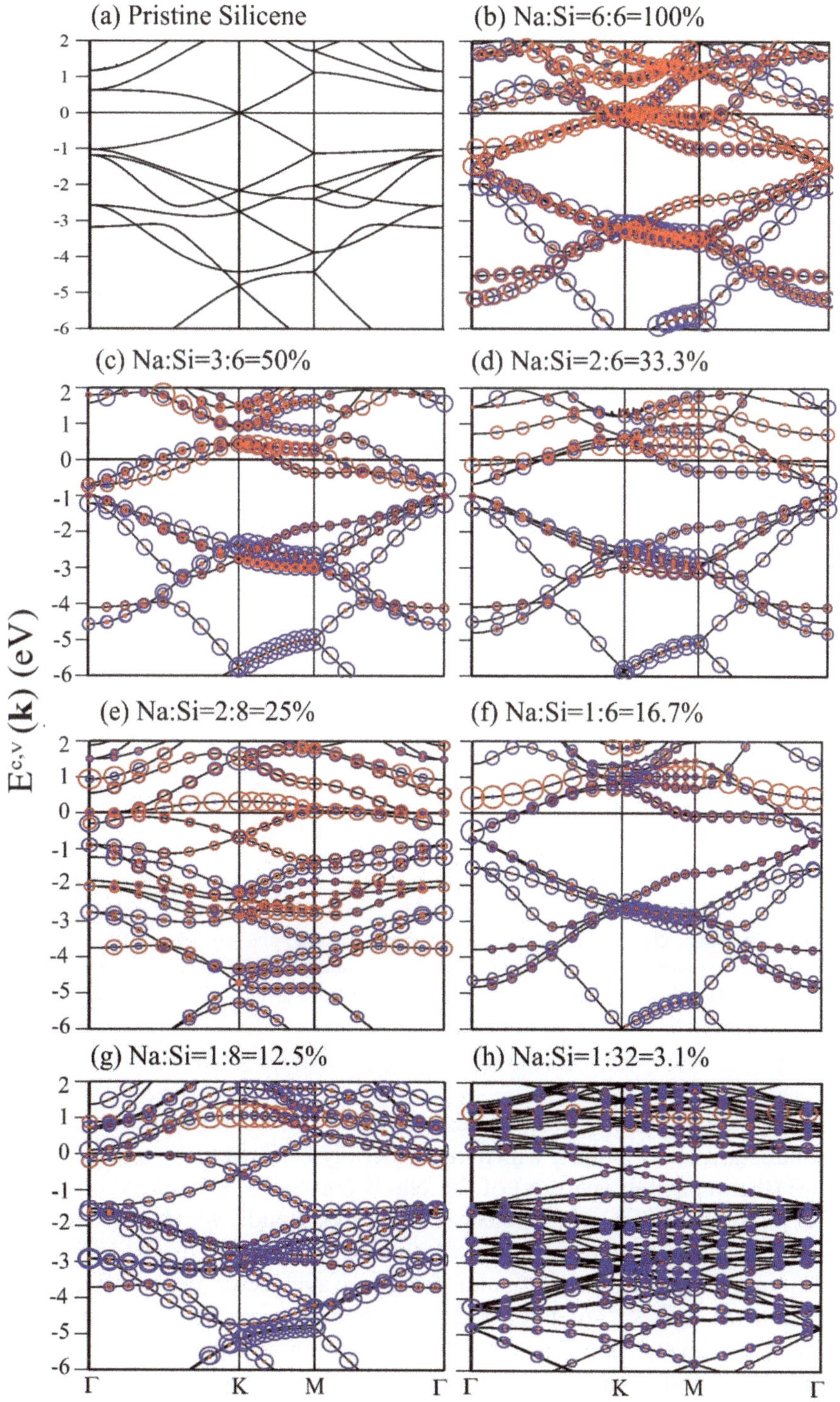

Figure 8.2. Various electronic energy spectra for different absorption with Na and Si contributions (the blue and red circles): (a) pristine silicone, and (b) 6:6, (c) 3:6, (d) 2:6, (e) 2:8, (f) 1:6, (g) 1:8 and (h) 1:32 Na concentrations.

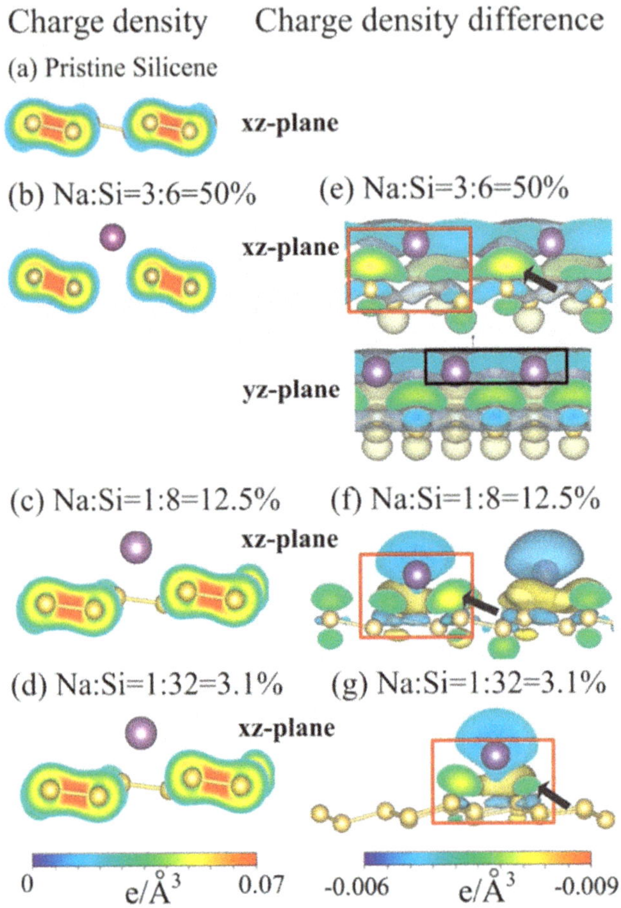

Figure 8.3. The spatial charge distribution for (a) pristine silicene, and (b) 3:6, (c) 1:8 and (d) 1:32 Na concentrations. Their corresponding variation after Na-chemisorption along the (x, z)- and (y, z)-planes are shown in (e)–(g). The π-bonding variation is indicated by the black arrows, whereas the Na–Na and Na–Si interactions are shown by the black and red rectangles, respectively.

result of the zone-folding effect. In general, there exists a separation of the valence and conduction Dirac points, which can generally be attributed to the different site energies (the ionization energies) of the Na-3s and Si-3p_z orbitals.

The well-behaved π- and σ-electronic energy bands which arise in sodium-adsorbed silicene systems can be examined efficiently using high-resolution ARPES measurements, in particular for the wave-vector dependence along the high-symmetry points. To date, such experiments have been very successful in identifying the diverse band structures of few-layer graphene systems with different layer numbers and stacking configurations, e.g. a gapless valence Dirac cone in monolayer graphene [28–30], two pairs of parabolic energy dispersions in AB-stacked bilayer systems [30, 31], coexistent linear and parabolic bands in twisted

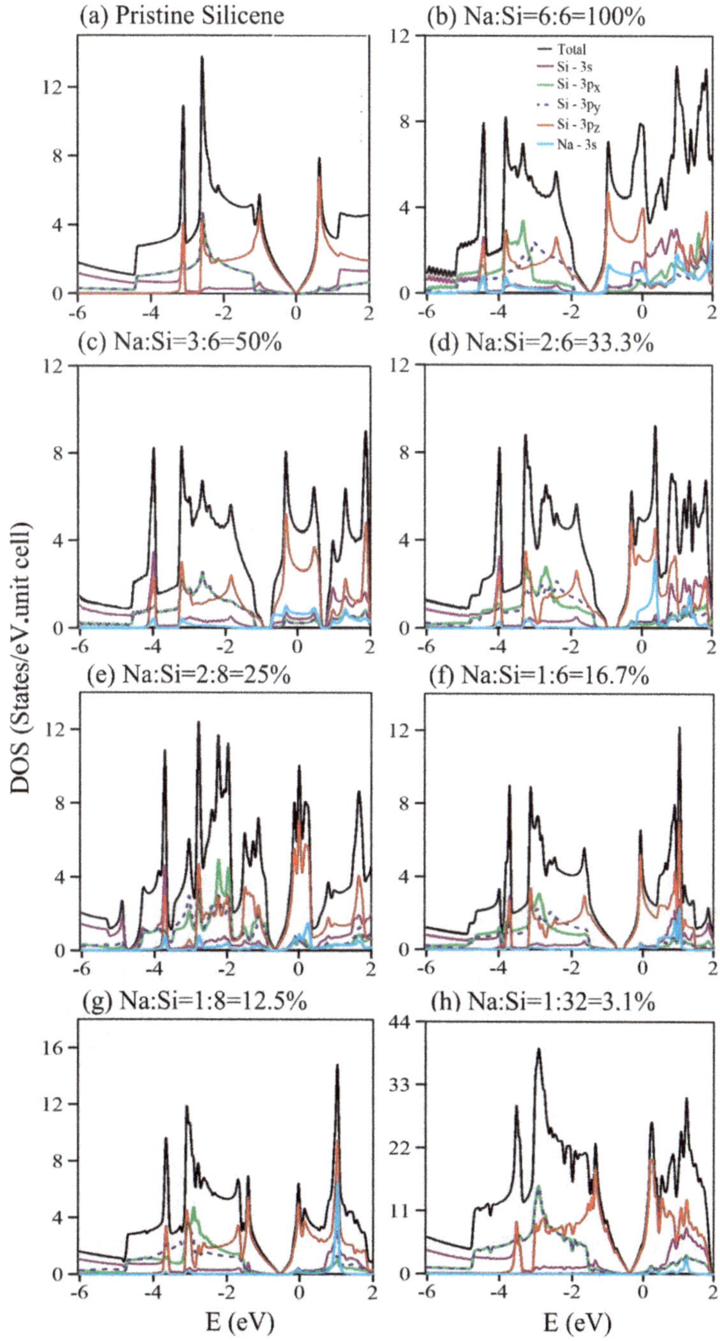

Figure 8.4. The atom- and orbital-decomposed density of states for Na-adsorbed silicene systems: (a) pristine silicene, and (b) 6:6, (c) 3:6, (d) 2:6, (e) 2:8, (f) 1:6, (g) 1:8 and (h) 1:32.

bilayer graphenes [32], linear and parabolic bands for trilayer ABA stacking [30, 33], and linear, partially flat and sombrero-shaped bands for ABC trilayer stacking [33], where the low-lying energy bands are initiated from the K/K′ valleys related to the planar honeycomb lattices. To date, Li, Na and K adatom adsorption [34–36] on graphene surfaces has been verified to present a blue shift of the Fermi level and the linear Dirac cone. Similar ARPES examinations could be performed for alkali-adsorbed silicenes to explore the smaller shift of E_F relative to the K or Γ valley, modifications of the Dirac cone (the creation of energy spacings and anisotropic spectra) and the splitting middle-energy valence bands near the M point. As a result, they are useful for estimating the partial charge transfer from guest to host atoms and understanding the modified π bonding.

The alkali-adsorbed graphenes and silicenes are worthy of closer examination (first-principles results for the former are available in [8]). There exist important similarities and differences between these two condensed-matter systems. After adatom chemisorption, the geometric structures of the pristine materials almost remain the same; thus the σ-electronic orbitals take little part in the interlayer chemical bonding. The σ-electronic states are affected only slightly by the zone-folding effects. The π chemical bonding, which behaves well under the $3p_z$- and $2p_z$-orbital bonding, is modified weakly by alkali–carbon and/or alkali–silicon interactions. This leads to the separation of the valence and conduction Dirac points and anisotropic energy dispersions. Obviously, electrons are transferred from the alkali adatoms to the host atoms. A lot of free conduction electrons exist between the Fermi level and the C-, Si- and alkali-dominated conduction states. However, E_F is, respectively, ~1.0 and 0.3 eV at low adatom concentrations. Its value even reaches 3.2 eV below the Dirac points for full-adsorption graphene systems. It should be noted that detailed studies have been performed successfully for alkali-adsorbed graphene systems regarding the densities of free carriers under any adatoms, concentrations and configurations. A specific conclusion has been reached: the dramatic transformation of the outermost s orbital of each alkali adatom into one free conduction electron (full charge transfer under various adsorption cases). Such layered materials are able to generate very high 2D free carrier densities and thus present great potential for applications in electronic devices [8, 34–36]. On the other hand, alkali-related silicenes exhibit only partial charge transfer, <0.5 e, according to the smaller magnitude of E_F and the k-space area covered from the conduction Dirac-point state to the Fermi-momentum states. In short, graphene can provide more active $2p_z$-orbital dangling bonds and thus very effective n-type doping in chemical modifications, compared to the the $3p_z$-orbital bonds of silicene.

The spatial charge distributions and their variation (ρ and $\Delta\rho$) after sodium adsorption can be used to examine the changes in π and σ chemical bonding on the silicene surface and coupling-induced orbital hybridizations. Similar bonding phenomena are present in other alkali-adsorbed silicene systems (not shown). Apparently, two kinds of orbital bonding in the honeycomb lattice can survive simultaneously under any sodium adsorption cases, as shown in figures 8.3(a)–(c). The σ-electronic orbital hybridizations, with a very high charge density between two silicon atoms, are affected little by sodium adsorption, corresponding to the almost vanishing $\Delta\rho$ in figures 8.3(d)–(f). This is responsible for the rigid red-shift of the σ energy bands and the absence of valence bands co-dominated by sodium and carbon atoms (figures 8.2(a)–(f)). On the other hand, significant modifications of the π bonding are observable through the charge variation between the sodium and carbon atoms (the red rectangles), in which part of the adatom electrons is transferred to the host-atom region (the black arrow). Furthermore, they are observed relatively easily under the high-concentration condition (figure 8.3(d)). These features suggest the existence of only a single 3s–3p$_z$ orbital hybridization in the sodium–silicon bonds, accounting for the Na- and Si-co-dominated conduction bands. Specifically, observable charge variation on the (y, z)-plane is revealed between two neighboring sodium atoms at sufficiently high concentrations (figure 8.3(e)), owing mainly to the creation of Na–Na bonds. In short, the critical orbital hybridization and free conduction electrons associated with alkali adsorption depend little on specific adatoms, distributions, concentrations, 2D and 1D dimensions, or even edge structures.

The 3s–3p$_z$ orbital hybridization in the Na–Si bond is able to create unusual van Hove singularities in the density of states. All the sodium-adsorbed silicenes, as shown in figures 8.4(b)–(h), present a strongly, moderately or lightly modified dip with, respectively, an observable, very narrow or negligible energy spacing below the Fermi level. This result depends strongly on the Na concentration, and it becomes vanishing at sufficiently low concentrations (<3.1% in figure 8.4(h)). The Dirac-cone-induced structure has an obvious red shift relative to E_F, illustrating clear n-type doping behavior, e.g. −0.5 eV for the specific 50% case (figure 8.3(c)). The area covered between the bottom of the conduction dip and the Fermi level determines the Na-created free electron density. Generally, the valence and conduction dips are dominated by the Si-3p$_z$ orbitals (the red curves). However, the Na-3s orbitals (the cyan curves) also make certain contributions to the conduction dips at very high adatom concentrations, >50% in figures 8.4(b) and (c), owing mainly to Na–Na interactions. That both the 2p$_z$ and 3s orbitals possess merged van Hove singularities indicates significant interlayer single-orbital chemical bonding. Most such structures appear in the conduction-state region away from the dip, while a few of them might arise in the deeper region for the >50% cases. The above-mentioned features suggest the existence of distorted π bonding on a buckled silicene surface. The valence π-band widths can be estimated approximately from the calculated density of states, since it is related closely to the initial valence dip, the prominent peak and two asymmetric peaks, as observed in pristine silicene (figure 8.4(a)). Their magnitudes are listed in table 8.1, and they are comparable to those in the absence of alkali adsorption (table 4.1). There exist splitting π and π^* prominent peaks, which are accompanied by extra minor structures, and are attributed to zone folding, interlayer atomic interactions and enhanced buckling.

For the σ electrons of silicon host atoms, the 3p$_x$ and 3p$_y$ orbitals (the green and dashed blue curves) display identical contributions in the presence of a symmetric environment and different contributions in the presence of an asymmetric environment under the (x, y)-plane projection (figure 8.1). The initial σ-electronic valence band creates obvious shoulder structures in the range -1.65 eV $< E < -1.20$ eV, depending on the Na concentration, which is deeper than for the pristine case (figure 8.4(a)). The whole band width includes the first shoulder, the symmetric peak and the second peak, which, respectively, correspond to the parabolic, saddle and parabolic band-edge states at the Γ, M and K points (figures 8.2(a)–(h)). Moreover, a few van Hove singularities, which are associated with the four orbitals of the Si atoms, occur simultaneously (the red, green, purple and dashed blue curves, figure 8.4), e.g. those at -4.0 eV $< E < -3.20$ eV related to the anti-crossing of the π and σ bands. In short, the modified π and σ chemical bonding could be identified approximately from the specific van Hove singularities, and such examinations are supported by the band structures and spatial charge densities.

Sodium-adsorption-enriched van Hove singularities can be verified directly using high-resolution STS measurements (details in section 3.2), as has been achieved successfully for carbon-based materials with well-characterized sp^2 bonding. For example, current experiments on layered graphene systems have identified an isotropic V-shaped structure vanishing at the gapless Dirac point for monolayer graphene [28–30], zone-folding-created peaks in the logarithmic form for twisted bilayer graphenes [32], a gate-voltage-induced energy gap for bilayer AB and trilayer ABC stacking [30, 31], a delta-function-like peak centered around E_F for any ABC stacking [33], a pair of asymmetric peaks relative to a dip structure at E_F for AAB trilayer stacking [30, 33], the semimetallic property for graphite [31, 33, 36], and its splitting π and π^* strong peaks at deeper or higher state energies [31, 33]. Further STS examinations for alkali-adsorbed silicene systems will focus mainly on the modified dip structure below the Fermi level, the split π and π^* peaks, and the adatom-generated extended structure.

8.2 Mg adatoms

Magnesium-adsorbed silicenes, shown in figures 8.5(a)–(f) and table 8.2, present hollow-site optimal positions. While such structures are similar to those of Na adsorption (table 8.1), the Mg–Si bond lengths and the adatom heights (~2.7 and 1.65 Å, respectively) are much shorter than the Na–Si ones (~3.15 Å). One hundred per cent Mg adsorption could survive under the 6:6 case (figure 8.5(a)), in which there exist sufficiently strong Mg–Mg bonds. The buckling of the honeycomb lattice and the height difference of the A and B sublattices are largest among the three kinds of metal adatom adsorption at high concentrations, such as the drastic change from 0.48 Å to 0.96, 0.84 and 0.79 Å under the high-concentration $6{:}6_d$, $3{:}6_s$ and $2{:}6_d$ cases. The enhanced buckling is expected to induce more important sp^3 bonding (non-orthogonality of π and σ bonding) and spin–orbit coupling in silicene. These geometric features suggest sufficiently strong chemical bonding between the magnesium adatoms and silicon atoms, which is associated with two 3s-valence orbitals of Mg. The π bonding, which is extended on the silicene surface, will be modified strongly or even disappear completely through critical interlayer atomic interactions. The non-uniform top- and side-view environments created by different chemical bonding can be verified using STM and TEM/LEED (see sections 3.1 and 3.2).

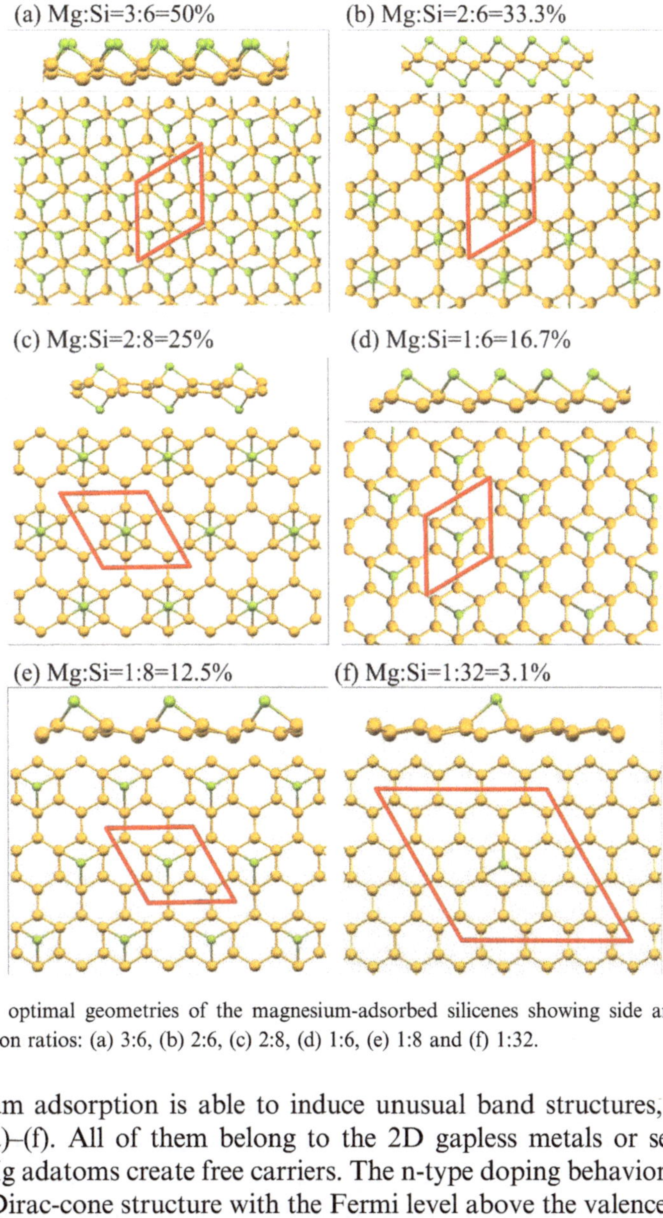

Figure 8.5. The optimal geometries of the magnesium-adsorbed silicenes showing side and top views for specific adsorption ratios: (a) 3:6, (b) 2:6, (c) 2:8, (d) 1:6, (e) 1:8 and (f) 1:32.

Magnesium adsorption is able to induce unusual band structures, as shown in figures 8.6(a)–(f). All of them belong to the 2D gapless metals or semimetals, in which the Mg adatoms create free carriers. The n-type doping behavior is defined by a modified Dirac-cone structure with the Fermi level above the valence Dirac point. Part of the Mg-3s orbitals are transferred from adatoms to Si host atoms. Generally, there exist drastically modified low-lying energy dispersions, owing mainly to the extremely non-uniform environment in an enlarged unit cell and more important or complicated interlayer orbital hybridizations. Detailed analyses are required to identify the critical mechanisms for the main features of electronic energy spectra. The strongly modified Dirac-cone structures (the large green circles), which appear at the Γ or K points, are characterized by highly anisotropic energy dispersions and zone-folding-induced split subbands with or without observable Dirac-point energy

Table 8.2. The calculated nearest passivated Si–Si and Mg–Si bond lengths, Mg-heights, total magnetic moments, buckle height and binding energies of Mg-absorbed silicene systems.

Mg:Si	Si–Si bond length (passivation) (Å)	Mg–Si bond length (Å)	Mg-height (Å)	Magnetic moment/ atom (μ_B)	Buckle height (Å)	E_b (eV)
6:6 = 100%	2.46	2.86	2.24	0.00	0.94	–0.77
3:6 = 50%	2.43	2.81	2.08	0.00	0.86	−1.66
2:6 = 33.3%	2.38	2.81	2.09	0.00	0.79	−1.33
2:8 = 25%	2.40	2.70	1.95	0.00	0.73	−1.67
1:6 = 16.7%	2.33	2.76	1.89	0.46	0.61	−0.92
1:8 = 12.5%	2.33	2.81	1.94	0.00	0.59	−0.98
1:32 = 3.1%	2.31	2.88	2.12	0.53	0.52	−0.82

spacing. These unusual bands are contributed mostly by the seriously distorted π bonding in the silicene surface (the blue circles). According to the main features of the low-energy spectra, the whole π-band width is very difficult to identify directly from electronic structures. Furthermore, initial σ valence energy subbands arise in the Γ valley using concave-downward dispersions in the range $E_v < -1.20$ eV (see figure 8.9 for the density of states, the solid green and dashed blue curves for the $3p_x$ and $3p_y$ orbitals, respectively). Interestingly, the Mg-induced and the Mg- and Si-induced energy bands (the red circles) present partially flat or oscillatory dispersions, which are very close to, or even merged with low-lying π- and σ-electronic structures. This phenomenon can survive under any Mg absorption case. Since the magnesium adatoms possess two 3s orbitals (the red circles), they are able to create Mg-dominated and Mg- and Si-dominated conduction and valence states. The conduction states are very obvious at sufficiently high concentrations, such as the electronic conduction energy spectra under the 6:6 and 3:6 cases (figures 8.6(a) and (b)). This unusual result is related closely to the significant Mg–Mg chemical bonding. However, with low concentrations, the Mg-induced energy bands are near the Fermi level, e.g. the band structures for the 1:8 and 1:32 conditions (figures 8.6(e) and (f)).

The spatial charge distributions and their variation after Mg-chemisorption, as shown in figures 8.7(a)–(f), can be used to determine the important orbital hybridizations in Si–Si, Mg–Si and Mg–Mg bonds. The π-electronic bonding (figures 8.7(a)–(c) on the (x, z)-plane) is modified strongly under chemical absorption, while the opposite is true for the σ-electronic bonding. There are significant and unusual charge density differences between the guest and host atoms. The $\Delta\rho$, with highly non-uniform variation, are reduced greatly near the guest atom and enhanced near the host atom through the Mg–Si bonds, such as those in figures 8.7(e)–(f) within the red rectangles by the red and blue region. The Mg-induced charge density variation is more obvious compared to the Na adsorption case. The main reason for this lies in the two 3s-valence electrons for each Mg atom. The strong orbital

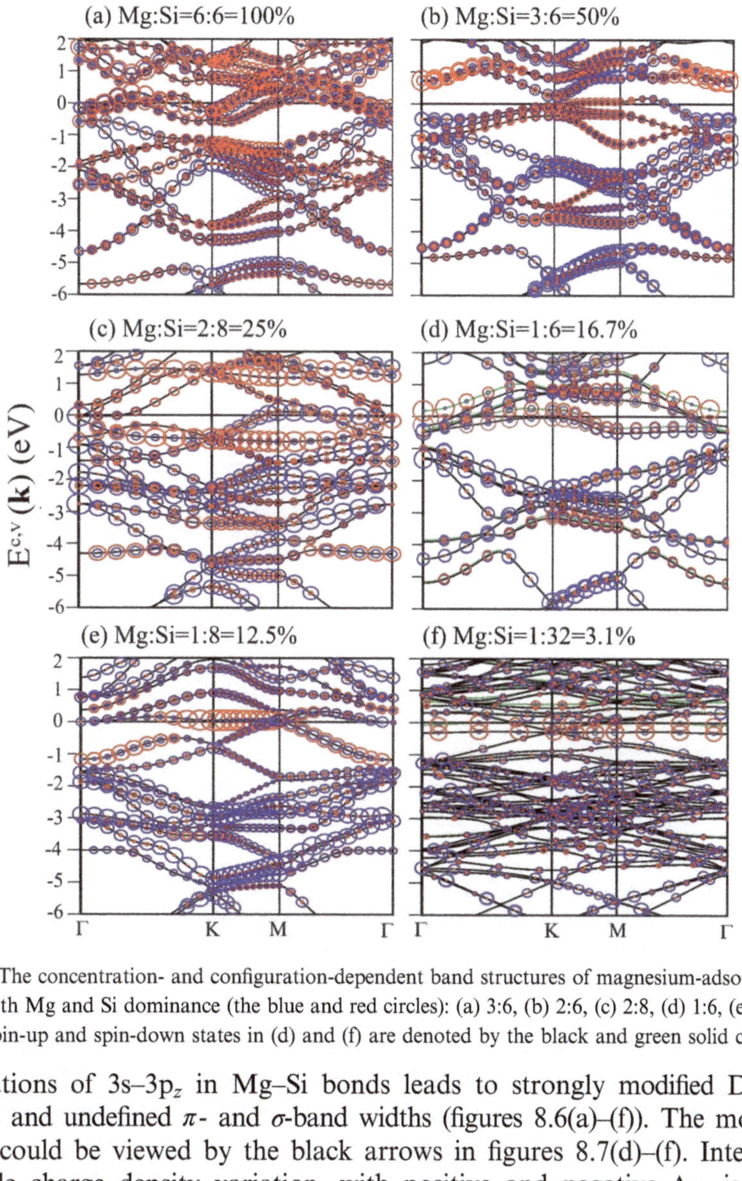

Figure 8.6. The concentration- and configuration-dependent band structures of magnesium-adsorbed silicene materials with Mg and Si dominance (the blue and red circles): (a) 3:6, (b) 2:6, (c) 2:8, (d) 1:6, (e) 1:8 and (f) 1:32. The spin-up and spin-down states in (d) and (f) are denoted by the black and green solid curves.

hybridizations of 3s–3p_z in Mg–Si bonds leads to strongly modified Dirac-cone structures and undefined π- and σ-band widths (figures 8.6(a)–(f)). The modified π-bonding could be viewed by the black arrows in figures 8.7(d)–(f). Interestingly, observable charge density variation, with positive and negative $\Delta\rho$, is revealed between two Mg atoms in high Mg-concentration cases, e.g. within the black rectangle in figure 8.7 on the (y, z)-plane for the 3:6 configuration. The strong 3s–3s orbital hybridizations in Mg–Mg bonds are responsible for the wide Mg-dominated valence and conduction energy spectra, which are consistent with the atom- and orbital-decomposed densities of states (see figures 8.9(a) and (b)). The above-mentioned results show that it might be difficult to simulate the first-principles band structures using the tight-binding model.

The ferromagnetic configurations, which are characterized by spin-split valence and conduction energy bands and the net magnetic moments (figure 8.6 and

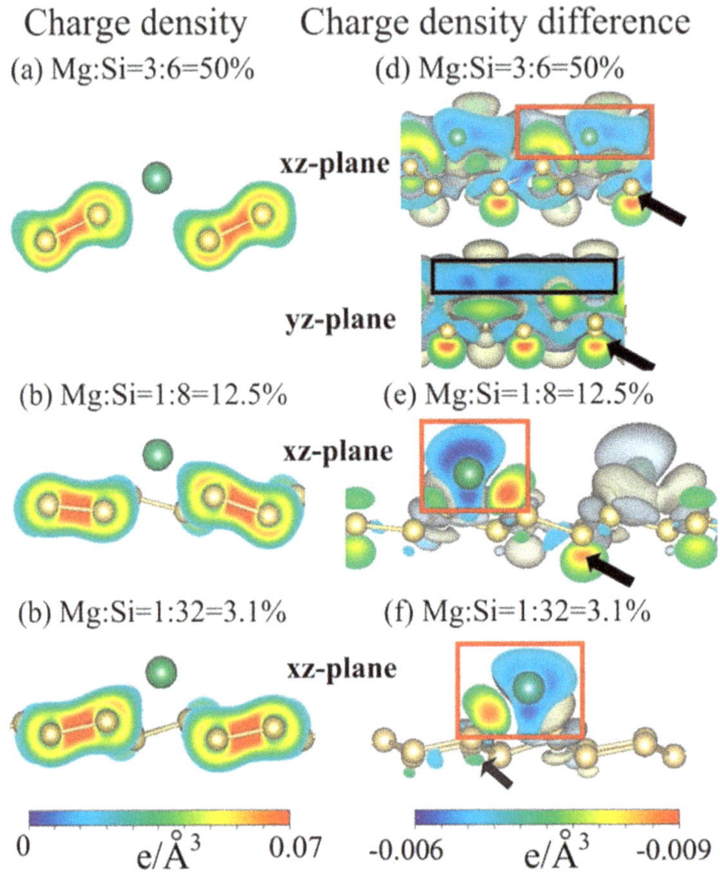

Figure 8.7. The spatial charge distributions (a)–(c) and variation after Mg adsorption (d)–(f) on the (x, z)- and (y, z)-planes at different Mg:Si ratios: (a) and (d) 3:6, (b) and (e) 1:8, and (c) and (f) 1:32.

table 8.2), are further examined using the spin density distributions of specific Mg adsorption cases. For example, the 1:6 and 1:32 conditions, as displayed in figures 8.8(a) and (b), exhibit unusual ferromagnetic phenomena. The Mg adatoms (the green spheres in figure 8.8(a) enclosed by a black parallelogram) of the 1:6 case seem to be able to create a dominant spin density with an extremely non-uniform distribution, where most of the spin-dependent volumes might be localized at the upper plane of $z > 0$. On the other hand, the host Si atoms (the yellow spheres) make a minor contribution, and only half of them (the B sublattice with longer Mg–Si bonds) provide a very weak magnetic moment through the rather small spin volumes in the lower plane. The Mg and Si dominance remain similar under the 1:32 case (figure 8.8(b)). However, the positive contributions from the guest and host atoms become larger and the Si-created spin distributions are extended lightly to the upper plane. The above-mentioned ferromagnetic distributions are associated with the two 3s-orbital electrons of Mg adatoms, which is also supported by Mg-dominated spin-split energy bands (figures 8.6(d) and (f)) and the atom-, spin- and orbital-

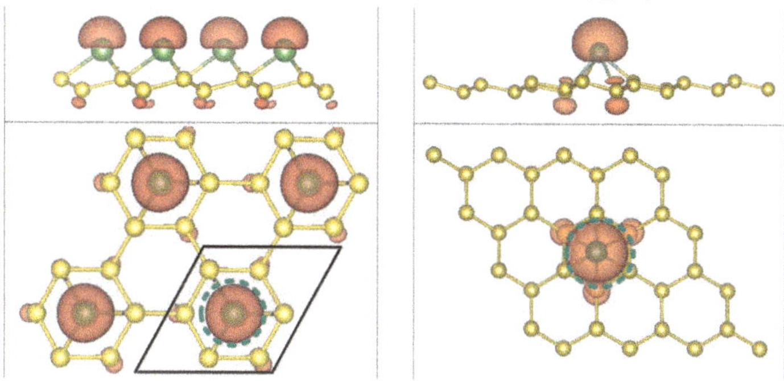

Figure 8.8. The spin density distributions for ferromagnetic (a) 1:6 and (b) 1:32 magnesium-adsorbed silicene materials are shown clearly on the (x, z)- and (x, y)-planes.

decomposed densities of states (see figures 8.9(d) and (f)). They are able to induce on-site Coulomb interactions in the intrinsic Hamiltonian [37]. As a result, the orbital hybridizations, spin-dependent many-particle interactions and spin–orbit coupling need to be included in the Hubbard tight-binding model [38]. It seems to be difficult to obtain the very complicated parameters for suitable phenomenological models. The theoretical predictions for Mg-adatom-generated ferromagnetism could be tested directly using the high-resolution SP-STM (see section 3.2).

The atom-, orbital- and spin-projected van Hove singularities' densities of states, as illustrated in figures 8.9(a)–(f), are available for identifying the critical orbital hybridizations in Mg-adsorbed silicenes and thus the complicated energy spectra and the existence of π and σ bonding. The density of states is finite at the Fermi level for all the Mg adsorption cases, in which its magnitude is very large for the partially flat bands and the oscillatory energy dispersions, e.g. that of the 1:8 case (figure 8.9(e)). It is almost impossible to estimate the real 2D conduction electron density as a result of the frequent band crossing and anti-crossing (figures 8.6(a)–(f)). Also, a detailed comparison of the metal-induced free carrier densities between the Na-, Mg- and Al-adsorbed silicene systems become meaningless. A seriously distorted dip structure, corresponding to the strongly modified Dirac cone, can be identified roughly only from the separated forms, such as those under the specific 1:6 and 1:32 conditions (figures 8.9(d) and (f)). The bottom of the conduction part presents a red shift relative to E_F. Below this structure, the van Hove singularities might be comparable for the Si-$3p_x$, -$3p_y$ and -$3p_z$ orbitals (the solid green, dashed blue and red curves), or they arise frequently. On the other hand, this characteristic becomes ambiguous in the presence of neighboring van Hove singularities (the other four conditions). Furthermore, the special structures (the red curves near $E = 0$), which arise mainly from the π- and π^*-electronic states (the valence and conduction states due to the $3p_z$ orbitals), are merged together. They might not be suitable for identifying n- or p-type doping. The above-mentioned features of the $3p_x$, $3p_y$ and $3p_z$ orbitals indicate that it would be very difficult to define the π-band width and thus the σ band width well.

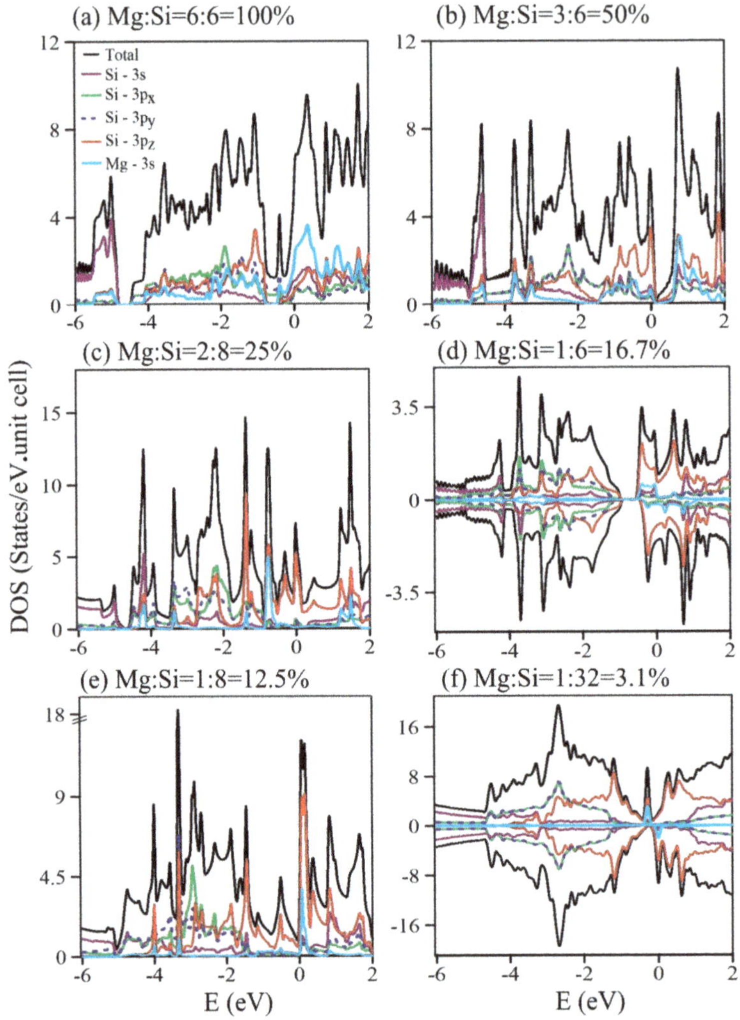

Figure 8.9. The atom-, orbital- and spin-projected van Hove singularities for Mg-adsorbed silicene materials for the (a) 6:6, (b) 3:6, (c) 2:8, (d) 1:6, (e) 1:8 and (f) 1:32 cases.

The Si-3p$_z$ orbitals appear to make important contributions to the density of states in the very wide range of -5.0 eV< E < 2.0 eV. Their van Hove singularities are accompanied simultaneously by those of Si-3s, -3p$_x$ and -3p$_y$ (the purple curves for the Si-3s curve). This result suggests enhanced sp^3 bonding. Interestingly, strongly merged special structures are revealed in the Si-3p$_z$ and Mg-3s orbitals (the red and light blue curves), where certain energy regions cover the valence and conduction states. The adsorption-induced 3s–3p$_z$ hybridizations in the Mg–Si bonds play a critical role in the fundamental electronic properties. They belong to multi-orbital bonding, which is completely different to the single-orbital bonding in the Na–Si bonds. Specifically, the Mg-3s orbitals have an obvious valence- and conduction-state region of -2.0 eV $< E < 2.0$ eV under very high concentrations (the 6:6 and 3:6 cases in figures 8.9(a) and (b)), directly reflecting significant Mg–Mg bonds and Mg-dominated conduction bands, respectively (figures 8.6(a) and (b)). The critical orbital hybridizations may be combined with the ferromagnetic configurations, which are characterized by the spin-split density of states. The spin-up and spin-down densities of states below the Fermi level determine the net magnetic moment. Most of the deep valence states cancel each other out and, furthermore, the main contributions arise from the unbalanced electronic energy spectrum across E_0, e.g. those for the 1:6 and 1:32 cases in figures 8.9(d) and (f), respectively. The first-principles results show that the Mg-3s and Si-3p_z orbitals dominate the spin-split distribution, in particular for the former. This agrees with the dominant and observable ferromagnetic spin configuration due to the Mg and Si atoms, as shown in figures 8.8(a) and (b).

The predicted features of the electronic properties in magnesium-adsorbed silicene materials could be examined using high-resolution ARPES, SP-ARPES, STS and SP-STS measurements (details in chapter 3). ARPES and SP-ARPES focus on the blue shift of the Fermi level, the complicated low-lying energy spectra with modified Dirac cones and Mg-dominated bands, the spin-split or spin-degenerate bands across or near E_F, the frequent crossing and anti-crossing, and the existence of complete π and σ bands along the high-symmetry points. STS and SP-STS focus on the finite density of states at E_F, the strongly modified dips with or without energy spacing, the n-type doping behavior, the numerous van Hove singularities due to the sp^3 and (Si, 2p_z)–(Mg,3s) bonding, and the Mg-3s- and Si-3p_z-induced spin splitting at low energy. Such experimental studies could provide full information about the critical orbital hybridizations in the Si–Si, Si–Mg and Mg–Mg bonds and in the extremely non-uniform environment of a moiré superlattice. Certain important similarities and differences have been revealed for Mg-adsorbed silicene and graphene systems. Both materials present a hollow-site optimal position, while graphene systems possess planar honeycomb lattices, but not great enhancement of buckled lattices. The guest–host atomic interactions are similar, however, the charge transfer is smaller for the 3s–2p_z orbital hybridization in the Mg–C bond, as identified from the spatial charge density difference (figures 8.7(d)–(f)). The non-buckling geometries and the weaker interlayer chemical bonding which arise under Mg adatom adsorption on the graphene surface create well-behaved electronic and magnetic properties. The Dirac-cone structure, with or without observable energy spacing, is identified easily from the uniform energy dispersion across the Fermi level. E_F is above the conduction Dirac point and free electrons exist between them. This demonstrates clear n-type doping, and the carrier density is higher compared to the silicene case. Interestingly, its real value is very difficult to evaluate precisely for both materials. Most importantly, simple band structures show well-defined π- and σ-electronic states and band widths, which are consistent with the sufficiently separated van Hove singularities. Finally, Mg adatoms and carbon host atoms are unable to generate specific spin configurations under any absorption case.

8.3 Al adatoms and the creation of diverse phenomena

The optimal aluminum adatom adsorption on the silicene surface, as shown in figures 8.10(a)–(f), is situated at the valley position (table 8.3). Also note that the valley and top positions possess different chemical environments, i.e. they are capable of creating diverse fundamental properties. The highest adatom adsorption can be achieved through the 3:6 case of single-sided adsorption with greatly enhanced buckling. However, 100% saturated adsorption with the double-sided configuration is not stable and thus disappears completely, owing mainly to the quite

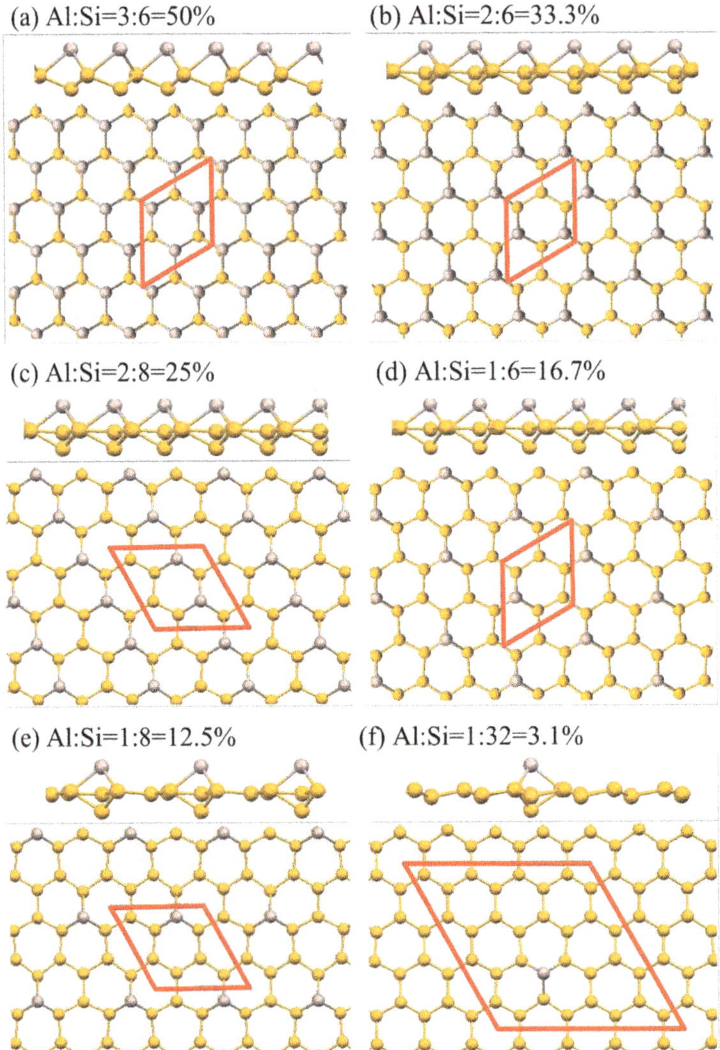

Figure 8.10. The optimal geometric structures for aluminum-adsorbed silicene systems with top and side views under various adsorption ratios: (a) 3:6, (b) 2:6, (c) 2:8, (d) 1:6, (e) 1:8 and (f) 1:32.

Table 8.3. The calculated nearest passivated Si–Si and Al–Si bond lengths, Al-heights, total magnetic moments, buckle height and binding energies of Al-absorbed silicene systems.

Al:Si	Si–Si bond length (passivation) (Å)	Al–Si bond length (Å)	Al-height (Å)	Magnetic moment/ atom (μ_B)	Buckle height (Å)	E_b (eV)
3:6 = 50%	2.47	2.89	2.14	0.00	0.74	−3.11
2:6 = 33.3%	2.47	2.71	1.93	0.00	0.70	−2.69
2:8 = 25%	2.47	2.72	2.05	0.34	0.75	−2.73
1:6 = 16.7%	2.39	2.61	1.90	0.00	0.75	−3.10
1:8 = 12.5%	2.34	2.60	1.89	0.73	0.64	−3.47
1:32 = 3.1%	2.32	2.58	1.84	0.00	0.49	−3.72

strong repulsive interactions between the neighboring aluminum adatoms. This appears to be very different to the Na and Mg adsorption cases. The Al–Si bond lengths, which depend on the adsorption condition, lie in the range of ~2.58–2.89 Å. They become longer with the increase of Al concentration, in particular at the highest concentration. In contrast, the optimal position of aluminum-adsorbed graphene systems corresponds to the hollow site on a planar honeycomb lattice. The concentration- and configuration-dependent Al–C bond lengths are ~2.54–2.57 Å (details in [13]), and the highest Al adatom concentration with a stable geometry was found be the 2:8 case under double-sided adsorption of 25% [13]. The geometric symmetry appears to be responsible for the critical chemical bonding and thus the electronic and magnetic properties. Important differences are expected to exist between the aluminum-adsorbed silicenes and graphenes, e.g. different geometries, that could be demonstrated using near-future STM, TEM and LEED studies (details in chapter 3).

The valley sites and the multi-orbital hybridizations in aluminum-adsorbed silicene systems can create rich and unique band structures (figures 8.11(a)–(f)), in sharp contrast to the Na and Mg adsorption cases (figures 8.2(b)–(h) and 8.6(a)–(f)). The low-lying band structures are very complicated under any adsorption case (even in the presence of low Al concentrations). Generally, it is very difficult to identify accurately the strongly modified Dirac-cone structure, the π and π^* bands and the σ bands at sufficiently high concentrations (figures 8.11(a)–(d)), since there exist certain conduction subbands across the Fermi level, non-uniform, oscillatory and partially flat energy dispersions, Si- and/or Al-dominated valence and conduction bands, and frequent crossing and anti-crossing behaviors. However, the separated Dirac-cone structures at the K point will be recovered gradually during the decrease Al concentration, e.g. those for the 1:8 and 1:32 cases (figures 8.11(e) and (f)). The above-mentioned features further lead to difficulties in evaluating the 2D free carrier densities. Interestingly, the Al adatoms play a critical role in the low-energy states across the Fermi level, in particular for the high-concentration cases, >25% in figures 8.11(a)–(c). Also, they make significant contributions to the deeper valence states in the range of -5.20 eV < E < -3.0 eV, in which their width is sensitive to the

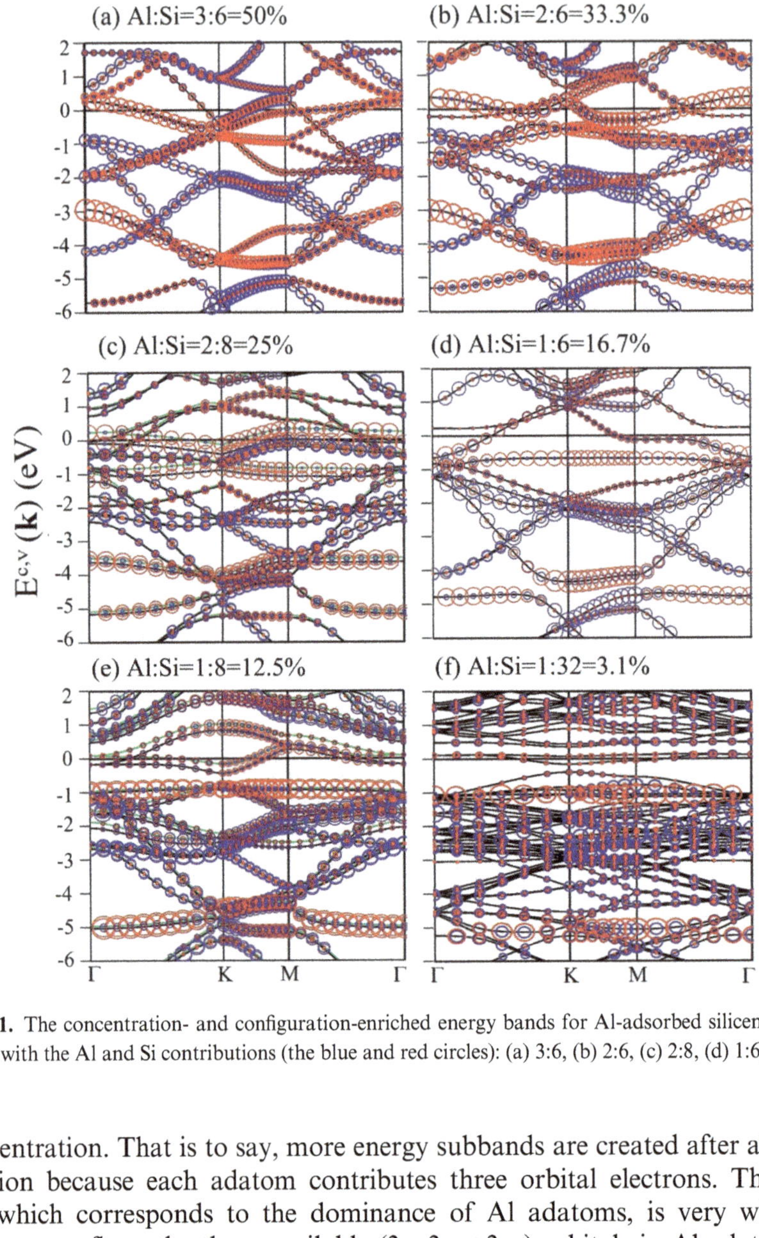

Figure 8.11. The concentration- and configuration-enriched energy bands for Al-adsorbed silicene materials, combined with the Al and Si contributions (the blue and red circles): (a) 3:6, (b) 2:6, (c) 2:8, (d) 1:6, (e) 1:8 and (f) 1:32.

Al concentration. That is to say, more energy subbands are created after aluminum adsorption because each adatom contributes three orbital electrons. The energy width, which corresponds to the dominance of Al adatoms, is very wide. This phenomenon reflects the three available (3s, $3p_x$+$3p_y$) orbitals in Al-related bonds directly. Most importantly, the characterizations of the π and σ valence bands in Si-induced states would become meaningless. Under certain Al adsorption cases, spin-split energy bands arise at low energies; furthermore, they might cross the Fermi level, such as the 2:8 and 1:8 cases in figures 8.11(c) and (e) shown by the black and green curves. These electronic energy spectra illustrate clearly the ferromagnetic configuration with an observable net magnetic moment (table 8.3; see figures 8.12(a) and (b)). They are expected to generate unusual spin-dependent transport properties and thus have high potential in nanoscale electronic devices [18, 39–41]. In addition, the spin-induced magnetic properties are completely absent in Al-adsorbed graphene systems [13, 16].

The aluminum adsorption on the silicene surface presents unusual spatial charge density distributions and differences (figures 8.13(a)–(f)), which are very different from the sodium- and magnesium-adsorbed cases (figures 8.3(a)–(f) and 8.7(a)–(f)). The critical factor is the strongest chemisorption due to three valence electrons in each aluminum atom. The optimal valley-site position is capable of generating multi-orbital hybridizations, in which one Al adatom presents significant chemical bonding with two silicon host atoms on the nearest-neighbor A and B sublattices of the same Si-dimer, as illustrated in figures 8.13(a)–(c). A sufficiently high carrier density can reach the intermediate region between the Al and Si atoms, resulting in obvious charge redistributions of Al-(3s, $3p_x$+$3p_y$) and Si-(3s, $3p_x$, $3p_y$, $3p_z$) orbitals. Prominent Al–Si bonding is also revealed through the drastic changes of carrier densities, where large charge transfers with positive and negative values appear simultaneously through the dark red and blue regions in figures 8.13(d)–(f), respectively. Furthermore, the prominent Al adsorption effects will strengthen the sp^3 bonding in the Si–Si bonds greatly or destroy the good characterization of the π and σ bonds. Specifically, sufficient important Al–Al bonds arise only under high Al concentrations, e.g. the 3:6 configuration with apparent charge variation between two adatoms (the black rectangle in the lower panel of figure 8.13(d)). In contrast, both the Na–Si and Mg–Si atomic interactions are related closely to the host atoms on the different dimers (figures 8.3(a)–(c) and figures 8.7(a)–(c)), where the carrier density is very dilute in the intermediate region close to the Na and Mg adatoms. The important difference to the Al adsorption case arises from the hollow-site absorption position, and the one 3s valence electron in Na and the two 3s valence electrons in Mg.

The carrier density, which is responsible for the fundamental electronic properties, deserves closer investigation. Among the three kinds of metal adatom adsorption on the silicene surface, the Al- and Na-adsorbed cases present,

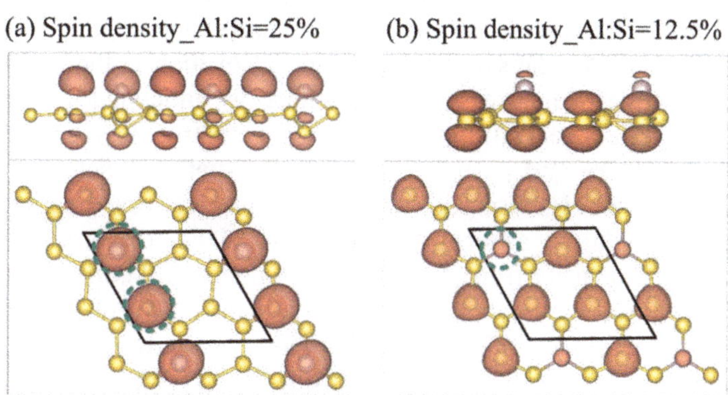

Figure 8.12. The spin density distributions for ferromagnetic (a) 2:8 and (b) 1:8 Al-adsorbed silicene materials are shown clearly on the (x, z)- and (x, y)-planes.

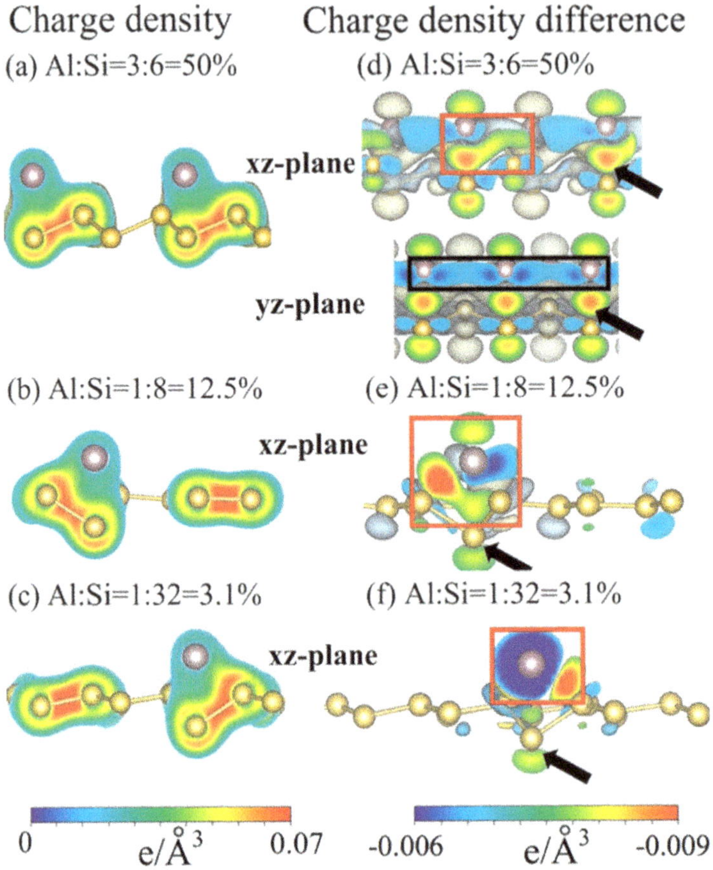

Figure 8.13. The spatial charge distributions (a), (b) and (c), and their variation after Al chemisorption (d), (e) and (f) along the (x, z)- and (y, z)-planes: (a) and (d) 3:6, (b) and (e) 1:8, and (c) and (f) 1:32 Al-concentrations.

respectively, the most complicated and simple charge distributions and the largest and smallest quantities in charge transfer, owing mainly to the number of valence electrons of (3s, $3p_x$+$3p_y$) and 3s orbitals in each adatom (three for Al and two for Na). As a result, for Al, the saturated adatom adsorption is higher than 50%, the modified Dirac-cone structure is absent, there is undefined n-type doping at sufficiently high Al concentrations, and there are meaningless π- and σ-electronic states, more oscillatory and partially flat energy dispersions near the Fermi level, and very complex van Hove singularities due to the electronic energy spectra. For Na, the saturated adatom adsorption is equal to 100%, the modified Dirac-cone structure is identified easily from the low-lying valence and conduction bands, and there are well-behaved doping behaviors under any chemisorption, slightly distorted π- and σ-electronic states, uniform band structures near the Fermi level and less complex van Hove singularities due to electronic energy spectra.

The ferromagnetic spin arrangements arise only in certain Al-adsorbed silicene systems, the 2:8 and 1:8 cases in figures 8.12(a) and (b), respectively, as observed in

Mg (figures 8.9(a) and (b)) but not the Na adsorption cases. The net moment seems to be related to the atomic configurations of the guest and host atoms (tables 8.1–8.3). Interestingly, the 2:8 and 1:8 Al-adsorbed silicene systems present completely different magnetic phenomena. For the 2:8 system, the Al and Si atoms (the green circles and yellow spheres in the black parallelogram) make, respectively, dominant contributions and minor or observable contributions to the extreme ferromagnetism, in which only two host atoms below the guest atoms can create significant spin distributions. However, the opposite ferromagnetic contribution is revealed clearly in the 1:8 system, i.e. the Si and Al atoms, respectively, play important and weak roles in creating the specific ferromagnetic configuration. Only three host atoms and one guest atom in the A and B sublattices provide the unusual spin arrangement; furthermore, the ferromagnetic configuration is almost symmetric around the (z = 0)-plane, as indicated by the (x, z)-plane side view. Generally, the atom- and spin-dominated magnetic behaviors are consistent with the density of states near Fermi level (figures 8.14(a) and (f)). The extremely anisotropic or non-homogeneous spin densities should match the physical, chemical and material environments of enlarged moiré superlattices. That is to say, the spin-configuration-induced many-particle Coulomb interactions [8] and the multi- and single-orbital hybridizations [8, 13, 42] compete and/or cooperate with each other to achieve the lowest ground state energy.

The significant orbital hybridizations, and the π and σ bonding, which could survive in Al-adsorbed silicene systems are investigated thoroughly through the atom-, orbital- and spin-projected van Hove singularities (figures 8.14(a)–(f)). All the absorption cases present a finite density of states at the Fermi level, indicating metallic or semimetallic behaviors directly. The seriously distorted V-shaped structure, corresponding to the strongly modified Dirac-cone structures, is dominated by the Si-3p$_z$ orbital (the red curve) and arises below E_F only under low Al concentrations, e.g. those for the 1:8 and 1:32 cases (figures 8.14(e) and (f)). This indicates the existence of the addressed π bonding and n-type doping. However, it is almost impossible to define the whole π-band width from the total or Si-3p$_z$ density of states (the black and red curves). On the other hand, the V-shaped bottom is very difficult to identify from the merged van Hove singularities close to E_F; therefore, the free conduction electrons or valence holes, and their 2D density and π bonding become meaningless. In general, the special structures due to the four orbitals of the Si host atoms could appear simultaneously within certain energy ranges of the valence and conduction spectra. Such a result suggests the significant and observable sp^3 bonding in the Si–Si bonds. Most importantly, the Si-(3s,3p$_x$,3p$_y$,3p$_z$) and Al-(3s, 3p$_x$+3p$_y$) orbitals (the purple, green, dashed blue, red, light blue and yellow curves) frequently present similar van Hove singularities at identical energies within the whole range except for very low Al concentrations (e.g. the 1:32 case in figure 8.14(f)). Their multi-orbital hybridizations appear to play critical roles in the varied electronic properties. Specifically, under very high Al concentrations, such as the 3:6 and 2:6 configurations, the effective distribution widths of Al-3s and Al-(3p$_x$+3p$_y$) are more than 5 eV, reflecting clearly the creation of Al-dominated valence and conduction bands (figures 8.11(a) and (b)). That is to say, the three

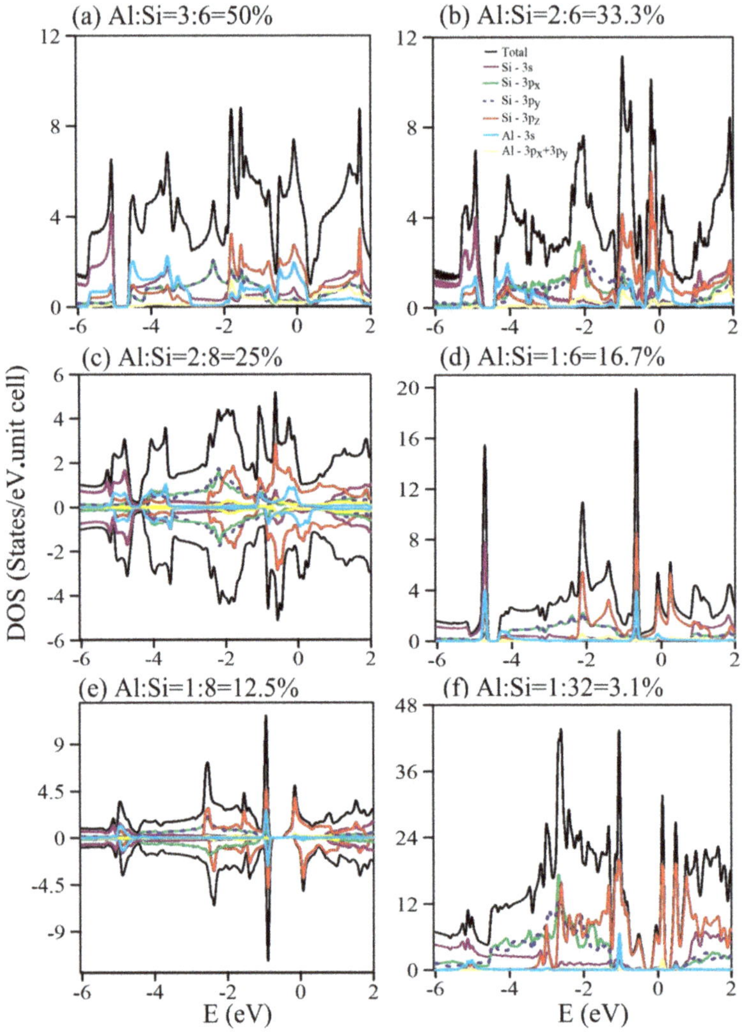

Figure 8.14. The atom-, orbital- and spin-projected densities of states for Al-adsorbed silicenes for specific absorption cases: (a) 3:6, (b) 2:6, (c) 2:8, (d) 1:6, (e) 1:8 and (f) 1:32.

orbital hybridizations survive in the Al–Al bonds. The spin-up and spin-down states are found only for the 2:8 and 1:8 cases (figure 8.14(c) and (e)), in which their unbalanced spectra lie in the low-energy range across the Fermi level. The former and the latter are, respectively, dominated by the Al-3s and Si-3p_z orbitals; therefore, the guest and host atoms under different conditions are responsible for the dominant spin distributions (figures 8.12(c) and (e)).

ARPES, STS and SP-STS measurements of Al-adsorbed silicene systems are very efficient for exploring the electronic energy spectra fully. The tested features include the irregular relationship between the Fermi-momentum states and the Al concentration, the anisotropic Dirac-cone structure with an observable energy spacing, the blue shift of E_F at the sufficiently low Al concentrations, the low-lying oscillatory or partially flat energy bands initiated from the stable K or Γ valley, and the Al- and Si-co-dominated valence bands which are accompanied by rich van Hove singularities in terms of semimetallic and metallic behavior at E_F, undefined π and σ band widths, a strongly modified V-shape below E_F, and merged special structures from the ($3s,3p_x+3p_y$)–($3s,3p_x+3p_y$), ($3s, 3p_x+3p_y$)–($3s,3p_x,3p_y,3p_z$) and sp^3 orbital hybridizations due to the Al–Al, Al–Si and Si–Si bonds, respectively. Such experimental studies are very helpful in clarifying the above-mentioned chemical modifications after aluminum adsorption on the silicene surface.

There exist certain important differences between the Al-adsorbed graphenes [13, 43, 44] and silicenes in terms of the fundamental properties, according to first-principles calculations. The main reasons for this arise from the chemical environments and critical orbital hybridizations. The Al-adsorbed graphene systems exhibit, as predicted in previous studies [13], hollow-site optimal positions, lower saturation adsorption (the 25% case), smaller binding energies, longer Al–C bonds, a blue shift of the Fermi level relative to the conduction Dirac point, a 2D density of free conduction electrons comparable to that in the alkali adsorption cases (almost full charge transfer), a slightly modified Dirac-cone structure, a quasi-red-shift of σ-electronic valence bands regardless of the adsorption effects, Al- and C-co-dominated valence and conduction bands corresponding, respectively, to the ($3s$, $2p_z$) and ($3p_x+3p_y$, $2p_z$) orbital mixing at the specific energies $E^c \sim 1.20$ eV and $E_v \tilde{}\; - 3.20$ eV, spin-degenerate states without ferromagnetic configurations under any adsorption conditions, a finite density of states characteristic of n-type doping, an obvious dip structure below E_F, merged van Hove singularities related to the above-mentioned adsorption bonding types, and observable widths of the π and σ valence bands. It seems to be relatively easy to characterize the π and σ bonding and the significant orbital hybridization in the Al–C, C–C and Al–Al bonds. Experimental studies can clarify these diverse phenomena due to chemical adsorption on different 2D surfaces.

The Na-, Mg- and Al-adsorbed silicene systems appear to differ from one another in terms of the fundamental properties. The important differences include the attractive interaction energies, the optimal absorption positions, the slightly or greatly enhanced buckling, the adatom–silicon band lengths, the metallic or semi-metallic behaviors (n-type or irregular doping), the slightly, strongly or completely modified Dirac-cone structures at low energy, the adatom-, silicon-dominated and adatom- and Si-co-dominated valence and/or conduction bands, the red quasi-rigid shift of the π and σ valence bands or their drastic and dramatic changes (well-behaved and undefined π- and σ-electronic states, respectively, due to the Na and Mg or Al adsorption effects), the diverse spatial charge densities associated with the single- or multi-orbital hybridization in Na–Si or Mg–Si and Al–Si bonds (the approximate separation of π and σ bonding or more complicated orbital hybridization), the number, energy and form of van Hove singularities enriched by moiré superlattices (zone-folding effects) and complex chemical bonding, and the presence or absence of ferromagnetic configurations (spin-split electronic states across the Fermi level, spin density distributions; net magnetic moments) in the Na-adsorbed or Mg- and Al-adsorbed silicene systems. The three kinds of metal adatom adsorption can create 2D free conduction electrons, where their carrier densities are very difficult to evaluate accurately from the calculated band structures because of the strongly anisotropic Dirac cones and the adatom-dominated or adatom- and Si-co-dominated low-lying energy bands with oscillatory and partially flat dispersion relations. The fundamental properties, which are varied by specific metal adatom chemisorption, are attributed to the extremely non-uniform chemical environments, the adatom-dependent orbital bonding, and the guest- and host-atom-induced spin arrangements. Most importantly, the Na–Si, Mg–Si and Al–Si bonds, which account for diverse physical phenomena, are deduced to possess 3s–3p_z, 3s–3p_z and (3s, 3p_x+3p_y)–(3s, 3p_x, 3p_y, 3p_z) orbital hybridization, respectively. Furthermore, only the 3s–3pz orbital hybridization in the Na–Si bond belong to the simple single-orbital condition.

References

[1] Pollak E *et al* 2010 The interaction of Li+ with single-layer and few-layer graphene *Nano Lett.* **10** 3386–8

[2] Tiwari A P, Shin S, Hwang E, Jung S G, Park T and Lee H 2017 Superconductivity at 7.4 K in few layer graphene by Li-intercalation *J. Phys.: Condens. Matter* **29** 445701

[3] Petrovic M *et al* 2013 The mechanism of caesium intercalation of graphene *Nat. Commun.* **4** 2772

[4] Lee E and Persson K A 2012 Li absorption and intercalation in single layer graphene and few layer graphene by first principles *Nano Lett.* **12** 4624–8

[5] Tang Y, Zhang H, Shen Z, Zhao M, Li Y and Dai X 2017 The electronic and diffusion properties of metal adatoms on graphene sheets: a first-principles study *RSC Adv.* **7** 33208–18

[6] Liu X *et al* 2016 Metal intercalation-induced selective adatom mass transport on graphene *Nano Res.* **9** 1434–41

[7] Bianchi M *et al* 2010 Electron–phonon coupling in potassium-doped graphene: angle-resolved photoemission spectroscopy *Phys. Rev.* B **81** 041403

[8] Tran N T T, Lin S Y, Lin C Y and Lin M F 2018 *Geometric and Electronic Properties of Graphene-related Systems: Chemical Bonding Schemes* (Boca Raton, FL: CRC Press)

[9] Jin K H, Choi S M and Jhi S H 2010 Crossover in the adsorption properties of alkali metals on graphene *Phys. Rev.* B **82** 033414

[10] Praveen C S, Piccinin S and Fabris S 2015 Adsorption of alkali adatoms on graphene supported by the Au/Ni (111) surface *Phys. Rev.* B **92** 075403

[11] Lin Y T, Lin S Y, Chiu Y H and Lin M F 2017 Alkali-created rich properties in graphene nanoribbons chemical bondings *Sci. Rep.* **7** 1722

[12] Uthaisar C, Barone V and Peralta J E 2009 Lithium adsorption on zigzag graphene nanoribbons *J. Appl. Phys.* **106** 113715

[13] Lin S Y, Lin Y T, Tran N T T, Su W P and Lin M F 2017 Feature-rich electronic properties of aluminum-absorbed graphenes *Carbon* **120** 209–18

[14] Chan K T, Neaton J B and Cohen M L 2008 First-principles study of metal adatom adsorption on graphene *Phys. Rev.* B **77** 235430

[15] Lin M C, Gong M, Lu B, Wu Y, Wang D Y, Guan M and Dai H 2015 An ultrafast rechargeable aluminium-ion battery *Nature* **520** 324

[16] Ao Z M and Peeters F M 2010 High-capacity hydrogen storage in Al-adsorbed graphene *Phys. Rev.* B **81** 205406

[17] Ao Z M, Jiang Q, Zhang R Q, Tan T T and Li S 2009 Al doped graphene: a promising material for hydrogen storage at room temperature *J. Appl. Phys.* **105** 074307

[18] Sivek J, Sahin H, Partoens B and Peeters F M 2013 Adsorption and absorption of boron, nitrogen, aluminum, and phosphorus on silicene: stability and electronic and phonon properties *Phys. Rev.* B **87** 085444

[19] Xu S, Fan X, Liu J, Jiang Q, Zheng W and Singh D J 2019 Adsorption of Na on silicene for potential anode for Na-ion batteries *Electrochim. Acta* **297** 497–503

[20] Olsson E, Chai G, Dove M and Cai Q 2019 Adsorption and migration of alkali metals (Li, Na, and K) on pristine and defective graphene surfaces *Nanoscale* **11** 5274–84

[21] Ozçelik V O, Kecik D, Durgun E and Ciraci S 2014 Adsorption of group IV elements on graphene, silicene, germanene, and stanene: dumbbell formation *J. Phys. Chem.* C **119** 845–53

[22] Sahin H and Peeters F M 2013 Adsorption of alkali, alkaline-earth, and 3D transition metal atoms on silicene *Phys. Rev.* B **78** 085423

[23] Xu Q, Yang G M, Fan X and Zheng W T 2019 Adsorption of metal atoms on silicene: stability and quantum capacitance of silicene-based electrode materials *Phys. Chem. Chem. Phys.* **21** 4276–85

[24] Quhe R *et al* 2012 Tunable and sizable band gap in silicene by surface adsorption *Sci. Rep.* **2** 853

[25] Hussain T, Chakraborty S and Ahuja R 2013 Metal-functionalized silicene for efficient hydrogen storage *Chem. Phys. Chem.* **14** 3463–6

[26] Siegel D A, Regan W, Fedorov A V, Zettl A and Lanzara A 2013 Charge-carrier screening in single-layer graphene *Phys. Rev. Lett.* **110** 146802

[27] Bostwick A, Ohta T, Seyller T, Horn K and Rotenberg E 2007 Quasiparticle dynamics in graphene *Nat. Phys.* **3** 3640

[28] Ohta T, Bostwick A, McChesney J L, Seyller T, Horn K and Rotenberg E 2007 Interlayer interaction and electronic screening in multilayer graphene investigated with angle-resolved photoemission spectroscopy *Phys. Rev. Lett.* **98** 206802

[29] Ohta T, Bostwick A, Seyller T, Horn K and Rotenberg E 2006 Controlling the electronic structure of bilayer graphene *Science* **313** 951954

[30] Zhou S Y, Siegel D A, Fedorov A V and Lanzara A 2008 Metal to insulator transition in epitaxial graphene induced by molecular doping *Phys. Rev. Lett.* **101** 086402

[31] Coletti C *et al* 2013 Revealing the electronic band structure of trilayer graphene on SiC: an angle-resolved photoemission study *Phys. Rev.* B **88** 155439

[32] Papagno M, Rusponi S, Sheverdyaeva P M, Vlaic S, Etzkorn M, Pacil D, Moras P, Carbone C and Brune H 2011 Large band gap opening between graphene Dirac cones induced by Na adsorption onto an Ir superlattice *ACS Nano* **6** 199204

[33] Virojanadara C, Watcharinyanon S, Zakharov A and Johansson L I 2010 Epitaxial graphene on 6 H–SiC and Li intercalation *Phys. Rev.* B **82** 205402

[34] Sugawara K, Kanetani K, Sato T and Takahashi T 2011 Fabrication of Li-intercalated bilayer graphene *AIP Adv.* **1** 022103

[35] Kim K S, Walter A L, Moreschini L, Seyller T, Horn K, Rotenberg E and Bostwick A 2013 Coexisting massive and massless Dirac fermions in symmetry-broken bilayer graphene *Nat. Mater.* **12** 887892

[36] Zhang X L, Liu L F and Liu W M 2013 Quantum anomalous Hall effect and tunable topological states in 3D transition metals doped silicene *Sci. Rep.* **3** 2908

[37] Jin H, Jeong H, Ozaki T and Yu J 2009 Anisotropic exchange interactions of spin–orbit-integrated states in Sr_2IrO_4 *Phys. Rev.* B **80** 075112

[38] Dong Y J, Wang X F, Vasilopoulos P, Zhai M X and Wu X M 2014 Half-metallicity in aluminum-doped zigzag silicene nanoribbons *J. Phys. D: Appl. Phys.* **47** 105304

[39] Deng X Q, Zhang Z H, Fan Z Q, Tang G P, Sun L and Li C X 2016 Spin-filtering and rectifying effects for Al-doped zigzag-edged silicene nanoribbons with asymmetric edge hydrogenation *Org. Electron.* **32** 41–6

[40] Das R, Chowdhury S, Majumdar A and Jana D 2015 Optical properties of P and Al doped silicene: a first principles study *RSC Adv.* **5** 41–50

[41] Tran N T T, Nguyen D K, Glukhova O E and Lin M F 2017 Coverage-dependent essential properties of halogenated graphene: a DFT study *Sci. Rep.* **7** 1–13

[42] Park H L and Chung Y C 2011 Metal (Li, Al, Ca and Ti) absorbed graphene with defects for hydrogen storage: first-principles calculations *J. Nanosci. Nanotechnol.* **11** 10624–8

[43] Rastegar S F, Peyghan A A and Hadipour N L 2013 Response of Si- and Al-doped graphenes toward HCN: a computational study *Appl. Surf. Sci.* **265** 412–7

[44] Huang Y K, Chen S C, Ho Y H, Lin C Y and Lin M F 2014 Feature-rich magnetic quantization in sliding bilayer graphenes *Sci. Rep.* **4** 7509

Chapter 9

Carbon-, boron- and nitrogen-substituted silicene compounds

Chemical substitution on layered materials is capable of creating completely different components or inducing significant modifications of pristine lattices through very strong host–guest multi-orbital hybridization. Using current experimental growth methods, ternary and binary compounds, characterized by $B_xC_yN_z$, have been synthesized successfully for 3D bulk systems [1–5], 2D planes [6–8], 1D cylindrical nanotubes [9–11], 1D nanoribbons [12–14] and 0D quantum dots [15, 16]. Their geometric structures vary from three to zero dimensions, as observed in carbon-related systems [17–20]. This indicates clearly that each atom possesses, at least, three half-filled orbitals. Similar syntheses have been performed for high-potential Si–C compounds [21]. Generally, it is easy to produce the above-mentioned compounds, while the opposite may be true for specific components. For example, in the high-technique (most modern method) experimental syntheses it is difficult to manipulate the ratio among and between the B, Ca and N atoms and the Si and C atoms. Such non-monoelement condensed-matter systems are predicted and through experiment have been been found to exhibit observable energy gaps or to belong to the significant semiconductors. The main reason lies in the different ionization energies of their components, which are consistent with the tight-binding model with non-vanishing diagonal Hamiltonian matrix elements (the sublattice-dependent site energies). Engineering of the gap could be achieved by transforming the strength relations in the π, σ and sp^3 bonding (p_z-, (s, p_x, p_y)- and (s, p_x, p_y, p_z)-orbital hybridization). B-, C- or N-substituted silicene systems are chosen to address a specific question: how can one explore the diverse fundamental properties of such systems using the critical roles of π, σ and sp^3 chemical bonding?

The chemisorption and substitution of silicon atoms on graphene are very interesting topics, since first-principles predictions are available for understanding the important differences between these two kinds of chemical modification [22, 23] and provide further information about other approaches for the formation of silicon carbide compounds [24]. Based on the viewpoint of a carbon-created honeycomb lattice, silicon atoms are regarded as adatoms and guest atoms under chemisorption and substitution, respectively. The adatoms present optimal positions at the bridge sites above the graphene surface [25], while the guest atoms show Si- and C-co-dominated A and B sublattices with deformed hexagons [26]. Both silicon and carbon atoms have quite active dangling bonds, leading to significant multi-orbital hybridization in the Si–C, Si–Si and C–C bonds. Most importantly, the planar bonding between the Si and C atoms is expected to be stronger and more complicated compared to the perpendicular configuration. That is to say, the π and σ bonding of graphene will be changed drastically by Si substitution [27]. This is reflected directly in the diverse properties, spatial charge distributions, atom-dominated energy bands, and atom- and orbital-projected densities of states. Furthermore, their gradual transformation could be achieved through a Si- or C-substituted graphene or silicene honeycomb lattice (details below and in chapter 10).

The non-mono-element materials with significant substitution effects have been explored in many experimental and theoretical studies. For example, both $B_xC_yN_z$ binary and ternary compounds [28, 29] and silicon carbide [21], which are grown frequently in experimental laboratories, are highly suitable for basic studies [30, 31] and various applications [32]. The former have been identified to present dimension-dependent symmetric structures, including 3D bulk structures [1, 2], 2D few-layer uniform and deformed lattices [6, 7], 1D cylindrical nanotubes [9], and 1D and 0D nanoribbons with open and passivated boundaries [15]. According to experimental measurements, such systems may be metals [1], semimetals [33], or finite- and zero-gap semiconductors [34, 35]. These diverse electronic properties lead to a variety of potential applications, e.g. electrical devices [36], optical sensors [37] and energy storage [38]. Moreover, following on from the first experimental synthesis using the Lely method in 1893, pure Si–C compounds can now be produced in large quantities. They belong to the light insulators with large band gaps. Compared to diamond, it is relatively easy to generate 3D silicon carbides artificially. Only few of these could survive on Earth, although they are found frequently in cosmology. As a result of their unique physical, chemical and material properties, their wide-ranging applications include growth substrates [39], bullet-protection materials [40], light diodes [41], field-effect transistors [42] and so on.

Chemical substitution (or doping) is an efficient method to vary the fundamental properties greatly [43] and enhance potential applications [25], as observed for chemisorption [44]. Currently, there are some theoretical [45–47] and a few experimental [48] studies on guest-atom-substituted silicene materials. Previous predictions show that Ni-doped monolayer silicene presents ferromagnetic, anti-ferromagnetic and non-magnetic spin configurations, metallic or semimetallic behaviors due to n-type doping, and spin-degenerate or spin-split energy bands, which are sensitive to the concentration and arrangement of guest atoms [49]. Also, single-atom substitution of B, N, Al or P might exhibit n- or p-type doping, but semiconducting behaviors are revealed only in more complicated ternary compounds [50]. Systematic investigations of the diverse fundamental properties and their critical mechanisms appear to have been absent until now. This work is focused on the diverse geometric, electronic and magnetic properties of B-, C- and N-substituted silicenes. A theoretical framework, based on multi-orbital hybridization and atom-created spin configurations, is further developed to analyse concentration- and configuration-dependent phenomena in detail. For example, the zero-, finite-gap and semimetallic behaviors, the doping-induced free electrons or holes, and the main features of the π and σ bands can be understood from the specific relationships among the hybridized π, σ and sp^3 chemical bonding. On the experimental side, a controllable synthesis method, using MTMS and hexane as precursors, is proposed to generate large-area graphene-based Si–C binary 2D compounds [51]. Furthermore, silicon carbide nanosheets have been synthesized successfully using a catalyst-free carbothermal method and post-sonication process [52], in which atomic force microscopy (AFM) measurements show an average thickness of ~2–3 nm and a size of ~2 μm.

B-, C- and N-substituted silicene systems are explored thoroughly using first-principles calculations with spin-dependent interactions, in particular for the relationships among the π, σ and sp^3 chemical bonding. Such guest atoms possess, respectively, three, four and five valence electrons in terms of an isolated atomic configuration [53–55], and are suitable for understanding fully the diverse fundamental properties that arise due to chemical substitution. The optimal geometries (the bond lengths of X–X, X–Si and Si–Si, and the planar and buckled honeycomb lattices), the host- and guest-atom-dominated energy bands, the p-type, semiconducting and n-type behaviors, the spatial charge distributions, the atom- and orbital-projected van Hove singularities, and the magnetic moments and spin densities are analysed in detail through the developed theoretical framework (details in section 2.1). That is, critical single- and multi-orbital hybridization and host- and guest-atom spin configurations will be proposed in order to comprehend the unusual results. The calculated properties are expected to depend strongly on the concentrations and configurations of the guest atoms, and the latter are classified into single, meta, ortho and para substitution configurations. The band properties near the Fermi level (the modified Dirac-cone structures or the multi-orbital-induced low-lying energy bands; carrier doping or finite-gap semiconductors), the well-behaved or undefined π and σ valence bands, the mixing and crossing of the π- and σ-electronic energy spectra, their whole band widths, the various 2D van Hove singularities due to substitution-induced band-edge states, and the presence or absence of spin configurations are the focuses of study. Interestingly, most of the free hole and electron densities will be estimated from the reduced and enhanced π bonding, respectively, with only a few from the σ bonding case. The significant differences between the different guest-atom systems (the B–, C– and N–Si binary compounds) are also compared in detail from the viewpoint of the electronic orbitals.

9.1 Carbon-substitution-enriched fundamental properties

Carbon-substituted silicene systems are able to present unusual geometries, as shown in table 9.1 and figures 9.1(a)–(l). Four types of typical C substitution configurations, which include the meta, ortho, para and single cases, are chosen for a model investigation. In general, the para and single cases, respectively, possess the lowest and highest ground state energies, i.e. the para configuration is most stable among them, and it is expected to be synthesized relatively easily using experimental growth methods. Carbon and silicon atoms possess four outer orbitals of, respectively, ($2p_x$, $2p_y$, $2p_z$, 2s) and ($3p_x$, $3p_y$, $3p_z$, 3s). In addition to the pure elements, substitution of carbon, as illustrated in figures 9.1(a)–(l), will create hybridized π and σ bonding ($2p_z$–$3p_z$ and (2s, $2p_x$, $2p_y$)–(3s, $3p_x$, $3p_y$)) in a silicene honeycomb lattice (the standard orbital hybridization cases are shown in figure 1.1). This appears to be responsible for the extremely non-uniform chemical and physical environments in an enlarged unit cell. As a result, the Si–Si and Si–C bond lengths and the height difference between the A and B sublattices might lie in specific ranges (table 9.1). Specifically, the full C substitution case, with the only uniform environment, is completely different to the pristine case, since they have, respectively, planar and buckled honeycomb lattices (figures 9.1(b) and (a)); δ_z = 0–0.48 Å. When the C concentration is sufficiently low, the buckled structures will recover, such as for a C: Si ratio lower than 15%. For the other conditions, the existence of buckling depends strongly on the concentration and configuration of any substitution.

The specific relationships between the π, σ and sp^3 bonding in pristine monolayer silicene account for the main features of the band structures. The electron–hole symmetry around the Fermi level is modified only weakly at low energies, as illustrated in figure 9.2(a). The first pair of valence and conduction bands which is nearest to $E_F = 0$ is initiated from the stable K and K′ valleys. A very narrow band gap of $E_g \sim 0.01$ meV arises in a slightly separated Dirac cone, owing mainly to weak spin–orbit coupling (single-particle interaction) This result is roughly consistent with the predictions under the tight-binding model [56]. There is a valence and conduction saddle-M-point structure at −1.02 eV and 0.59 eV, respectively. Finally, the π band ends at the stable Γ valley, where the π-band energy width for the valence states (the energy spacing between the initial K point and the final Γ point) is about 3.2 eV. Such electronic spectra appear to originate from the π bonding of pure $3p_z$–$3p_z$ orbital hybridization in a buckled honeycomb lattice. On the other hand, the four-fold degenerate σ bands, which arise mainly from $3p_x$- and $3p_y$-orbital bonding, present parabolic energy dispersion relations from the stable Γ valley at $E^v \sim -1.04$ eV. The σ and π valence bands have obvious or observable mixing phenomena along the ΓK and ΓM directions, since there exists weak, but significant, sp^3 bonding. Moreover, the band width of the first σ-electronic states along ΓMK is about 3.45 eV. Generally, all the π- and σ-band (three-band) widths can be well defined for monolayer silicene. In addition, all the k-states in energy bands are doubly degenerate in terms of the spin degree of freedom, since the spin-up- and spin-down-dominated states have identical energy spectra. Similar results are found in chapter 4.

Table 9.1. The optimal geometric structures of carbon-substituted silicene systems under various concentrations and configurations with the C–C, C–Si and Si–Si bond lengths, the band gaps, the ground state energies per unit cell, and the height differences between the A and **B** sublattices.

Number of atoms/unit cell	Percentage	The ratio of C–Si	C–C bond length (Å)	C–Si bond length (Å)	Nearest Si–Si (Å)	Second nearest Si–Si (Å)	E_g^d (eV) / metal	Optimization E_0 (eV)	Δ (Å)
2	—	Pristine	—	—	2.250	—	0.001	—	0.48
	100%	1:1	—	1.785	—	—	2.560	−14.054	0
6	50%	2:4 meta	—	1.809	2.263	—	0.9	−36.704	0.4
	50%	2:4 ortho/para	1.430	1.836	2.261	—	0	−36.157	0
	20%	1:5	—	1.845	2.261	—	0	−32.562	0.35
8	33.3%	2:6 meta	—	1.810–1.853	2.225	—	0.4	−45.033	0.25
	33.3%	2:6 meta	1.466	1.911	2.237	2.186	0	−41.109	0
	33.3%	2:6 para	—	1.812	2.258	—	0	−45.894	0
	14.3%	1:7	—	1.866	2.280	2.196	0.09	−41.300	0.40
18	12.5%	2:16 meta	—	1.864 −1.886	2.300	2.252–2.269	0.14	−92.058	0.31 −0.75
	12.5%	2.16 ortho	1.480	1.955	2.302 −2.322	2.242–2.267	0	−91.046	0.38 −0.75
	12.5%	2.16 para	—	1.866	2.306 −2.329	2.238–2.255	0	−92.619	0.3–0.62
	5.8%	1:17	—	1.864	2.310	2.264–2.288	0	−89.163	0.25 −0.57

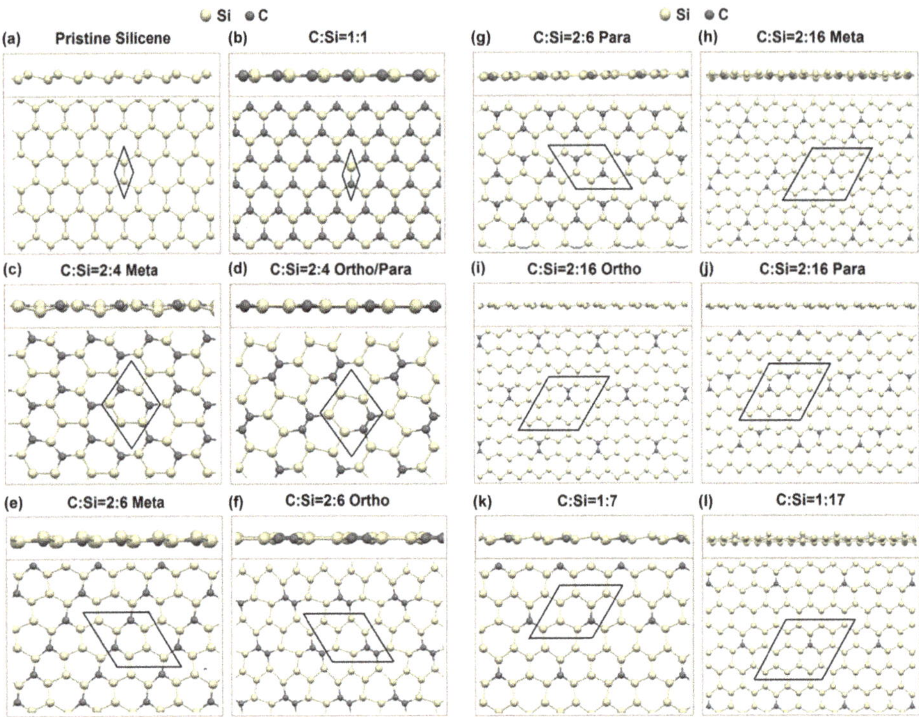

Figure 9.1. The top-view optimal geometric structures for C-substituted silicene systems for the (a) pristine, (b) full, (c) 50% meta, (d) 50% ortho/para, (e) 33.3% meta, (f) 33% ortho, (g) 33% para, (h) 12.5% meta, (i) 12.5% ortho, (j) 12.5% para, (k) 14.3% single and (l) 5.8% single substitution cases. Also shown in the narrow horizontal panels are the side views for the buckled structures.

The full C substitution silicene system, the silicon–carbon binary compound [57], presents an unusual electronic energy spectrum in figure 9.2(b), which is completely different to the pristine case (figure 9.2(a)). The occupied valence bands are highly asymmetric to the unoccupied conduction bands around $E_F = 0$. Such a material belongs to the direct-gap insulators with a band gap larger than 2.56 eV at the K/K′ point, as observed in a wide-gap monolayer boron–nitride system [58]. This is related closely to the strongly modified Dirac-cone structure across the Fermi level with highly anisotropic energy dispersions (e.g. the partially flat conduction band along KM). Both the valence and conduction bands which are nearest to E_F are, respectively, π and π^* bands due to the $2p_z$–$3p_z$ impure orbital hybridization in the Si–C bonds. The higher ionization energy of the $2p_z$ orbitals and the lower ionization energy of the $3p_z$ orbitals further lead to the creation of a band gap in the honeycomb lattice, and the C and Si dominance of the π and π^* bands, respectively (the red and blue circles). The first pair of energy bands is highly anisotropic. For example, the valence π states and conduction π^* states which are initiated at the K valley show different group velocities along the KM and KΓ directions, in particular for, respectively, the saddle point and the partially flat dispersion related to the M point. Furthermore, the π-band energy width is $E_w \sim 3.71$ eV (according to the

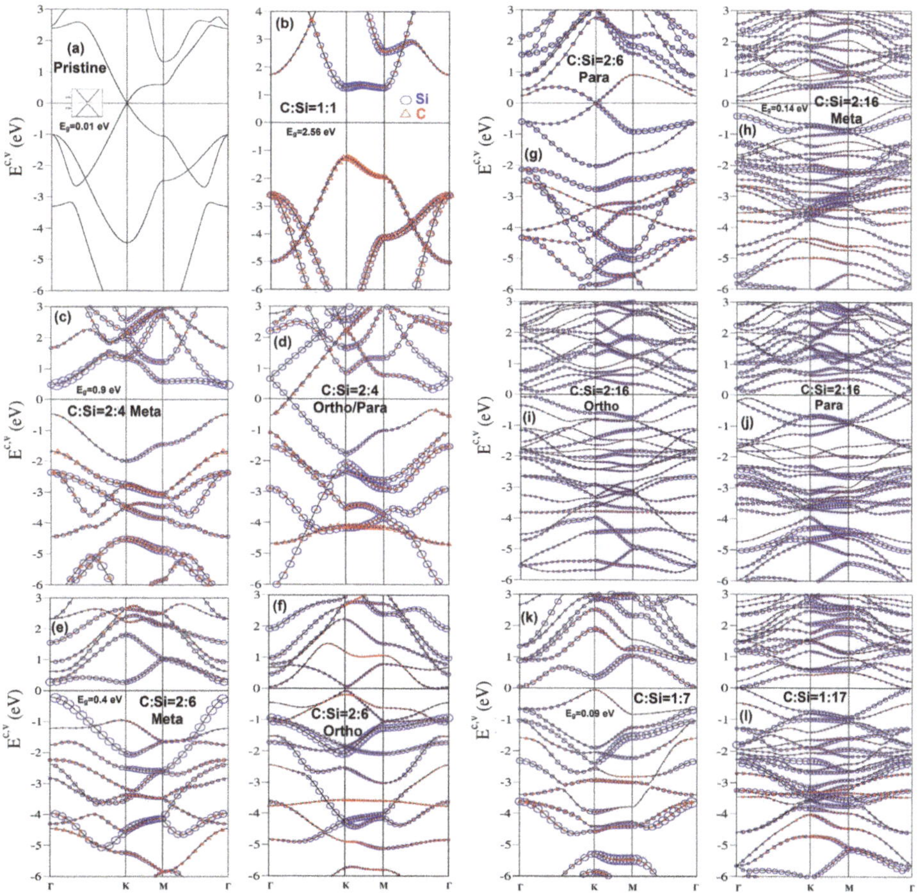

Figure 9.2. Electronic structures, with the dominance circles of the silicon (blue) and carbon atoms (red), for C-substituted silicene systems for the (a) pristine, (b) 1:1, (c) 2:4 meta, (d) 2:4 ortho/para, (e) 2:6 meta, (f) 2:6 ortho, (g) 2:6 para, (h) 2:16 meta, (i) 2:16 ortho, (j) 2:16 para, (k) 1:7 single and (l) 1:17 para substitution configurations.

difference between the K- and Γ-point energies −1.50 and −5.20 eV). Compared to the pristine $3p_z$–$3p_z$ case ($E_w \sim 3.20$ eV in figure 9.2(a)), a larger band width might indicate stronger $2p_z$–$3p_z$ bonding. The two σ bands, which arise from ($3p_x$, $3p_y$)–($2p_x$, $2p_y$) hybridization, appear at $E^v \sim -2.60$ eV initiated from the Γ point, but remain as four-fold degenerate electronic states there. Furthermore, the contributions from different orbitals are comparable in the C- and Si-co-dominated energy bands. The concave-downward parabolic valley near the Γ point changes dramatically into concave-upward valleys or saddle-point forms under the ΓK and ΓM directions, respectively. Therefore, the first σ-band width is about 4.30 eV. There exists direct crossing of the π and σ valence bands along any direction, e.g. the MΓ and KΓ directions. This result indicates a clear lack of evidence of sp^3 bonding [59]. As a result, the fundamental properties of the 1:1 Si–C compound are dominated by well-behaved π and σ bonding in the absence of sp^3 bonding.

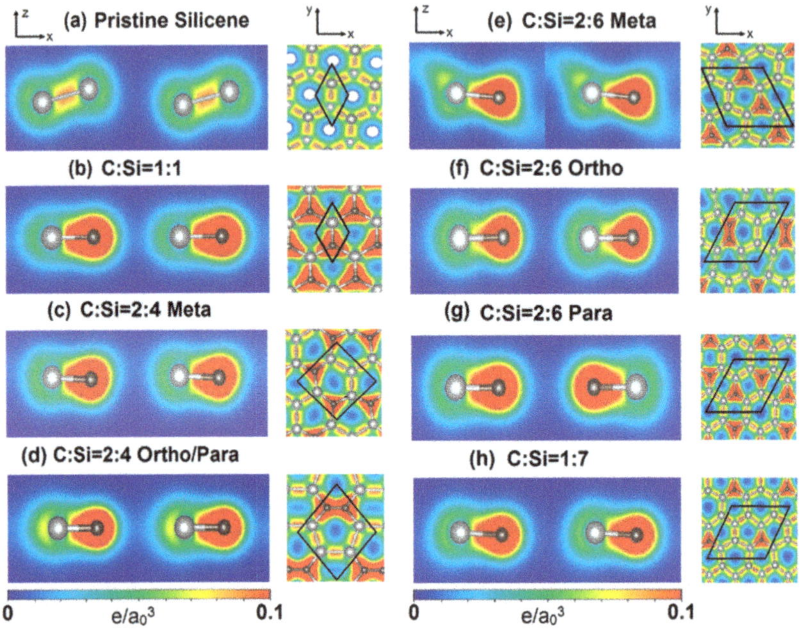

Figure 9.3. The spatial charge densities of carbon-substituted silicene materials for various concentrations and configurations: (a) the absence of guest atoms and the (b) 1:1, (c) 2:4 meta, (d) 2:4 ortho/para, (e) 2:6 meta, (f) 2:6 ortho, (g) 2:6 para and (h) 1:7 conditions. They are shown on the (x, z)- and (x, y)-planes, illustrating clear π and σ chemical bonding and their orthogonality or non-orthogonality.

Electronic energy spectra are varied greatly through the modulation of the concentration and distribution configuration. The 2:4 meta and 2:4 ortho/para cases, as shown in figures 9.2(c) and (d), present finite- and zero-gap behaviors, respectively, corresponding to the highest occupied state and the lowest one occupied state at the Γ point ($E_g \sim 0.90$ eV) and the gapless Dirac-cone structure along the ΓK direction (but not the ΓM direction). That the low-lying valence and conduction bands are initiated from the stable Γ valley is associated with the zone-folding effect. According to the pristine and full substitution configurations (figures 9.2(a) and (b)) the valence and conduction bands in the energy range of $|E^{c,v} < 1.30|$ eV, including the first pair, arise mainly from the 2p$_z$- and 3p$_z$-orbital hybridization. Such bonding is responsible for the low-energy fundamental physical properties. This result is also confirmed from the orbital-projected density of states (see figure 9.4(b)). It should be noted that the 2:4 ortho/para case (figure 9.2(d)) belongs only to the zero-gap semiconductors because of the vanishing density of states at the Fermi level. The low-lying energy bands are dominated by the 3p$_z$ orbitals of the silicon-host atoms because of their higher weight. More energy subbands arise under enlarged unit cells; therefore, band crossing and anti-crossing will happen frequently. In addition, it might be possible to define the π-band energy widths through the ΓK → KM → MΓ direction (through examination of the 3p$_z$-projected density of states in figures 9.4(c) and (d)), corresponding to 4.0 and 4.70 eV for the meta and ortho configurations, respectively. It would be very difficult

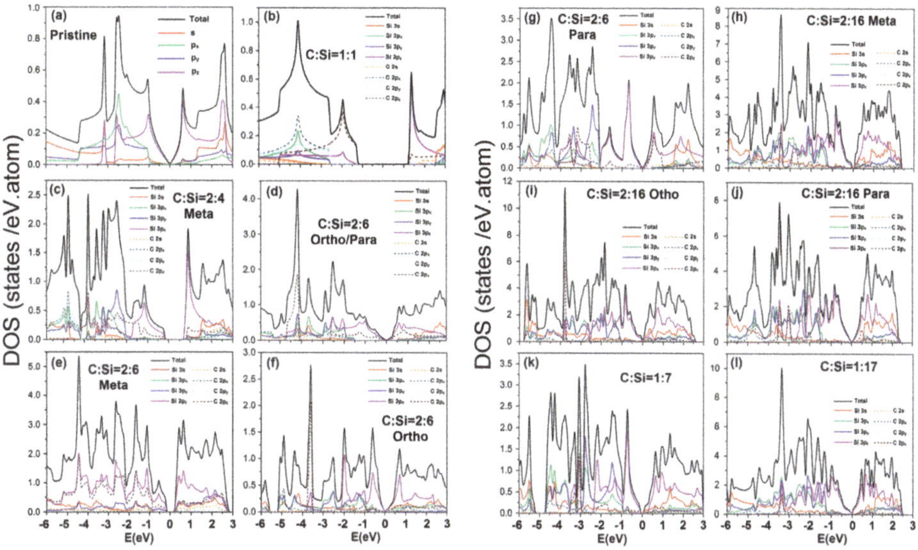

Figure 9.4. Various van Hove singularities in atom- and orbital-decomposed densities of states for carbon-substituted silicene systems under various cases: (a) pristine, (b) 1:1, (c) 2:4 meta, (d) 2:4 ortho/para, (e) 2:6 meta, (f) 2:6 ortho, (g) 2:6 para, (h) 2:16 meta, (i) 2:16 ortho, (j) 2:16 para, (k) 1:7 single and (l) 1:17 single configurations, which are illustrated by, respectively, the solid red, green, blue and purple curves for the 3s, $3p_x$, $3p_y$ and $3p_z$ orbitals and the dashed red, green, blue and purple curves for the 2s, $2p_x$, $2p_y$ and $2p_z$ contributions.

(or become meaningless) to characterize band widths with a further decrease of guest-atom concentration, as a result of more complicated valence subbands. The first σ valence bands are identified roughly from the initial Γ-states at $E^v \sim -1.80$ eV for the meta case (figure 9.2(c)), while very strong zone-folding effects prevent their characterization under the ortho condition (figure 9.2(d)).

Interestingly, two kinds of band properties are also revealed in the 2:6 cases, as shown in figures 9.2(e)–(g). The meta, ortho and para configurations exhibit, respectively, small-, narrow- and zero-gap behaviors (finite- and zero-gap semiconductors) according to the first pair of energy bands nearest to the Fermi level. The meta case has an indirect band gap of $E_g \sim 0.31$ eV, which is determined by the specific energy spacing between the highest occupied state at the K point and the lowest unoccupied state at the Γ point (figure 9.2(e)). Furthermore, the first valence band has oscillatory energy dispersions. As a result, two stable valleys dominate the low-energy physics simultaneously. Significant overlaps of the valence and conduction bands, which arise along any wave-vector direction, appear under the ortho configuration in figure 9.2(f). There are five energy subbands with weak and oscillatory dispersion relations near the Γ point, and three energy subbands near the the K and M points. They are responsible for the formation of partially flat bands and thus the high density of states near the Fermi level (see figure 9.4(f)). The para case, in figure 9.2(g), shows a gapless Dirac-cone structure at the K/K′ valley only with a single Fermi-momentum state (a zero density of states at E_F in figure 9.4(g)). There are more finite energy spacings between the valence and conduction subbands in the Γ valley. For any guest-atom configuration, most electronic energy spectra are dominated by silicon atoms, while the opposite is true for carbon atoms.

Obviously, more complicated electronic energy spectra arise during the decrease of guest-atom concentration (under a reduced number of Si–C bonds and an increased number of C–C and Si–Si bonds). Three typical kinds of distribution configurations are revealed for the 2:16 condition, namely the meta, ortho and para configurations in figures 9.2(h)–(j). Their low-lying band structures present strongly modified Dirac cones with a direct gap of $E_g \sim 0.14$ eV due to parabolic valence and conduction dispersions near the Γ valley for the meta configuration, a gapless and crossing structure along the ΓM direction and finite energy spacing along the ΓK direction for the ortho configuration, and similar zero-gap behavior for the para configuration. In addition, the single 1:7 substitution in figure 9.2(l) creates an $E_g \sim 0.09$ eV indirect band gap associated with the valence K state and conduction Γ state. It should be noted that the second and third electronic structures belong only to the zero-gap semiconductors (the vanishing density of states at E_F in figures 9.2(i) and (j)). The moiré superlattices create a lot of energy subbands within the smaller first Brillouin zones, leading to frequent band crossing and anti-crossing. As a result, it is very difficult to characterize the π-band width (and similarly for the σ-band widths). However, the orbital-projected density of states might be useful in examining the relationships among the π, σ and sp^3 bonding. Interestingly, the C-substituted silicenes will recover gradually to the pristine case (figure 9.2(a)) at very low guest-atom concentrations, e.g. the appearance of an almost gapless Dirac cone under the 1:17 case in figure 9.2(l). In short, the energy subbands, with $|E^{c,v} < 1.5|$ eV, are dominated by the $2p_z$- and $3p_z$-orbital hybridization, and the other deeper- or higher-energy electronic states might be related closely to the the $2p_x$, $2p_y$, 2s, $3p_x$, $3p_y$ and 3s orbitals or the eight orbitals in Si and C atoms. Carbon substitution in silicene can induce only modified Dirac-cone structures, but not free carriers (n- or p-type doping).

High-resolution angle-resolved photoemission spectroscopy (ARPES) measurements provide the only way to examine and identify the occupied electronic energy spectra along specific 2D and 1D wave-vector directions (details in section 3.3). This method has been used to identify a lot of unusual low-lying energy dispersions in graphene-related sp^2-bonding materials, such as the linear valence Dirac cone in monolayer graphene [60], the two pairs of parabolic bands in bilayer AB stacking [61], the partially flat, sombrero-shaped and linear energy bands in trilayer ABC stacking [61], the bilayer- and monolayer-like energy bands of Bernal graphite at the K and H corners, respectively (the AB-stacked system at $k_z = 0/k_z = \pi$) [62], and the 1D parabolic energy subbands of graphene nanoribbons [63]. Detailed and careful ARPES studies are required for the predicted band structures of C-substituted silicene materials. That is, they include the zero- and finite-gap behaviors with modified Dirac-cone structures across the Fermi level or without carrier doping, the π bands or subbands with an initial K or Γ valley, the intermediate saddle-M-point form and the final Γ valley, the degenerate σ bands initiated at deeper energies from the Γ valley, their band crossing and anti-crossing along the KΓ and MΓ directions, and the spin-degenerate electronic states, which are all very sensitive to the configurations and concentrations of guest atoms (figures 9.2(a)–(l)). Full information about verified band structures is very useful in understanding the substitution-enriched relationships between the π, σ and sp^3 bonding in Si–Si, Si–C and C–C chemical bonds, and the non-magnetic property.

According to first-principles calculations, the tight-binding model might be capable of simulating the low-lying spin-degenerate energy dispersions in carbon-substituted silicene systems. The site energies of different orbitals, impure or pure multi-orbital hybridization, non-uniform environments (the optimal positions of host and guest atoms), and buckled or planar structures need to be taken into account simultaneously. For example, full C substitution, which includes uniform Si–C bonds in the smallest unit cell, possesses different orbital energies of 3s, $3p_x$, 2s and $2p_x$), and impure π ($3p_z$–$2p_z$) and σ ((3s, $3p_x$, $3p_y$)–(2s, $2p_x$, $2p_y$)) chemical bonding. In general, it is difficult to determine the various hopping integrals due to multi-orbital hybridization, in particular for those of moiré superlattices with many atoms in large unit cells. Harrison's rule [64] may be an efficient method for solving the non-uniform chemical bonds under dilute substitution. When suitable parameters are achieved under numerical fitting with first-principles calculations, the generalized tight-binding models in the presence of external electric and magnetic fields [65, 66] are very powerful in obtaining a detailed understanding of the rich magnetic quantization phenomena of Si–C-based 2D compounds. Similar simulation methods could be used for the more complicated B- and N-substituted silicene compounds. The spatial charge density distributions are able to provide certain observable evidence about the existence of impure or pure π and σ chemical bonding and their non-orthogonalities or orthogonalities. Pristine 2D silicene, as shown in figure 9.3(a) with (x, z)- and (x, y)-plane projections, shows well-defined $3p_z$–$3p_z$ and (3s, $3p_x$, $3p_y$)–(3s, $3p_x$, $3p_y$) orbital hybridization, in particular for the latter projection. ρ is a symmetric distribution around the Si–Si bond center. Most of the charge density is accumulated between two silicon atoms (the red region), corresponding to the very strong σ bonding of three orbitals. The neighboring Si atoms are also attracted together through the parallel $3p_z$ orbitals perpendicular to the (x, y)-plane, in which the π bonding appears in the outer region shown by the light green and blue color. As a result of buckling, there is weak, but significant sp^3 hybridization under the non-orthogonality of π and σ chemical bonding (this is very difficult to observe directly in the current figure; details in section 4.1). When the Si–Si bonds become Si–C bonds under full substitution, ρ presents a dramatic transformation, illustrated in figure 9.3(b). The charge density is highly asymmetric to the C–Si bond center, owing mainly to the different electron affinities of the guest and host atoms. There exist more carriers around the guest C atoms (the red region); that is, electrons are transferred from silicon to carbon atoms. All the C–Si bonds consist of a planar honeycomb lattice with a stronger σ bonding, compared to those of the pristine system (figure 9.3(a)). The σ and π bonding can be defined roughly under impure $2p_z$–$3p_z$ and (2s, $2p_x$, $2p_y$)–(3s, $3p_x$, $3p_y$) orbital hybridization, respectively. They are reflected directly in the π- and σ-electronic valence subbands with direct crossing (figure 9.2(b)). Most importantly, the different types of chemical bonding are orthogonal to each other; therefore, the sp^3 orbital hybridization is negligible in the 1:1 case. Each Si–C bond has an identical chemical environment, i.e. only one chemical bond in a unit cell. This illustrates clearly that it is relatively easy to simulate the first-principles band structure using the tight-binding model [67].

With decreasing carbon concentration (figures 9.3(c)–(h)), the presence of π and σ chemical bonding is found relatively easily from the spatial charge densities. There exist C–Si, Si–Si and even C–C bonds, in which the latter are stable under specific ortho cases (table 9.1). Almost symmetric carrier distributions are revealed between two silicon atoms for any concentration and configuration, in which they possess the lowest charge density among the three kinds of chemical bonds. Similar phenomena appear in the C–C bonds with the highest carrier densities, e.g. ρ in figure 9.3(d)–(f) in the 2:4 and 2:6 ortho conditions. Generally, the spatial charge densities in different chemical bonds are not sensitive to the changes in various carbon substitution cases. These further illustrate that the σ and π chemical bonding might be well separated from each other; therefore, they could be defined roughly as carbon-substituted silicene systems. As a result, the low-energy fundamental properties are expected to be dominated by the π-electronic states due to the modified Dirac-cone structures. In contrast, different chemical bonds lead to complicated orbital hybridization and thus increase the difficulties in obtaining suitable phenomenological models. When the first-principles electronic energy spectra along the high-symmetry paths are simulated successfully by the tight-binding model with non-uniform and multi- or single-orbital hopping integrals [68] (details in section 2.2), the varied fundamental properties could be explored fully in future studies, such as the rich and unique magnetic quantization phenomena [66] predicted and observed in layered graphene systems [69].

There are five categories of van Hove singularities in carbon-substituted silicene systems and four in pristine systems, as illustrated in figures 9.4(a)–(l). The orbital- and atom-decomposed densities of states are very useful in understanding bonding-induced special structures fully. The critical points (the band-edge states) in the energy–wave-vector space include the linear Dirac-cone structure, local minima or maxima of parabolic energy dispersions, saddle points, constant-energy loops (related closely to band anti-crossing) and partially flat bands (figures 9.2(a)–(l)). Their densities of states generate, respectively, the V-shaped form, discontinuous shoulders, logarithmically divergent peaks, asymmetric peaks in the square root divergence and delta-function-like peaks. The pure monolayer silicene in figure 9.3(a) presents an almost linear E-dependence across the Fermi level with a vanishing density of states (a quasi-V-shape in the range of -0.5 meV $\leqslant E \leqslant 0.5$ meV; the purple curve of the $3p_z$ orbitals), a logarithmically symmetric π peak at -1.02 eV and π^* peak at 0.59 eV, the π- and σ-mixing-created square root asymmetric peaks at -2.20, -2.65 and -3.15 eV (the red, green and blue curves of the $3p_x$, $3p_x$ and $3p_y$ orbitals, respectively), the Γ-valley π shoulder at -3.20 eV, the initial σ-$3p_x$ and -$3p_y$ shoulder at -1.03 eV at the Γ point, their M-saddle-point symmetric peak at -2.47 eV and the K valley shoulder at -4.40 eV. As a result, the π- and σ-band widths are $\sim$3.15 and 3.36 eV, respectively. It should be noted that the 3s orbitals (the red curve) are accompanied frequently by the partners of $3p_x$ and $3p_y$

orbitals, but their contributions become significant at deeper or higher energies, e.g. the density of states within $E < -3$ eV. The above-mentioned features of the van Hove singularities further illustrate well-behaved π and σ chemical bonding and their weak, but important, hybridization.

The structure, energy and number of van Hove singularities appear to display dramatic transformations under full carbon substitution, as indicated in figure 9.4(b). The special structures include the vanishing density of states within a band gap of $E_g \sim 2.56$ eV at the K/K′ valley, the initial π shoulder at −1.26 eV and the delta-function-like π^* peak at 1.26 eV (also see the low-lying π and π^* bands in figure 9.2(b)), the −1.95 eV symmetric π peak in the logarithmic divergence, the final π shoulder at −5.00 eV, the first σ shoulder at −2.75 eV, the −4.10 eV logarithmic peak and the second discontinuous structure at −7.01 eV. Also note that the first valence structure is dominated by the C-2p$_z$ orbitals and the first conduction structure is dominated by the Si-3p$_z$ orbitals due to an obvious difference of ionization energy. The widths of the π and the first σ bands are, respectively, ~3.74 and 4.26 eV. Moreover, there is no evidence of π–σ band mixing, owing mainly to the absence of simultaneous four-orbital structures. The absence of sp^s bonding agrees with direct π–σ subband crossing (figure 9.2(b)).

The van Hove singularities in a whole energy range become more complicated during the decrease of the guest C atom concentration, as a result of zone-folding effects and significant (3s, 3p$_x$, 3p$_y$, 3p$_z$)–(2s, 2p$_x$, 2p$_y$, 2p$_z$) multi-orbital hybridization. For example, the π-band widths in the 2:4 meta and 2:4 ortho configurations respectively, as illustrated in figures 9.4(c) and (d), exhibit diverse low-energy van Hove singularities due to the dominant 3p$_z$–2p$_z$ chemical bonding. The 2:4 meta configuration has a π-electronic zero density of states in the 0.90 eV gap region (the Γ valley in figure 9.2(c)), a threshold valence shoulder at −0.45 eV (the first composite conduction-state structure due to the discontinuous shoulder and delta-function-like peak at 0.45 eV), a strong logarithmic-form peak at −1.41 eV, and second and third step structures at −1.69 and −2.71 eV, respectively. Concerning the 2:4 ortho configuration, the 3p$_z$–2p$_z$-varied van Hove singularities show a gapless V-shaped form (the Dirac-cone energy spectrum in figure 9.2(d)), the first valence shoulder at −0.5 eV (the initial conduction shoulder at 0.30 eV), a logarithm-step composite structure at −1.01 eV, a symmetric peak arising from two opposite shoulders at −1.32 eV, a similar peak at −2.10 eV and a −2.65 eV asymmetric peak in the square root form. The π-band widths in the former and latter are, respectively, estimated to be ~4.05 and 4.70 eV. Furthermore, the first σ-band widths are wider than 4.5 eV. Obviously, the low-energy physics are dominated by the 3p$_z$ orbitals of the Si host atoms.

As the guest-atom concentration declines further, there are more van Hove singularities. Generally, it is possible to characterize the π-band width from the effective distributions of the 3p$_z$ and 2p$_z$ orbitals under any concentration and configuration (figure 9.4(a)–(l)). However, the opposite is true for the first σ-band width except for the full substitution and pristine cases (figures 9.4(b) and (a)). Certain van Hove singularities, which are close to the Fermi level, are very useful to comprehend the band-edge states in the first pair of valence and conduction bands.

The main features include the vanishing density of states at or across the Fermi level (figures 9.4(c)–(e) and (d)–(g)), the electron–hole asymmetry about the first pair of obvious shoulders, the prominent π peak in the logarithmic form, the second and third shoulders and so on. Finally, the dip structure at E_F, with a very small gap comparable to the broadening factor, will arise for a sufficiently low concentration, e.g. the almost gapless behavior in the 1:17 case (figure 9.4(l)). Also note that the single-particle interactions of spin–orbit coupling cannot create any significant effects on the electronic properties and thus the other fundamental properties.

High-resolution STS measurements are an efficient way to identify the valence and conduction band-edge states (the van Hove singularities of the densities of states; details in section 3.2). To date, STS has confirmed the rich and unique densities of states in dimension-, stacking- and layer-number-dependent graphene-based materials, such as the width- and edge-structure-dominated or chirality- and radius-related band gaps and a lot of asymmetric prominent peaks in 1D semiconducting graphene nanoribbons and carbon nanotubes [70, 71], a V-shaped structure with a dip at the Fermi level in monolayer graphene [72], a Dirac-point red shift thorough alkali adatom chemisorption (n-type doping) [73], a gate-voltage-created band gap in bilayer AB-stacked graphene [74], a surface-state peak only at E_F in trilayer ABC-stacked graphene [75], semimetallic behavior (a finite density of states at E_F), and strongly symmetric π and π^* peaks at the middle energies for Bernal graphite [76]. Similar STS studies can be conducted on various van Hove singularities in carbon-substituted silicene systems, including the zero-, finite-gap and semimetallic band properties, the π and σ band widths, and five kinds of configuration- and concentration-dependent special structures (V-shaped across the Fermi level, shoulders, symmetric peaks in the logarithmic form, delta-function-like peaks and square root asymmetric peaks in figures 9.4(a)–(l)). Both experimental STS and ARPES results could provide sufficient information on the strength relationships between π, σ and sp^3 bonding during the chemical substitution of guest atoms.

9.2 Varied phenomena in B-, C- and N-substituted silicenes

When a boron-substituted silicene system is under a specific case (figure 9.5), all the intrinsic interactions are included in the detailed calculations of first-principles optimal processes. However, the B- and Si-dependent spin, configurations and interactions are suppressed fully by the very strong and complicated chemical bonding. That is to say, the spin-induced magnetic configuration is absent in any B-substituted silicenes, in contrast to that found in the C- and N-related systems (figures 9.1 and 9.9). Compared to pristine silicene, an enlarged unit cell is determined mainly by the B substitution concentration and configuration. The uniform chemical bonding in a moiré superlattice could survive only under the 1:1 condition with the three nearest-neighboring B–Si bonds in a planar honeycomb lattice (table 9.2), in which their lengths are ~1.95 Å. With decreasing B concentration, the other modulated chemical environments include B–Si and Si–Si bonds for any case, as well as B–B bonds only for ortho cases. Interestingly, the original Si–Si and B–Si lengths can be utilized to estimate the length of the B–B bond, or the

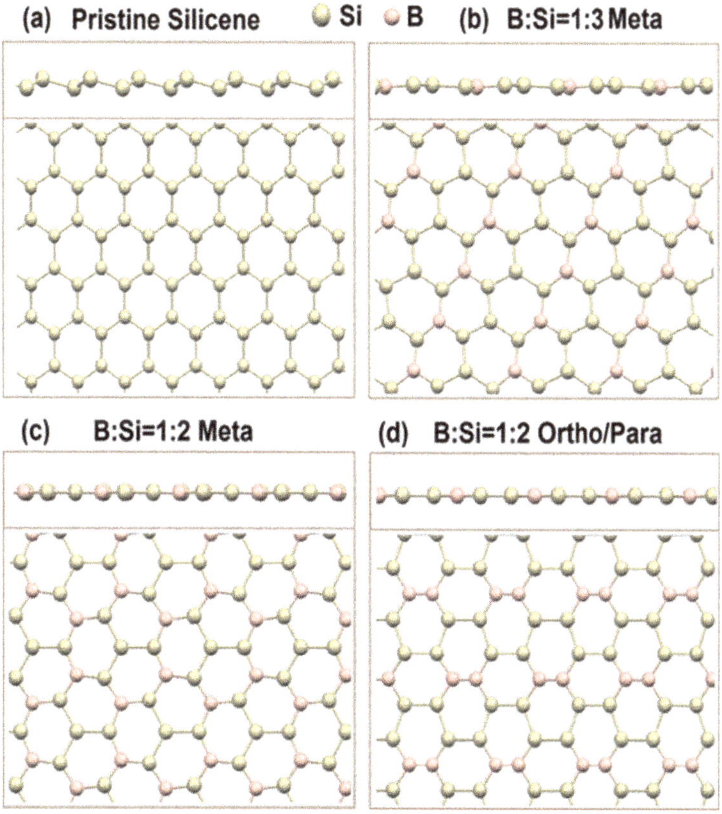

Figure 9.5. The top- and side-view optimal geometric structures for boron-substituted silicene systems for the (a) pristine, (b) 1:3 meta, (c) 1:2 meta and (d) 1:2 ortho/para configurations.

average superpositions of any two bonds correspond roughly to the third bond. A suitable and reliable estimation should be attributed to well-separated or -defined π and σ chemical bonding. This indicates clearly the negligible contributions due to sp^3 bonding. Compared to the carbon substitution cases, the boron-related bond lengths are longer in the current cases. This result is consistent with the higher ground state energy, as shown in tables 9.1 and 9.2. That is to say, the B-substituted silicene systems possess weaker chemical bonding. Most importantly, two or three kinds of chemical bonds lead to a non-uniform chemical environment in an enlarged unit cell, which is illustrated by the fact that the B–Si and Si–Si bond lengths and even the height differences of the A and B sublattices present observable fluctuation ranges (table 9.2). This behavior is expected to create more complicated hopping integrals and orbital hybridization in the phenomenological models [77, 78].

Boron-substituted silicene systems present rich and unique band structures, owing mainly to the three valence electrons of the B atom in terms of the 2s, $2p_x$, $2p_y$ and $2p_z$ orbitals. A 1:1 2D boron–silicon compound, with a planar honeycomb lattice, is an unusual 2D metal or semimetal, as shown in figure 9.6(b). The dominance of the boron and silicon atoms (the red and blue circles) are comparable to each other for

Table 9.2. The optimal geometric structures of boron-substituted silicene systems under various concentrations and configurations with the B–B, B–Si and Si–Si bond lengths, band gaps, ground state energies per unit cell, and height differences between the A and B sublattices.

Number of atoms/unit cell	Percentage	Ratio of B–Si	B–B bond length (Å)	B–Si bond length (Å)	Nearest Si–Si (Å)	Second nearest Si–Si (Å)	E_g^d (eV) / metal	Optimization E_0 (eV)	Δ (Å)
2	—	Pristine	—	—	2.250	—	0.001	—	0.48
	100%	1:1	—	1.953	—	—	M	−10.427	0
6	50%	2:4 meta	—	1.963	2.290	—	M	−30.640	0
	50%	2:4 ortho/para	1.639	1.965	2.295	—	M	−30.560	0
	20%	1:5	—	1.923	2.262	—	M	−29.831	0
8	33.3%	2:6 meta	—	1.943–1.992	2.239	2.280	M	−40.022	0.07
	33.3%	2:6 meta	1.717	1.997	2.294	2.256	M	−39.643	0.34
	33.3%	2:6 para	—	1.945	2.281	—	M	−40.365	0
	14.3%	1:7	—	1.969	2.268	2.233	M	−39.217	0
18	12.5%	2:16 meta	—	1.969 −2.002	2.279	2.259–2.268	M	−87.851	0.24 −0.64
	12.5%	2.16 ortho	1.717	2.011	2.284 −2.313	2.269–2.307	M	−87.463	0.28–0.6
	12.5%	2.16 para	—	1.947	2.266 −2.292	2.236–2.256	M	−88.161	0.07 −0.15
	5.8%	1:17	—	1.961	2.283	2.268–2.280	M	−86.982	0.2–0.6

all the valence and conduction bands. The asymmetric energy spectrum of the valence and conduction bands about the Fermi level is enhanced greatly by boron substitution, since E_F intersects with the former. Plenty of free valence holes are created simultaneously near the stable K and Γ valleys. The π or π^* and the first σ bands can be well defined according to, respectively, a modified Dirac-cone structure and concave-downward parabolic energy dispersions. The π- and π^*-electronic energy spectra are initiated from the K point, respectively, corresponding to E^v = 1.70 eV and E^c = 2.00 eV with an observable Dirac-point spacing of ~0.30 eV. Compared to that of pristine silicene (figure 9.6(a)), the Dirac-cone band structure is modulated strongly by impure hybridization of $2p_z$–$3p_z$ orbitals. Clearly, the π-band energy width grows from 3.20 to 4.00 eV. On the other hand, the hybridized σ bonding of (2s, $2p_x$, $2p_y$)–(3s, $3p_x$, $3p_y$) orbitals will affect their energy dispersions, e.g. the energy and the number of band-edge states. The double-degenerate σ-band energy at the Γ point is higher than the Fermi level, such that the electronic states close to it become free valence holes. Interestingly, the first and second σ-band widths, which survive along the ΓMK direction, are identified easily to be ~4.30 and 6.20 eV, respectively. The anti-crossing and crossing behaviors of the σ and π bands are, respectively, vanishing and present, further indicating that their chemical bonding is orthogonal to each other in a planar honeycomb B–Si lattice. That is to say, the anti-crossing-induced van Hove singularities will be absent in the orbital and atom-projected densities of states (see figure 9.8(a)). Most importantly, the π and σ energy bands are characterized well in a full boron substitution, as revealed in monolayer silicene [78] and [79] graphene.

Obviously, the concentrations and configurations of boron guest atoms play a critical role in the varied electronic properties, as shown in figures 9.6(a)–(l). They are able to dominate and/or specify the rich multi-orbital hybridization of various types of chemical bonds and thus create unusual band structures. The main features of the band structures are very sensitive to the B substitution case, in which the strongly modified properties include asymmetry of the hole and electron energy spectra around the Fermi level, the non-uniform dominance of boron and silicon atoms, initial valence and conduction states from the K and Γ valleys, the energy spacing of separated Dirac points, the π- and σ-band widths (the initial and final state energies), the number of band-edge states and the 2D free hole densities. Interestingly, the almost gapless or perfect Dirac-cone structure in pristine silicene (figure 9.6(a)) becomes seriously deformed under various B substitution cases. The separated Dirac-cone energy spacing might depend on the different ionization energies (the different site energies) of the B-$2p_z$ and Si-$3p_z$ orbitals [80]. Whether modified Dirac-cone structures arise at the K or Γ point is associated with the zone-folding effect (an enlarged unit cell) [81]. That is to say, the original hexagonal first Brillouin zone is folded into the smaller zone according to the ratio of the enlarged superlattice versus the pristine superlattice. As a result, a pristine corner point might correspond to the same point or the origin in the reduced first Brillouin zone. The π and π^* valence and conduction energy subbands present a saddle-point structure near the M point with a prominent density of states. The valence and conduction Dirac cones are, respectively, dominated by boron and silicon atoms for the 1:1 case

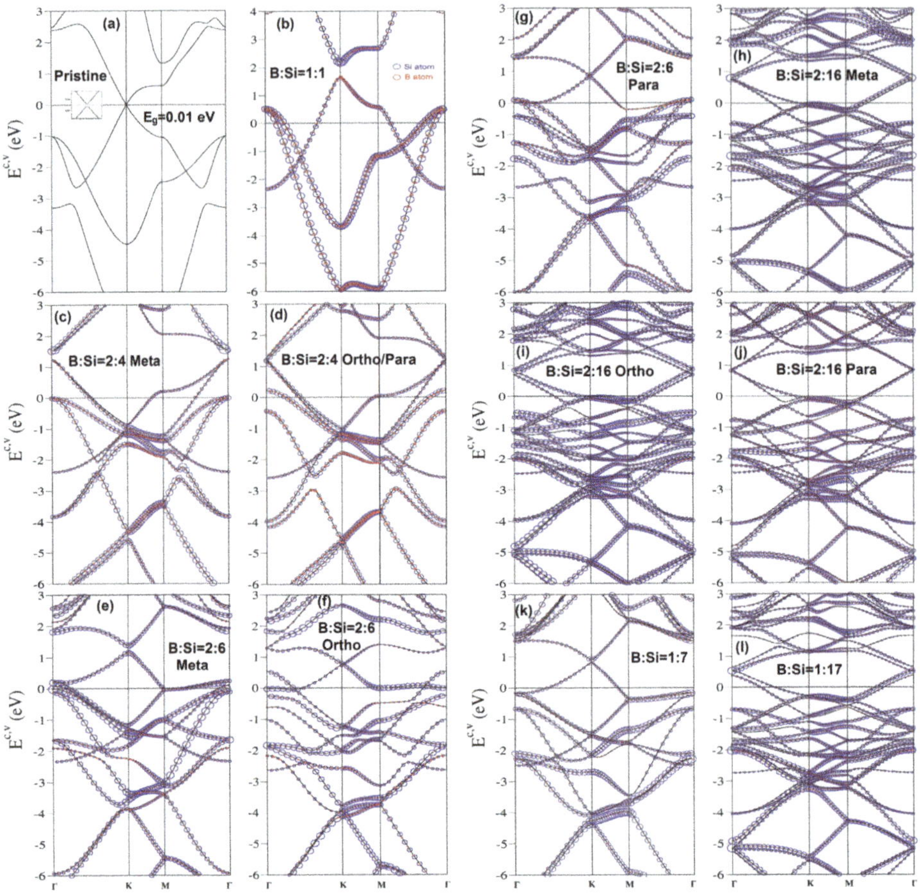

Figure 9.6. Band structures of B-substituted silicene materials with Si and B dominance (blue and red circles) under various cases: (a) without substitution, and the (b) 1:1, (c) 2:4 meta, (d) 2:4 ortho/para, (e) 2:6 meta, (f) 2:6 ortho, (g) 2:6 para, (h) 2:16 meta, (i) 2:16 ortho, (j) 2:16 para, (k) 1:7 single and (l) 1:17 single substitution configurations.

(figure 9.6(b)), while their dominance suddenly belongs to the silicon atom during the decrease of concentration (figures 9.6(c)–(l); also see the density of states in figures 9.8(b)–(l)). With decreasing guest-atom concentrations, there are more valence and conduction subbands (π- and σ-electronic bands), leading to frequent subband crossing. However, few anti-crossing behaviors, which are dominated by the σ valence bands, could survive at $E^v < -2.3$ eV except for in the 1:1 case (figure 9.6(b)). Such phenomena, which might rely on the hybridized σ bonding of (2s, $2p_x$, $2p_y$)–(3s, $3p_x$, $3p_y$) orbitals, could create 1D constant-energy loops (quasi-1D parabolic dispersions) in energy–wave-vector space. Of course, the boron concentration is lower and so is the 2D free hole density, as observed in the reduced first Brillouin zone. In short, the unusual non-uniform physico-chemical environment in a moiré superlattice, which originates mainly from the impure and pure π- and σ-electronic orbital hybridization in specific chemical bonds, are responsible for the diverse electronic energy spectra. High-resolution ARPES measurements on the main features of the occupied states are very useful in examining the B guest-atom substitution effects, such as the concentration- and configuration-dependent π and σ valence bands, the modified Dirac-cone structures with anisotropic energy dispersions and Fermi momenta, and the unoccupied hole states near the stable K or Γ valleys.

Interestingly, all the B-substituted silicene systems are predicted to exhibit *p*-type doping phenomena, in which the 2D free hole density is deduced to be the same as that of the guest B atoms. Boron and silicon atoms, respectively, possess three and four valence electrons, and they create complex π- and σ-chemical bonding through the $2p_z$–$3p_z$/$2p_z$–$2p_z$/$3p_z$–$3p_z$ and (2s, $2p_x$, $2p_y$)–(3s, $3p_x$, $3p_y$)/(2s, $2p_x$, $2p_y$)–(2s, $2p_x$, $2p_y$)/(3s, $3p_x$, $3p_y$)–(3s, $3p_x$, $3p_y$) orbital hybridization. Well-separated π and σ states have been confirmed from band structures (figures 9.6(a)–(l)). Generally, the pure or mixed π- and σ-electronic behaviors require four valence orbitals in each atom. This will lead to unusual p-type doping because there are only three valence orbitals in boron. According to the conservation of electronic states, the *k*-space area of the unoccupied valence states is equal to the boron-atom number per unit cell. For example, compared to a pristine system (figure 9.6(a)), 1:1 B-substituted silicene (figure 9.6(b)) presents the same number of valence and conduction bands, but different Fermi levels. E_F, as measured from the middle of the separated Dirac-cone structure, is 0 eV for the valence band and 1.80 eV for the conduction band. Such a result should be independent of the distribution and concentration of boron guest atoms. Based on the lower-energy valence bands in figures 9.6(b)–(l), most of the free holes are provided by the π band near the K or Γ valley (hybridized or pure π bonding), and part of them originate from the σ bonds close to the Γ valley (the σ bonding due to the hybridization of the $2p_x$, $2p_y$, $3p_x$ and $3p_y$ orbitals), as further confirmed from the atom- and orbital-projected densities of states in figure 9.8. p-type doping phenomena could be verified directly from the ARPES measurements on the Fermi-momentum valence states near the K and Γ valleys and from STS measurements on the finite density of states in the valence spectrum at the Fermi level. Most importantly, they indicate greatly reduced π bonding (slightly reduced σ bonding) in boron-substituted environments compared to carbon-related environments.

Certain important features of the electronic energy spectra in boron-substituted silicene systems can be understood from their charge density distributions, as shown in figures 9.7(a)–(h) with (*x*, *z*)- and (*x*, *y*)-plane projections. For the full substitution, only the B–Si bonds which create a planar honeycomb lattice present an unusual carrier density (figure 9.7(b)). The σ bonding between the guest and host atoms, the (2s, $2p_x$, $2p_y$)–(3s, $3p_x$, $3p_y$) orbital hybridization, is revealed clearly in the red region. Furthermore, it is somewhat modified into an asymmetric distribution relative to the bond center. The higher carrier density is localized around the B guest atoms, indicating the charge transfer from silicon to boron atoms. The σ carrier density is accompanied by observable π bonding ($2p_z$–$3p_z$ orbital hybridization) in the outer region under an almost orthogonal condition. Generally, the spatial charge distribution of the full B substitution is highly anisotropic even in the

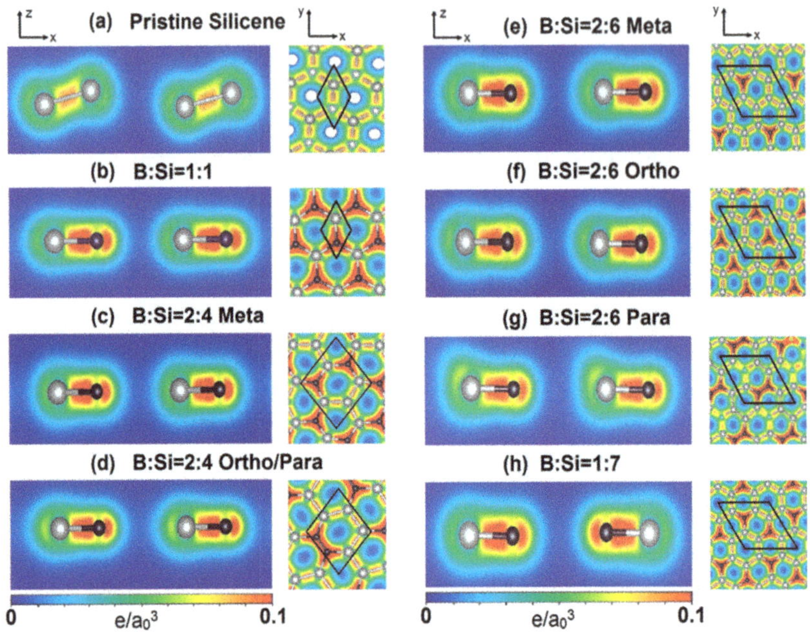

Figure 9.7. The spatial charge distributions in boron-substituted silicene systems under various concentrations and configurations: (a) without guest atoms, and the (b) 1:1, (c) 2:4 meta, (d) 2:4 ortho/para, (e) 2:6 meta, (f) 2:6 ortho, (g) 2:6 para and (h) 1:7 cases, with (x, z)- and (x, y)-plane projections due to well-defined π and σ chemical bonding.

presence of only B–Si bands, where its density declines quickly during variation from the hexagonal sides to the center of the honeycomb lattice. The above-mentioned features lead to changes in the σ- and π-electronic energy spectra (figure 9.6(b)), such as a modified Dirac-cone structure and direct crossing of the π and σ valence subbands. However, the red shift of the Fermi level, corresponding to the valence Dirac point (the creation of free holes in valence holes) are very difficult to identify directly from the current charge density distribution.

Most importantly, the π and σ bonding can be characterized well under various guest-atom concentrations and configurations, as illustrated in figures 9.7(c)–(h). For non-full boron substitution, there exist the B–Si, Si–Si and B–B bonds, which show obvious σ chemical bonding along the (x, z)- and (x, y)-planes. That is, pure and impure σ-electronic hybridization, which induce high density distributions between two atoms, arise in the different bonds. Furthermore, similar phenomena can also be found in the rich π bonding. As a result, well-defined π and σ chemical bonding depends little on the boron substitution case. Interestingly, this phenomenon indicates that the tight-binding model might be reliable in simulating first-principles band structures (figures 9.6(a)–(l)) [82] in which the atom- and orbital-dependent hopping integrals and site energy model of the former can be obtained from well-fitted energy spectra along the high-symmetry points (details in section 2.2). The consistency of these two methods is very useful in further studies on the other fundamental properties, e.g. the optical [83], magnetic [84] and transport properties of guest-atom-substituted silicenes and germanenes.

The atom- and orbital-decomposed densities of states, as illustrated in figures 9.8(a)–(l), are capable of providing more information about p-type doping phenomena and the π- and σ-band widths of boron-substituted silicene materials. The 2D energy spectra can create five kinds of van Hove singularities: distorted V-shapes with or without observable energy spacings above the Fermi level, shoulders, symmetric peaks in the logarithmic form, square root form asymmetric peaks and delta-function-like peaks corresponding to, respectively, the modified Dirac cones, extreme points (local minima or maxima), saddle points, constant-energy loops (quasi-1D parabolic dispersions), and partially flat energy dispersions in the energy–wave-vector space. The low-lying density of states, which is characterized by a modified Dirac-cone structure across the Fermi level, appears to be determined mainly by the Si-3p_z and B-2p_z orbitals with merged van Hove singularities under any substitution case (the solid purple curves and the dashed burgundy curves). This indicates the dominance of the pure and hybridized π bonding on the low-energy fundamental properties, e.g. its creation of the rich magnetic quantization in the B-, C- and N-substituted silicene systems [86]. The B-3p_z orbitals make larger contributions to the valence Dirac-cone only under the specific 1:1 case, while the opposite is true for the other cases. Furthermore, the Si-3p_z orbitals dominate the conduction-cone energy spectrum for all the substitution cases. The pure and hybridized π bonding can create a pair of sharp shoulders with an observable energy spacing or an asymmetric V-shaped structure with different slopes related to the valence and conduction Dirac points, owing mainly to the obviously modified Dirac cone (figures 9.6(b)–(l)). In general, the energy and spacing of the valence and conduction Dirac points decline quickly with a decrease of concentration. The special left-hand structure corresponds to the initial π-electronic valence states at the K or Γ valley. Then, at deeper energies along the KMΓ or ΓMK path, a prominent symmetric peak or peaks arise in the logarithmic form. Such structures are due to the split saddle-point valence states; furthermore, their positions might be above, near or below the Fermi level, depending on the critical case of 1:3. As for the final π-band shoulder structures, such van Hove singularities appear at $E \sim -2.5$ eV. Consequently, the total π-band width, which is wider than 3 eV, grows with increasing concentration. Three kinds of special structures can characterize the π-electronic energy subbands, illustrating concise orbital hybridization of 3p_z–3p_z, 3p_z–2p_z and 2p_z–2p_z. In addition to the strong π peak, there also exists a prominent π^* peak above the conduction Dirac-cone structure of ~0.5–0.7 eV, as a result of the M-saddle-point states. Interestingly, the 2D free hole density is proportional to the total area which is enclosed by the energy-dependent density of states between the valence Dirac-point structure and the Fermi level. When the real magnitude per unit cell is taken into account for the 2D electronic state density, such an area is deduced to be the same as the boron concentration under any substitution case. Generally, the σ orbitals of (2s, 2p_x, 2p_y) and (3s, 3p_x, 3p_y) make significant contributions to the densities of states at deeper and higher energies. The initial σ-band valence states, which arise mainly from the four p-orbitals (the solid green and blue curves and the dashed cyan and dark cyan curves), can create an obvious shoulder structure very close to the symmetric π peak in the range of -0.7 eV$< E < 0.2$ eV. That is, the initial σ-electronic valence states exhibit a dramatic transformation from an unoccupied to an occupied state as the B concentration decreases from the 1:1 full compound (figures 9.6(b)–(l)). These two structures might be merged together at a higher concentration, e.g. the 1:1 case in figure 9.8(b), and could be well separated at a lower concentration (the 2:16 cases in figures 9.8(h)–(j)). However, the σ and π bands do not have any anti-crossing behavior, which indicates the absence of sp^3 bonding. With a further decline of the state energy ($E < -0.7$ eV), many van Hove singularities, which consist of shoulders, symmetric peaks and anti-symmetric peaks, represent the emergence, saddle-point structures and anti-crossing of the σ-valence energy subbands. It might be very difficult to characterize the σ-subband widths from high-resolution STS measurements on their van Hove singularities, but not the initial σ-electronic energy. On the other hand, such experimental studies are available to provide sufficient information on the p-type (free hole) activity, the three and two energy-dependent van Hove singularities of the π- and π^*-electronic states, respectively (the initial shoulder, intermediate peak and final shoulder) and the first σ-electronic shoulder.

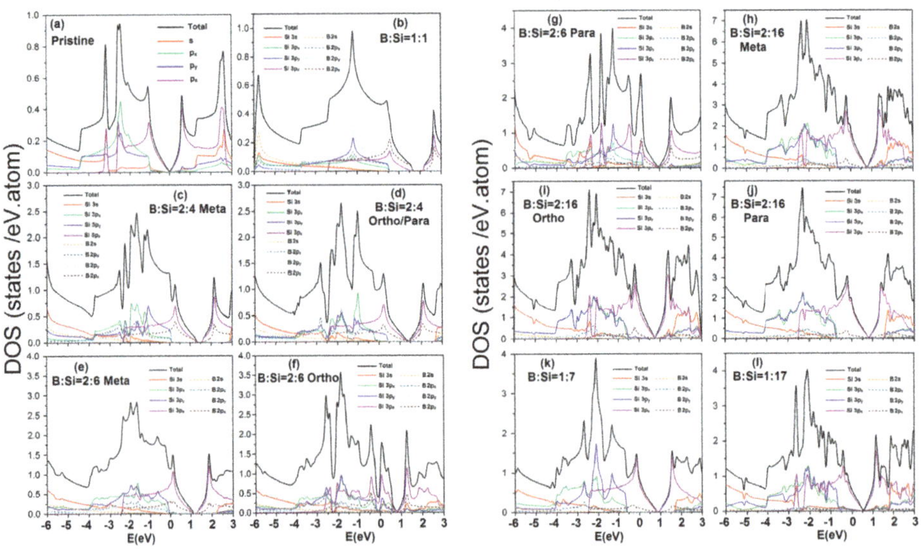

Figure 9.8. The significant van Hove singularities in the B-, Si- and orbital-projected densities of states for boron-substituted silicene materials under various cases: (a) without guest atoms, and the (b) 1:1, (c) 2:4 meta, (d) 2:4 ortho/para, (e) 2:6 meta, (f) 2:6 ortho, (g) 2:6 para, (h) 2:16 meta, (i) 2:16 ortho, (j) 2:16 para, (k) 1:7 single and (l) 1:17 single configurations, where the solid red, green, blue and purple curves correspond to the 3s, $3p_x$, $3p_y$ and $3p_z$ orbital contributions, and the dashed red, green, blue and purple curves to the 2s, $2p_x$, $2p_y$ and $2p_z$ orbital contributions.

9.3 Nitrogen-substituted silicene systems and the important differences between them

Nitrogen-substituted silicene systems exhibit unique geometries (figure 9.9 and table 9.3) compared to the carbon- and boron-based systems (figures 9.1 and 9.5, tables 9.1 and 9.2). To achieve optimal geometric structures, the spin-dependent many- and single-particle interactions, which correspond to, respectively, the on-site Coulomb and spin–orbit interactions [87], need to be included in the first-principles calculations. Extremely non-uniform bond lengths arise in the enlarged unit cell except in the pristine and 1:1 cases (figures 9.9(a) and (b)). The N–Si bonds are comparable in length to the C–Si bonds and shorter than the B–Si bonds. The N–N bonds, which are ~1.42 Å in length (table 9.3), are stable under the 2:4 ortho case (figure 9.9(d)). However, the opposite is true for the 2:6 and 2:16 ortho cases, in which the too large bond lengths correspond only to the separated distances. Furthermore, buckled geometries are found under various concentrations and configurations. Generally, the maximum height difference between the A and B sublattices is larger than that in the absence of nitrogen substitution. Also note that among the 1:1 nitrogen–, boron– and carbon–silicon compounds, only the first two possess buckled honeycomb lattices. Such features might be related closely to the multi-orbital hybridization of non-well-defined π and σ chemical bonding (discussed

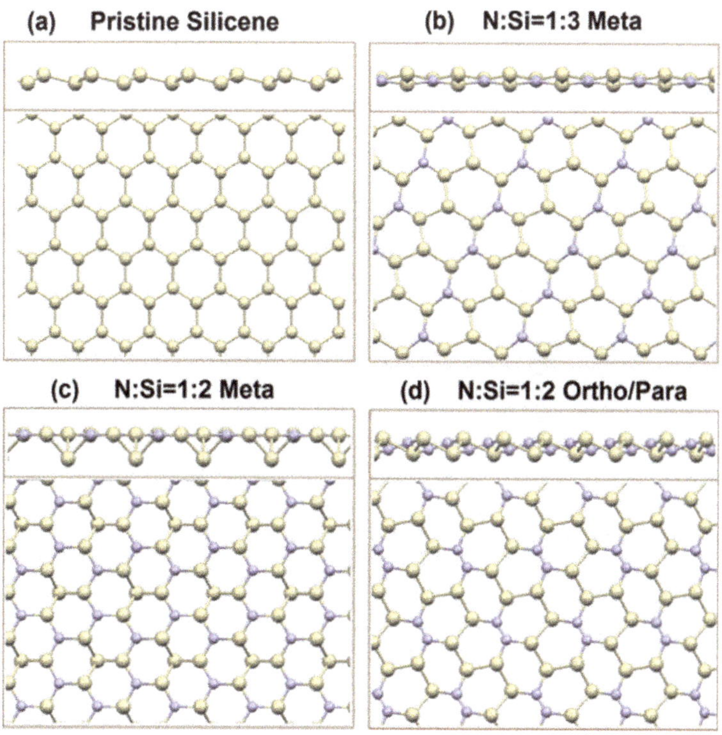

Figure 9.9. Top and side views of the optimal geometric structures in nitrogen-substituted silicene systems under the (a) pristine, (b) 1:3 meta, (c) 1:2 meta and (d) 1:2 ortho/para configurations.

Table 9.3. The optimal geometric structures of nitrogen-substituted silicene systems with net magnetic moments. Also note that the non-reasonable N–N bond lengths in the 2:6 and 2:16 ortho configurations correspond only to the distance of two host atoms.

Number atoms/unit cell	Percentage	The ratio of N–Si	N–N bond length (Å)	N–Si bond length (Å)	Nearest Si–Si (Å)	Second nearest Si–Si (Å)	Magnetic moment (μ_n)/ magnetism	E_g^d (eV) / metal	Optimization E_0 (eV)	Δ (Å)
2	—	Pristine	—	—	2.250	—		0.001	—	0.48
	100%	1:1	—	1.783	—	—	0.25	2.268	−13.783	0.57
6	50%	2:4 meta	—	1.800	2.367	—	1.06	SM	−36.794	0.9
	50%	2:4 ortho/para	1.418	1.783	2.307–2.365	—	0	SM	−34.874	0.65
	20%	1:5	—	1.781	2.306	—	0.51	0.087	−32.540	0.7
8	33.3%	2:6 meta	—	1.765–1.926	2.241	2.425	0	SM	−44.510	0.45
	33.3%	2:6 meta	3.103	1.654	2.349	2.300	0	SM	−45.017	0.88
	33.3%	2:6 para	—	1.809	2.373	—	0	SM	−46.041	0.41
	14.3%	1:7	—	1.819	2.273	2.201	0.11		−41.076	0.63
18	12.5%	2:16 meta	—	1.738–2.301	2.290–2.436	2.247–2.448	0	0.414	−91.682	0.38–0.68
	12.5%	2:16 ortho	3.491	1.672	2.324–2.426	2.262–2.325	0	0	−92.179	0.44–0.78
	12.5%	2:16 para	—	1.836	2.266–2.292	2.289–2.426	0	0	−92.883	0.2–0.41
	5.8%	1:17	—	1.830	2.315	2.264–2.329	0	SM	−88.810	0.35–0.65

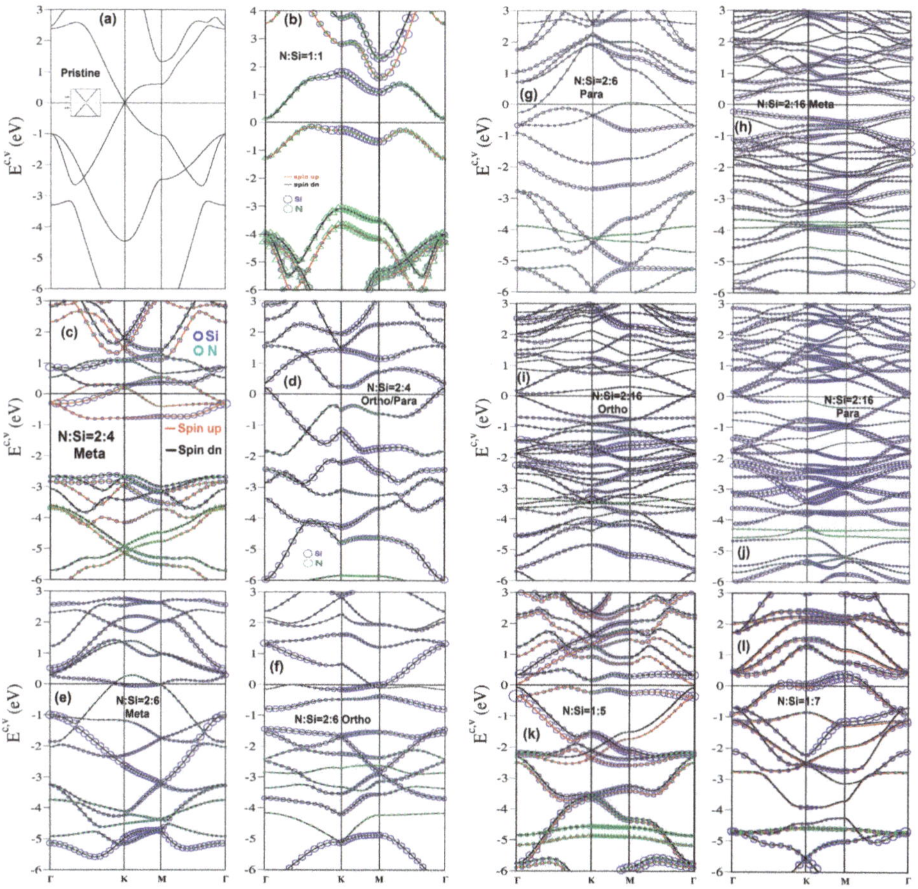

Figure 9.10. Electronic band structures of nitrogen-substituted silicene systems, with the dominance circles of the N and Si atoms (red and blue) under different substitution cases: (a) no substitution and the (b) 1:1, (c) 2:4 meta, (d) 2:4 ortho/para, (e) 2:6 meta, (f) 2:6 ortho, (g) 2:6 para, (h) 2:16 meta, (i) 2:16 ortho, (j) 2:16 para, (k) 1:5 single and (l) 1:7 single configurations.

below in relation to spatial charge densities). Highly asymmetric chemical environments, which are characterized by observable fluctuation ranges of N–Si and Si–Si bond lengths and buckled structures, are expected to create complicated charge distributions and thus vary the other fundamental properties [88].

Nitrogen atoms, with five valence electrons of (2s, $2p_x$, $2p_y$, $2p_z$), can create the unusual substitution effects in silicene-based electronic states (figures 9.10(a)–(l)), owing mainly to the more complicated multi-orbital hybridization in N–Si, Si–Si and N–N bonds, the enhanced buckling of the two sublattices and the ferromagnetic spin configuration. Electronic structures are very sensitive to changes in N concentration. For the 1:1 case (figure 9.10(b)), the uniform 2D nitrogen–silicon compound is an indirect-gap semiconductor, with a unique ferromagnetic spin configuration. The lowest unoccupied conduction state corresponds to the Γ point, while the highest occupied valence state is found between the Γ and K points. The indirect band gap is $E^i_g \sim 0.25$ eV. The occupied and unoccupied electronic energy spectra are highly asymmetric around the Fermi level, illustrating complex hopping integrals and site energies as a result of the strong interactions of atom-dependent orbitals. Clearly, the modified Dirac-cone structure near E_F is completely absent. This indicates that the original π bonding in pristine silicene presents a dramatic transformation (see figure 9.11(b) for charge distributions). It is almost impossible to define the initial, intermediate and final π-band states well along the high-symmetry paths; that is, the full investigation of the π-band widths becomes meaningless. Similar phenomena can be found in the σ-electronic energy subbands, illustrating clearly the absence of well-behaved or modified π and σ chemical bonding. The electronic energy spectrum shows only non-uniform wave-vector dependence, in which the band-edge states might appear between two high-symmetry points. Such a result will be reflected in the rich van Hove singularities of the density of states (see figure 9.11(b)) and optical excitation spectra [89]. Most importantly, the spin degeneracy is destroyed by the spin-induced many-body effects (the on-site Coulomb interactions in the Hubbard-like model) [90], where the ferromagnetic spin distributions are associated with the host and guest atoms (the spin density distribution in figure 9.12(a)). The spin-up and spin-down energy subbands have observable spacings of $E_s \sim 0.1$–2.0 eV, and can be examined using spin-polarized ARPES and STS measurements (details in sections 3.3 and 3.2). In general, the spin-

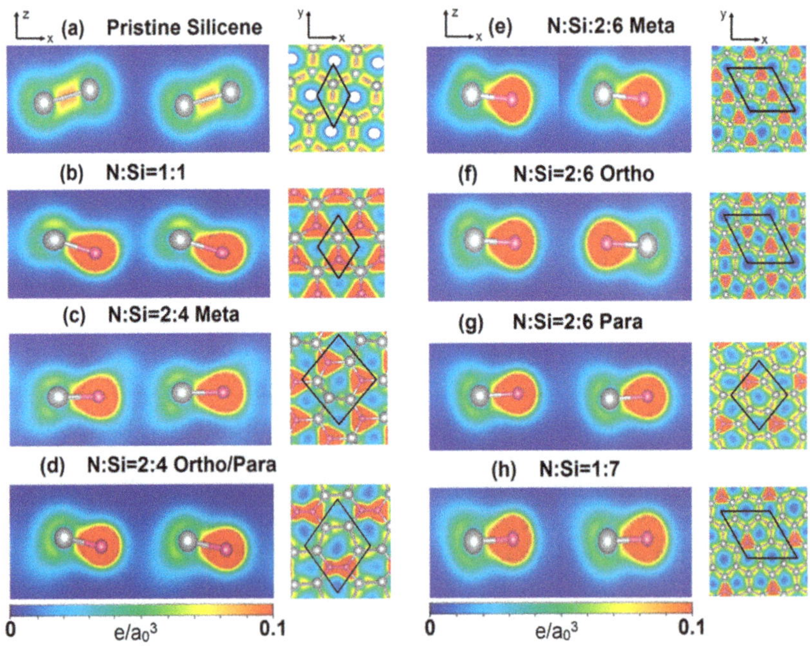

Figure 9.11. The spatial charge distributions of nitrogen-substituted silicene materials for various concentrations and configurations: (a) without guest atoms and the (b) 1:1, (c) 2:4 meta, (d) 2:4 ortho/para, (e) 2:6 meta, (f) 2:6 para, (g) 1:5 single and (h) 1:7 conditions. Two projections along the (x, z)- and (x, y)-planes indicate clear multi-orbital chemical bonding.

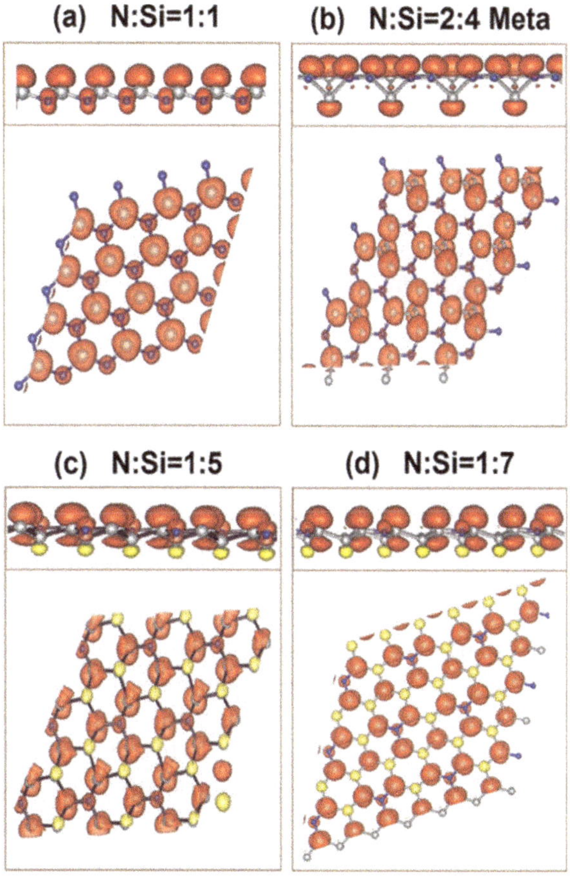

Figure 9.12. The spin charge distributions in nitrogen-substituted silicene systems for certain cases: (a) 1:1, (b) 2:4 meta, (c) 1:5 and (d) 1:7, with top and side views.

spilt energy subbands, which are more than 1.50 eV away from the Fermi level, belong to the occupied or unoccupied states simultaneously. As a result, such pairs do not make any contributions to the net magnetic moments. On the other hand, the first valence and conduction bands, as measured from the Fermi level arise from, respectively, the spin-up and spin-down configurations (the black and red curves). The spin-up configuration can create a ferromagnetic configuration in the buckled honeycomb lattice (table 9.3) and thus a large magnetic moment per unit cell. Interestingly, according to the atom dominance, each energy subband is co-dominated by silicon and nitrogen atoms (the blue and green circles) for the silicon and nitrogen contributions, respectively. It is expected to be very difficult to distinguish the N- and Si-dominated energy subbands from experimental studies [91]. The magnetic property is co-determined by the host and guest atoms. This will further lead to a very high simulation barrier using the tight-binding model, i.e. the multi-orbital-dependent hopping integrals, ionization energies and on-site Coulomb interactions make the phenomenological models very complex [92].

The ferromagnetic band structures deserve closer examination. They also arise in certain nitrogen substitution cases with sufficiently high concentrations, such as those of the 2:4 meta, 1:5 and 1:7 conditions (figures 9.10(c), (k) and (l)). The 2:4 meta, and the 1:7, 1:5 and 1:1 nitrogen substitution cases present semimetallic (or metallic) behaviors, in which their ferromagnetic configurations are dominated by, respectively, two pairs and one pair of spin-split valence and conduction bands nearest to the Fermi level. Whether such energy subbands, with non-uniform or strong wave-vector dependence, cross E_F will determine the existence of free carriers and their densities. Furthermore, they are related closely to the different contributions of the host and guest atoms. As a result, the silicon and nitrogen atoms might have specific weights and even opposite directions in the spin density distributions (see figures 9.12(a)–(d)). Only the nitrogen-substituted silicene systems are able to display ferromagnetic spin configurations, compared to the carbon- and boron-related systems [50].

Whether nitrogen-substituted silicenes present spin-split energy subbands, as well as finite-gap or gapless behavior, depends strongly on the concentration and arrangement of the guest atoms. Non-magnetic band structures also arise under high-concentration conditions, e.g. the 2:4 ortho/para, 2:6 meta, 2:6 ortho and 2:6 para cases with spin-degenerate electronic states, as shown in figures 9.10(d)–(g). This result suggests that the spin-dependent on-site Coulomb interactions are suppressed thoroughly by the complicated multi-orbital hybridization in the Si–Si, N–Si and N–N chemical bonds. Most importantly, the band property (the semiconducting or semimetallic/metallic behavior) is determined mainly by the low-lying energy subbands with complicated wave-vector dependence. For example, three 33.3% cases exhibit partial flat energy subbands across the Fermi level, and are thus expected to have a large density of states at E_F (a sufficiently high free carrier density; see figures 9.13(e)–(g)). It is also noted that it is almost impossible to distinguish n- or p-type doping in the nitrogen-substituted silicene systems from the complicated low-lying energy subbands, as has been achieved for boron-related systems (figures 9.6(a)–(l)). That is to say, the viewpoint of the valence electron number cannot be utilized to realize substitution-induced free carriers. The absence of ferromagnetic configurations appears frequently at sufficiently low concentrations, and is illustrated by the 2:16 meta, 2:16 ortho, 2:16 para and 1:17 para configurations in figures 9.10(h), (i), (j) and (m), respectively. Interestingly, the modified linear valence and conduction subbands (modified Dirac-cone structures), which are further studied using the van Hove singularities close to E_F (figures 9.13(h)–(j) and (l)), are initiated from the Γ point. They should be dominated by pure and hybridized π bonding. However, the band structures under any case are too complex to characterize the π- and σ-band widths clearly. Well-behaved chemical bonding is deduced to be completely absent in nitrogen-substituted silicenes.

The calculated band structures appear to be very complicated for nitrogen-substituted silicene systems. It might be difficult to simulate them using the tight-binding model with complex multi-orbital hopping integrals and site energies [87], and even Hubbard-like spin-induced on-site Coulomb interactions [90]. That is,

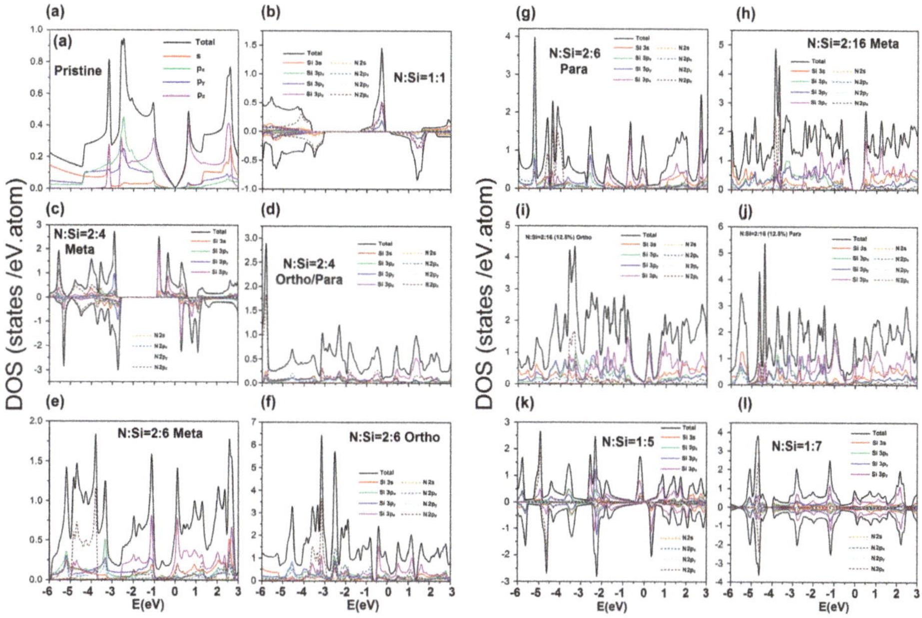

Figure 9.13. The atom- and orbital-decomposed densities of states for nitrogen-substituted silicene materials under different cases: (a) non-substitution, and the (b) 1:1, (c) 2:4 meta, (d) 2:4 ortho/para, (e) 2:6 meta, (f) 2:6 ortho, (g) 2:6 para, (h) 2:16 meta, (i) 2:16 ortho, (j) 2:16 para, (k) 1:5 single and (l) 1:7 single configurations. The solid red, green, blue and purple curves represent the 3s, $3p_x$, $3p_y$ and $3p_z$ orbitals and the dashed red, green, blue and purple curves represent the 2s, $2p_x$, $2p_y$ and $2p_z$ contributions.

phenomenological models might be unable to provide the critical mechanisms of and concise pictures for their fundamental properties. For example, the generalized tight-binding model [66], with magnetic-field effects, cannot be utilized to explore the unique magnetic quantization phenomena. As a result, first-principles calculations become the only way to understand the varied phenomena fully. The predicted electronic energy spectra can be verified through sensitive studies using ARPES and SP-ARPES (details in section 3.3). ARPES is very successful in identifying the electronic structures of layered materials, e.g. the dimension-varied energy bands in bulk graphites [93], graphenes [94], graphene nanoribbons [95] and substrate-enriched few-layer group-IV systems [96]. SP-ARPES has confirmed the spin-resolved occupied energy subbands of topological insulators [97], ferromagnetic materials [43] and superconductors [98]. The experimental study of nitrogen-substituted silicene systems is focused on the oscillatory energy dispersions near the Fermi level (the anisotropic Fermi momenta), the non-magnetic or ferromagnetic electronic states, the gapless or finite-gap band properties, and modified or absent Dirac-cone band structure at, respectively, sufficiently low or high guest-atom concentrations. The measured results could provide certain important information related closely to the complex relationships between the π, sp^2 and sp^3 bonding, and the strong competition and cooperation between the multi-orbital hybridization and spin configurations.

Nitrogen-substituted silicene systems exhibit unique charge density distributions, compared to pristine, carbon- and boron-substituted systems. For the 1:1 configuration (figure 9.11(b)), the N–Si bonds create a buckled honeycomb lattice, which is completely different to those of the C–Si and B–Si systems. The carrier density is highly asymmetric around the N–Si bond center on the (x, y)-, (x, z)- and (y, z)-plane projections, in which most of its distribution is localized around the nitrogen atoms, and even affects the formation of π bonding. The N–Si bonds are estimated broadly to have significant and complicated (2s, $2p_x$, $2p_y$, $2p_z$)–(3s, $3p_x$, $3p_y$, $3p_z$) sp^3-orbital hybridization. That is to say, it is very difficult to characterize the σ- and π-electronic bonding and states, respectively. There are also Si–Si and N–N bonds as the nitrogen concentration decreases (figures 9.11(c)–(h)). They seem to present seriously distorted charge distributions between two identical atoms, compared to the pristine cases. For example, a rectangular σ-electronic distribution, which survives in pristine silicene (the (x, y)-plane projection in figure 9.11(a)), changes dramatically into a triangular distribution under any nitrogen substitution except for the full substitution case. Serious distortions are also revealed in π-electronic bonding. The pure orbital hybridization is affected strongly by the non-uniform chemical environment. The pure and impure four-orbital hybridization indicate that it would be difficult to simulate the VASP band structures through the tight-binding model with orbital-induced hopping integrals and site energies [65] and spin-dependent interactions [99]. The above-mentioned results are predicted to create varied fundamental properties and important differences among the three kinds of guest-atom substitution. In short, the unusual features of charge density distributions might be related closely to the strong affinity and five valence electrons of the nitrogen atoms.

Ferromagnetic configurations, which appear in certain high-concentration nitrogen-substituted silicene systems, are revealed clearly in the spatial spin density distributions, as indicated in figures 9.12(a)–(d) with the top and side views. The net magnetic moment is very sensitive to the nitrogen concentration (table 9.3). Generally, the spatial spin distributions present are highly non-uniform and anisotropic. For the 1:1 case, only the spin-up distribution can survive in the stable honeycomb lattice of a uniform 2D nitrogen–silicon compound (the red regions in figure 9.12(a)). Interestingly, the spin-up configuration is revealed only in the upper plane of $z > 0$ for the host atoms (in gray), while its distribution is nearly symmetric about $z = 0$ around the guest atoms (in blue). The highly asymmetric spin distribution should play an important role in the creation of ferromagnetism and the spin-dependent on-site Coulomb interactions [100]. The nitrogen-induced red volumes are comparable to those associated with silicon atoms; that is, both atoms make the important contribution to the magnetic property. This result is consistent with the Ni and Si dominance of the first spin-split valence band below the Fermi level (figure 9.9(b)) or the comparable spin-up densities of states due to the $2p_z$ and $3p_z$ orbitals (see figure 9.13(b)). Similar magnetic phenomena have been found in the 2:4 meta case (figure 9.12(b)). However, the spin-up distribution is initiated gradually from the lower plane of $z < 0$, and the weight related to the nitrogen atoms declines quickly. With a further decreasing nitrogen concentration, the spin-up and spin-down states (in red and yellow) arise simultaneously, and both of them originate mainly from the neighboring silicon atoms, e.g. the 1:5 and 1:7 cases in figures 9.12(c) and (d), respectively. Also note that the guest N atoms present a weak spin-up configuration. Most importantly, quite strong competition between the opposite spin configurations leads to a weakened magnetic moment and thus reduces ferromagnetism greatly. The theoretical predictions for the spatial ferromagnetic configurations and the spin-created magnetic moments can be verified using, respectively, high-resolution SP-STM and SQUID measurements (details in sections 3.2 and 3.4). Specifically, non-magnetic spin configurations occur frequently in low-concentration cases ($\lesssim$13% for the ratio of guest and host atoms in table 9.3). Furthermore, the spin-degenerate energy subbands correspond to the modified Dirac-cone structures (figures 9.10(h)–(j) and (l)). It is deduced that atom-created spin configurations are suppressed effectively by well-behaved π bonding ($3p_z$–$3p_z$, $3p_z$–$2p_z$ and $2p_z$–$2p_z$ orbital hybridization, in particular for the $3p_z$–$3p_z$), i.e. the spin-induced on-site Coulomb potentials [101] are negligible in reducing the total ground state energy, compared to the charge-dependent atomic interactions [102].

The atom-, orbital- and spin-decomposed densities of states are able to provide more information about the complicated orbital hybridization and spin configurations in nitrogen-substituted silicene materials, as illustrated in figures 9.13(a)–(k) under various substitution cases. There are a lot of special structures in the density of states, owing mainly to the band-edge states of parabolic and oscillatory energy dispersions (figures 9.10(a)–(l)). Such special critical points, which survive in the energy–wave-vector space, are related closely to the extreme states in parabolic subbands or modified Dirac-cone structures, the partial flat bands along any direction or a specific direction (quasi-oD or 1D parabolic bands), and the saddle-point structures near the high-symmetry point. As a result, they create the van Hove singularities in terms of, respectively, shoulders or asymmetric V-shapes, delta-function-like peaks or divergent peaks in the square root forms, and logarithmically prominent peaks. Interestingly, the modified V-shaped structures, which are centered in the range -0.5 eV $< E <$ 0 eV, appear at sufficiently low concentrations, such as the 2:16 meta, 2:16 ortho, 2:16 para and 1:17 single cases (figures 9.13(i)–(l)). Such results indicate a gradual recovery of the Dirac-cone band structure during the decrease of N concentration. The density of states across the Fermi level is available to distinguish semiconducting and semimetallic or metallic behaviors. Only the 1:1 and 2:16 meta cases belong to the observable semiconductors (figures 9.13(b) and (h)), while the other cases are gapless conductors or semiconductors. Specifically, certain nitrogen-substituted silicene systems might have sufficiently high free carriers under the crossing of the partial flat subbands with the Fermi level, which is illustrated clearly by the large density of states at E_F in the 2:6 meta, 2:6 ortho, 2:6 para and 1:7 single configurations (figures 9.13(d)–(f) and (l)).

The atom-, orbital- and spin-dependent interactions play critical roles in the creation of various van Hove singularities. Generally, within the specific energy range of -1.0 eV $< E <$ 1.0 eV, the density of states is dominated by the Si-$3p_z$ and N-$2p_z$ orbitals (the solid and dashed purple curves), in which the 3s, $3p_x$, $3p_y$, 2s, $2p_x$ and $2p_y$ orbitals also make significant contributions. Furthermore, a series of extended van Hove singularities, associated with the π-electronic orbitals, are very difficult to identify from the rich special structures. As a result, the π and σ energy subbands cannot survive in any nitrogen-substituted silicene systems. It should be noted that the van Hove singularities due to the eight orbitals of the guest and host atoms might be merged together at identical energies. This result further illustrates impure or pure four-orbital hybridization in any chemical bonds, as observed earlier in the spatial charge distributions (figure 9.11). Concerning the spin-split density of states, the ferromagnetic configuration is determined mainly by the low-energy part across the Fermi level, where the spin-up and spin-down configurations, respectively, appear at $E < 0$ and $E > 0$, such as those of the 1:1, 2:4 meta, 1:5 single and 1:7 single conditions (figures 9.13(b), (c), (k) and (l)). The occupied electronic states appear to account for the net magnetic moment. On the other hand, at deeper or higher energies, two kinds of spin arrangements arise simultaneously, and their identical areas enclosed by the density of states (DOS) curve lead to the absence of ferromagnetism. Of course, they might possess different energies for the van Hove singularities.

High-resolution STS and SP-STS measurements are the most efficient and accurate way to explore the van Hove singularities of layered materials fully. For example, such experiments are highly suitable for identifying rich special structures in the density of states due to the various layer numbers and stacking configurations of few-layer graphene systems, such as those in monolayer graphene [103], AB bilayer and trilayer stacking [104], twisted bilayer graphenes [105], and ABC- and AAB-stacked trilayer graphene systems [106, 107]. Interestingly, the direct combination of SP-STM and SP-STS is able to examine the spin-induced magnetic properties thoroughly. To date, SP-STM has been successful in observing magnetic domains and domain-wall structures under the high spatial resolution of 10 nm [108]; furthermore, SP-STS can distinguish metallic and semiconducting ferromagnetism [109]. These methods are available for verifying the unusual van Hove singularities in nitrogen-substituted silicene systems, in which the main features include the finite or vanishing density of states at the Fermi level, many special structures due to complicated multi-orbital hybridization, a modified V-shape near E_F at sufficiently low concentrations, undefined π- and σ-band widths, and low-lying spin-degenerate or spin-split energy spectra. Such studies are able to provide more information on the multi-orbital- and spin-dependent intrinsic interactions.

There exist a lot of important differences between the carbon-, boron- and nitrogen-substituted silicene materials in terms of fundamental properties. They include the bond lengths, the total ground state energies, net or vanishing magnetic moments, the band properties across the Fermi level (the semiconducting, semi-metallic and p-type phenomena), modified Dirac-cone structures or the low-lying oscillatory dispersions, well-defined π and π^* energy subbands or undefined π- and σ-electronic states (the π- and σ-orbital bonding and the four-orbital hybridization), rich and complicated σ valence subbands, spin-degenerate or spin-split band structures, frequent crossing or little anti-crossing along the ΓM and KM paths, the atom-, orbital- and spin-dependent van Hove singularities, and highly non-homogeneous spin density distributions only under certain Ni substitution cases. The C–, B– and N–Si bonds lengths can be estimated from the original lengths, indicating clearly the characteristics of covalent bonds in silicon-based few-layer compounds. The C–Si bonds are the shortest and thus generate the lowest ground state energies. The C-substituted silicene compounds are synthesized relatively easily in experimental laboratories. Specifically, the N–N bonds are unstable under ortho configurations except for the 2:4 case. The hybridized π and σ chemical bonding in Si–C materials enlarges band gaps only at higher guest-atom concentrations, since both atoms possess four half occupied orbitals. Generally, the Fermi level is situated at the middle of the modified Dirac-cone structures. Furthermore, the initial σ valence subbands can be defined roughly from the zone-folding effects. Well-behaved π- and σ-electronic states are found in boron substitution. Interestingly, only the B-substituted silicenes are able to create valence holes, corresponding to the red shift of the Fermi level (relative to the Dirac points). The free carrier densities are identical to the boron concentrations according to the conservation of electronic states ($2p_z$ orbitals). As a result, such B–Si-based systems are very useful in potential applications, e.g. as electrical conductors [110]. On the other hand, the low-lying non-uniform energy subbands of the nitrogen-substituted silicenes demonstrate complicated ($2s$, $2p_x$, $2p_y$, $2p_z$)–($3s$, $3p_x$, $3p_y$, $3p_z$) orbital hybridization in N–Si bond and ($3s$, $3p_x$, $3p_y$, $3p_z$)–($3s$, $3p_x$, $3p_y$, $3p_z$) in Si–Si bonds. It is almost impossible to characterize the π and σ chemical bonding clearly. Also, the spin-split energy subbands, as well as the net moments and the spatial ferromagnetic configurations, can survive under different cases in which the spin states are associated with the guest and host atoms simultaneously. There exist ferromagnetic semiconductors or semimetals; furthermore, it is almost impossible to define the blue or red shift of the Fermi level in the absence of a modified Dirac-cone structure. Also note that these materials cannot be regarded as n-type materials, even with an extra valence electron in each nitrogen atom.

Most importantly, the substitution-enriched electronic structures are expected to present diverse fundamental properties. First-principles calculations show that the low-energy electronic states can be approximated by modified Dirac cones (π chemical bonding) for carbon- and boron-substituted silicene systems, but not for nitrogen-related systems. For the carbon- and boron-related systems, the generalized tight-binding model, accompanied with single- and many-particle theories, is available for the thorough investigation of the unique magnetic quantizations [111], the rich magneto-optical absorption spectra, the unusual quantum Hall conductivities [112] and the various magneto-plasmon modes [113]. Interestingly, the modulation of free hole densities, which is achieved easily through variation of boron concentration, is capable of creating diverse fundamental properties [114]. However, for nitrogen-related systems, the multi-orbital hopping integrals might be too complicated to obtain reasonable values, since the tight-binding model cannot fit the VASP band structures well along the high-symmetry points [115]. Consequently, it will be very difficult to explore the above-mentioned magnetic quantization phenomena fully.

9.4 Conclusion

Boron, carbon and nitrogen substitution can vary the geometric, electronic and magnetic properties of silicene-related binary compounds greatly, since the different guest atoms are able to provide three, four and five valence electrons, respectively. For the B and C binary silicene compounds, the magnetic spin configurations disappear completely through the roughly well-defined π and σ chemical bonding. On the other hand, the N–Si systems present complicated four-orbital hybridization and semimetallic or semiconducting ferromagnetic configurations under certain nitrogen substitution cases. The moiré superlattices can create planar or buckled structures, non-uniform bond lengths in an enlarged unit cell (co-existent X–X, X–Si and Si–Si bonds), and the typical meta, ortho (the highest ground state energy), para (the lowest ground state energy) and single configurations. Specifically, the N–N bonds are unstable except for the 2:4 ortho case. Among the three kinds guest atoms, the carbon-related bond lengths are the shortest and the boron-related bond lengths are the longest, thus corresponding to the lowest and highest ground state energies, respectively (the most stable systems). Interestingly, C, B and N guest-atom substitution generates diverse band properties across the Fermi level associated with, respectively, semiconducting, p-type and semiconducting/semimetallic behaviors. n-type metals cannot be created by such guest atoms. The largest band gap arises in the full carbon-atom substitution, e.g. $E_g \sim 2.53$ eV in a pure 100% SiC compound. When the π and π^* energy subbands dominate the low-lying electronic structures, they are initiated from the K/K′ or Γ valleys, become saddle structure centered at the M point and then end at the Γ valley. The initial symmetry points are determined mainly by the zone-folding effect. Their energy widths are identified clearly for the C- and B-substituted systems under any case, illustrating well-behaved π chemical bonding. Similar results are obtained from the first σ-electronic valence subbands. However, it is very difficult to characterize the π- and σ-electronic states well in the nitrogen substitution cases. The energy spectra of the electrons and holes are highly asymmetric for the boron and nitrogen atom substitution cases, owing mainly to the red shift of the Fermi level and the great reduction of π-electronic chemical bonding for the former, and the four-four impure or pure orbital hybridization for the latter. For the C-substituted silicenes, the modified Dirac-cone structure (the V-shaped van Hove singularity across the Fermi level) is recovered gradually for a sufficiently low concentration (approximately lower than 12.5%). The degenerate threshold σ bands, which are deeper than -1.0 eV, appear at the Γ point. The π and σ energy subbands show frequent crossing and little anti-crossing along the KΓ and MΓ directions, except for the full substitution case which presents only crossing behavior and thus pure sp^2 bonding (non-sp^3 and non-hybridized sp^2 bonding). Generally, the electronic states of boron-related silicenes can be regarded as zone-folding states with p-type doping, e.g. the Fermi level below the modified Dirac points, and the initial π and σ energy subbands at, respectively, the K/Γ and Γ points. On the other hand, the nitrogen-substituted silicenes display quite complicated band structures, such as strongly oscillatory wave-vector dependence at low energy, undefined π and σ subbands, and spin-split energy spectra (the semimetallic or semiconducting ferromagnetic configurations) under specific cases. Obtaining the multi-orbital hopping integrals and the spin-dependent many-particle interactions in the tight-binding model would become almost impossible for the nitrogen-substituted silicene systems [50]. Five kinds of van Hove singularities—the V-shaped structure, shoulders, logarithmically symmetric peaks, delta-function-like peaks and square root asymmetric peaks—correspond to, respectively, the linear Dirac cone, extreme band-edge states of parabolic dispersions, saddle points, partially flat bands and constant-energy loops. Whether the vanishing or finite density of states can survive within an energy range covering the Fermi level determines the above-mentioned three types of band properties. High-resolution ARPES, SP-ARPES, STS and SP-STS measurements are capable of examining the wave-vector-dependent π- and initial σ-electronic energy spectra, and the band crossing and anti-crossing phenomena. That is, they can provide much useful information on the complex relationships for the π, σ, sp^3 and multi-orbital bonding. There exist certain important differences between the B-, C- and N-substituted silicene systems, as a result of the different electron affinities, atomic configurations, ionization energies and orbital hybridization. They are suitable for exploring the diverse physical, chemical and material phenomena fully, e.g. the unusual magnetic quantization close to the Fermi level [111] and the E_F-enriched optical absorption spectra and Coulomb excitations [116]. Most importantly, boron substitution is predicted to create very high 2D free hole densities, the same as those of the guest atoms, leading to potential applications in nanoscale electronic devices [117]. Whether suitable tight-binding models, related to the multi-orbital hopping integrals of (2s, $2p_x$, $2p_y$, $2p_z$) and (3s, $3p_x$, $3p_y$, $3p_z$), can simulate the first-principles results is worthy of further careful investigation.

The differences and similarities between the chemisorption and substitution effects are revealed clearly in silicon-doped graphene systems, providing full information about single- and multi-orbital hybridization. Significant chemical bonding is proposed to account for the Si-induced chemical modifications in graphene systems. The calculated results show clearly that free carriers can be created using chemical adsorption and band gaps using identical methods and substitution. Specific modifications of silicon guest atoms are able to modulate the semiconducting, metallic and semimetallic behaviors, in a way which was difficult to observe for the other kinds of guest atoms [118, 119]. For example, the alkali adatoms can generate a high free electron density under a rigid band shift of graphene energy bands [120]. The Si-adsorbed and Si-substituted graphenes could serve as potential candidates in nanoscale applications [121]. The chemisorption and substitution of Si-guest atoms on monolayer graphene present unusual geometric properties, which are reflected directly in the spatial charge distributions. Si-adsorbed graphene is a non-buckled honeycomb lattice with an optimal bridge-site position. The planar structure indicates a very small variation in the σ bonding of C-2s, -$2p_x$ and $2p_y$ orbitals; therefore, they take little part in the Si–C bonds. This adsorption configuration is similar to that of graphene oxide [122], although there is a slight buckling in the latter. The Si–C bond length is about 2.1–2.5 Å, leading to significant multi-orbital hybridization. Furthermore, the C–C bond lengths almost remain the same. According to detailed analyses of the 3D charge density distributions, their spatial variation, and the atom- and orbital-projected densities of states, the sp^3–p multi-orbital hybridization is deduced to determine the chemical Si–C bonds. The theoretical predictions for the bridge site and Si–C and C–C bond lengths could be studied using high-resolution STM, TEM and LEED measurements. Also, Si-substituted graphene systems display planar geometries under the multi-orbital chemical bonding in the Si–C, C–C and Si–Si bonds. The Si–C bond length is about 1.62–1.83 Å, which is shorter than those in the Si-adsorbed systems. This means that four orbitals in the host and guest atoms are hybridized strongly with one another. STM experiments are available for examining the predicted Si–C, C–C and Si–Si bond lengths. Any chemisorption or substitution of Si-guest atoms alters the unusual band structure of monolayer graphene completely, in particular for the zero-gap semiconducting behavior and linear Dirac cone due to the π bonding of the $2p_z$ orbitals. In the 100% double- and single-sided adsorption cases, the Si-adsorbed graphene systems are semimetals with free conduction electrons and valence holes. Low-concentration systems belong to p-type metals with only free valence holes. In general, the Dirac-cone structure near the stable valley is distorted seriously after Si chemisorption. There are more valence and conduction energy subbands, which are accompanied by various band-edge states within a whole energy spectrum, e.g. the emergent low-lying energy bands along the KΓ and MΓ directions. However, the σ bands, which arise from the $2p_x$ and $2p_y$ orbitals of carbon atoms, exhibit a rigid red shift of ~1 and ~0.5 eV for the 100% and lower-concentration cases, respectively. The above-mentioned important results reveal the critical mechanisms directly, namely the multi-orbital hybridization of sp^3–p in Si–C bonds, sp^3–sp^3 in Si–Si bonds, and sp^2–sp^2 or π in C–C bonds. Such types of chemical bonding arise mainly from the four Si and C orbitals, as a result of the bridge-site adsorption positions, and the binding energies and charge distributions of the separated orbitals. High-resolution ARPES measurements are very useful in verifying the low-lying valence bands near or crossing the Fermi level along ΓK and ΓM and the rigid σ bands initiated from the Γ valley. On the other hand, all substitution cases result in semiconducting behavior with a finite or zero band gap. The Dirac-cone structure presents a deviation from the Γ point, a strong distortion, or even complete destruction. The number of valence and conduction energy subbands remains the same after chemical substitution; furthermore, they are co-dominated by the Si-guest and C-host atoms. The main features of band structures in Si-substituted graphene systems appear to be dominated by the sp^2–sp^2 and p–p orbital hybridization in the Si–C, C–C and Si–Si bonds. Many unusual van Hove singularities in the atom- and orbital-decomposed densities of states are created under Si-guest-atom chemisorption and substitution. Only pristine monolayer graphene presents a V-shaped structure with a vanishing value at the Fermi level, logarithmically symmetric π and π^* peaks at −2.41 and 1.81 eV, respectively, and a shoulder structure at −3.01 eV, of which the first two correspond to the π bonding of carbon atoms and the last one corresponds to the σ bonding of carbon atoms. The strong π-bonding evidence is destroyed completely by any chemisorption except for the very dilute Si-adatoms. They are replaced by a finite density of states at $E_F = 0$, many shoulders and peak structures. Furthermore, they are co-dominated by the four Si-3s, -3p_x, -3p_y and 3p_z orbitals and the single C-2p_z orbital, since their contributions are merged together. Specifically, the σ-band shoulder arises at −4.21, −4.11 or −3.5 eV, and the ~0.5–1.0 red shift is related closely to the different ionization energies of the Si and C atoms. The above-mentioned significant features further support and illustrate the sp^3–sp^3 and sp^2–sp^2 multi-orbital hybridization in Si–C bonds and C–C bonds, respectively. Concerning the substitution cases, the density of states is zero within a finite energy range centered at the Fermi level, except for a few systems with deformed V-shaped structures at E_F. All the C-substituted silicene systems are finite- or zero-gap semiconductors. The special structures which originate from the Si-3p_z and C-2p_z orbitals appear simultaneously. Also, similar behavior is found in the Si-3p_x and -3p_y and C-2p_x and -2p_y orbitals. These further illustrate the critical single- and multi-orbital hybridization in the Si–C bonds. The predicted van Hove singularities could be verified using high-resolution STS experiments [123].

References

[1] Terrones M, Grobert N and Terrones H 2002 Synthetic routes to nanoscale $B_xC_yN_z$ architectures *Carbon* **40** 1665–84

[2] Deng X, Kousaka H, Tokoroyama T and Umehara N 2014 Deposition and tribological behaviors of ternary BCN coatings at elevated temperatures *Surf. Coat. Technol.* **259** 2–6

[3] Qiu Y and Gao L 2003 Chemical synthesis of turbostratic carbon nitride, containing CN crystallites, at atmospheric pressure *Chem. Commun.* **18** 2378–9

[4] Bojdys M J, Muller J-O, Antonietti M and Thomas A 2008 Ionothermal synthesis of crystalline, condensed, graphitic carbon nitride *Chemistry—Eur. J.* **4** 8177–82

[5] Wang Y, Zhang J, Wang X, Antonietti M and Li H 2010 Boron- and fluorine-containing mesoporous carbon nitride polymers: metal-free catalysts for cyclohexane oxidation *Angew. Chem.* **49** 3356–9

[6] Prakash A, Sundaram K B and Campiglia A D 2016 Photoluminescence studies on BCN thin films synthesized by RF magnetron sputtering *Mater. Lett.* **183** 355–8

[7] Kester D, Ailey K, Davis R and More K 1993 Phase evolution in boron nitride thin films *J. Mater. Res.* **8** 1213–6

[8] Mirkarimi P B, McCarty K F and Medlin D L 1997 Review of advances in cubic boron nitride film synthesis *Mater. Sci. Eng.* **21** 47–100

[9] Terrones M *et al* 1996 Pyrolytically grown $B_xC_yN_z$ nanomaterials: nanofibres and nanotubes *Chem. Phys. Lett.* **257** 576–82

[10] Han W, Bando Y, Kurashima K and Sato T 1998 Synthesis of boron nitride nanotubes from carbon nanotubes by a substitution reaction *Appl. Phys. Lett.* **73** 3085

[11] Lourie O R, Jones C R, Bartlett B M, Gibbons P C, Ruoff R S and Buhro W E 2000 CVD growth of boron nitride nanotubes *Chem. Mater.* **12** 1808–10

[12] Erickson K J, Gibb A L, Sinitskii A, Rousseas M, Alem N, Tour J M and Zettl A K 2011 Longitudinal splitting of boron nitride nanotubes for thc facilc synthesis of high quality boron nitride nanoribbons *Nano Lett.* **11** 3221–6

[13] Chen W, Li Y, Yu G, Li C-Z, Zhang S B, Zhou Z and Chen Z-F 2010 Hydrogenation: a simple approach to realize semiconductor–half-metal–metal transition in boron nitride nanoribbons *J. Am. Chem. Soc.* **132** 1699–705

[14] Wei X, Wang M-S, Bando Y and Golberg D 2011 Electron-beam-induced substitutional carbon doping of boron nitride nanosheets nanoribbons, and nanotubes *ACS Nano* **5** 2916–22

[15] Zhang L, Zhang Z-Y, Liang R-P, Li Y-H and Qiu J-D 2014 Boron-doped graphene quantum dots for selective glucose sensing based on the 'abnormal' aggregation-induced photoluminescence enhancement *Anal. Chem.* **86** 4423–30

[16] Fei H *et al* 2014 Boron- and nitrogen-doped graphene quantum dots/graphene hybrid nanoplatelets as efficient electrocatalysts for oxygen reduction *ACS Nano* **8** 10837–43

[17] Zhu M and Diao G 2011 Review on the progress in synthesis and application of magnetic carbon nanocomposites *Nanoscale* **3** 2748

[18] Chen L-F *et al* 2012 Synthesis of nitrogen-doped porous carbon nanofibers as an efficient electrode material for supercapacitors *ACS Nano* **6** 7092–102

[19] Zhang H-F, Dohnalkova A C, Wang C-M, Young J S, Buck E C and Wang L-S 2001 Lithium-assisted self-assembly of aluminum carbide nanowires and nanoribbons *Am. Chem. Soc.* **2** 105–8

[20] Osikoya A O, Wankasi D, Vala R M K, Afolabi A S and Dikio E D 2014 Synthesis, characterization and adsorption studies of fluorine-doped carbon nanotubes *J. Nanomater. Biostruct.* **9** 1187–97

[21] Chabi S, Chang H, Xia Y and Zhu Y 2016 From graphene to siliconcarbide: ultrathin silicon carbide flakes *Nanotechnology* **7** 075602

[22] Zheng H and Duley W 2008 First-principles study of edge chemical modifications in graphene nanodots *Phys. Rev.* B **78** 045421

[23] Boukhvalov D W, Rudenko A N, Prishchenko D A, Mazurenkoc V G and Katsnelson M I 2015 Chemical modifications and stability of phosphorene with impurities: a first principles study *Phys. Chem. Chem. Phys.* **17** 15209

[24] Bekaroglu E, Topsakal M, Cahangirov S and Ciraci S 2010 First-principles study of defects and adatoms in silicon carbide honeycomb structures *Phys. Rev.* B **81** 075433

[25] Tran N T T, Dahal D, Gumbs G and Lin M-F 2017 Adatom doping-enriched geometric and electronic properties of pristine graphene: a method to modify the band gap *Struct. Chem.* **28** 1311–8

[26] Nguyen D K, Tran N T T, Chiu Y-H and Lin M-F 2017 Concentration-diversified magnetic and electronic properties of halogen-adsorbed silicene *Sci. Rep.* **9** 13746

[27] Jose D and Datta A 2014 Structures and chemical properties of silicene: unlike graphene *Acc. Chem. Res.* **47** 593–602

[28] Kim D H, Byon E, Lee S, Kim J-K and Ruh H 2004 Characterization of ternary boron carbon nitride films synthesized by RF magnetron sputtering *Thin Solid Films* **447–448** 192–6

[29] Md A M, Noguchi H, Kida T, Nagano M, Hirao N and Baba Y 2010 Growth and characterization of stoichiometric BCN films on highly oriented pyrolytic graphite by radiofrequency plasma enhanced chemical vapor deposition *Thin Solid Films* **518** 4163–9

[30] Souza T A, Silva M R A and Carvalho A C M 2010 Theoretical study of atomic arrangement in $B_xC_yN_z$ nanotubular structures *Phys. Proc.* **28** 84–9

[31] Mazzoni M S C, Nunes R W, Srgio A and Chacham H 2006 Electronic structure and energetics of $B_xC_yN_z$ layered structures *Phys. Rev.* B **73** 073108

[32] Xiao J L, Wang C B, Shen Q and Zhang L M 2015 Influence of nitrogen pressure on bonding structure and mechanical properties of pulsed laser deposited BCN thin films *Surf. Coat. Technol.* **215** 141–4

[33] Xiao X, Li H, Tie J and Lu J 2017 Half metallic behavior in H-terminated zigzag BC_2N nanoribbons *Appl. Mech. Mater.* **864** 30–5

[34] Venu K, Kanuri S, Raidongia K, Hembram K P S S, Waghmare U V and Datta R 2010 Band gap and chemically ordered domain structure of a graphene analogue $B_xC_yN_z$ *Solid State Commun.* **150** 2262–5

[35] Blase X, Charlier J C, De Vita A and Car R 2017 Structural and electronic properties of composite $B_xC_yN_z$ nanotubes and heterojunctions *Appl. Phys.* A **68** 293–300

[36] Han S-J *et al* 2017 High-speed logic integrated circuits with solution processed self-assembled carbon nanotubes *Nat. Nanotechnol.* **12** 861–5

[37] Chowdhury R and Adhikari S 2011 Boron-nitride nanotubes as zeptogram-scale bionanosensors: theoretical investigations *IEEE Trans. Nanotechnol.* **10** 659–67

[38] Songfeng E, Geng R, Zhu Z, Xie L, Lu W and Li C 2018 Large-scale fabrication of boron nitride nanotubes and their application in thermoplastic polyurethane based composite for improved thermal conductivity *Ceram. Int.* **44** 22794–9

[39] Wickenden D K, Faulkner K R and Brander R W 1971 Growth of epitaxial layers of galliumnitride on silicon carbide and corundum substrates *J. Cryst. Growth* **9** 158–64

[40] Bhatnagar A and Lang D 2006 Military and law enforcement applications of lightweight ballistic materials *Lightweight Ballistic Composites* (Amsterdam: Elsevier) pp 364–97

[41] Edmond J A 1990 Blue light emitting diode formed in silicone carbide *United States Patent* 4,918,497

[42] Baliga B J and Bhatnagar M 1990 Silicon carbide field effect transistor *United States Patent* 5,338,945

[43] Tran N T T, Nguyen D K, Glukhova O E and Lin M-F 2017 Coverage-dependent essential properties of halogenated graphene: a DFT study *Sci. Rep.* **7** 17858

[44] Tran N T T, Lin S-Y, Lin C-Y and Lin M-F 2017 *Geometric and Electronic Properties of Graphene-Related Systems* (London: Taylor and Francis)

[45] Hussain T, Kaewmaraya T, Chakraborty S and Ahuja R 2016 Defect and substitution-induced silicene sensor to probe toxic gases *J. Phys. Chem.* C **120** 25256–62

[46] Yin N, Dai Y, Wei W and Huang B 2018 Electronic structure engineering in silicene via atom substitution and a new two-dimensional Dirac structure Si_3C *Physica* E **98** 39–44

[47] Li S, Ren J-C, Ao Z and Liu W 2018 Enhanced stability and induced magnetic moments of silicene by substitutional doping of nickel *Chem. Phys. Lett.* **706** 202–7

[48] Chabi S, Chang H, Xia Y and Zhu Y 2016 From graphene to silicon carbide: ultrathin silicon carbide flakes *Nanotechnology* **27** 075602

[49] Manjanath A, Kumarcd V and Singh A K 2014 Mechanical and electronic properties of pristine and Ni-doped Si, Ge, and Sn sheets *Phys. Chem. Chem. Phys.* **16** 1667

[50] Sivek J, Sahin H, Partoens B and Peeters F M 2013 C: stability and electronic and phonon properties *Phys. Rev.* B **87** 085444

[51] Lv R, dos Santos M C, Antonelli C, Feng S, Fujisawa K and Berkdemir A 2014 Large-area Si-doped graphene: controllable synthesis and enhanced molecular sensing *Adv. Mater.* **26** 7593–9

[52] Sun L, Han C, Wu N, Wang B and Wang Y 2018 High temperature gas sensing performances of silicon carbide nanosheets with an n–p conductivity transition *RSC Adv.* **8** 13697
[53] Tang H and Ismail-Beigi S 2007 Novel precursors for boron nanotubes: the competition of two-center and three-center bonding in boron sheets *Phys. Rev. Lett.* **99** 115501
[54] Sawides 1985 Four-fold to three-fold transition in diamond-like amorphous carbon films: a study of optical and electrical properties *J. Appl. Phys.* **58** 518
[55] Jalili S and Vaziri R 2011 Study of the electronic properties of Li-intercalated nitrogen doped graphite *Mol. Phys.* **109** 687–94
[56] Liu C-C, Jiang H and Yao Y 2011 Low-energy effective Hamiltonian involving spin–orbit coupling in silicene and two-dimensional germanium and tin *Phys. Rev.* B **84** 195430
[57] Badzian A 2002 Stability of silicon carbon nitride phases *J. Am. Ceram. Soc.* **85** 16–20
[58] Elias C *et al* 2019 Direct band-gap crossover in epitaxial monolayer boron nitride *Nat. Commun.* **10** 2639
[59] Sahin H and Peeters F M 2013 Adsorption of alkali, alkaline-earth, and 3D transition metal atoms on silicene *Phys. Rev.* B **87** 085423
[60] Wallace P R 1947 The band theory of graphite *Phys. Rev.* **71** 622
[61] Chang C P, Lu C L, Shyu F L, Chen R B, Huang Y C and Lin M F 2013 Adsorption of alkali, alkaline-earth, and 3D transition metal atoms on silicene *Phys. Rev.* B **87** 085423
[62] Chang C P, Chiu C W, Shyu F L, Chen R B and Lin M F 2002 Magnetoband structures of AB-stacked zigzag nanographite ribbons *Phys. Lett.* A **306** 137–43
[63] Chen S C, Wang T S, Lee C H and Lin M F 2008 Magneto-electronic properties of graphene nanoribbons in the spatially modulated electric field *Phys. Lett.* A **372** 5999–6002
[64] Harrison W A and Tersoff J 1986 Tight binding theory of heterojunction band lineups and interface dipoles *J. Vac. Sci. Technol.* B **4** 1068
[65] Shih P-H, Do T-N, Gumbs G, Pham H D and Lin M-F 2019 Electric-field-diversified optical properties of bilayer silicene *Opt. Lett.* **44** 4721–4
[66] Shih P-H, Do T-N, Gumbs G, Huang D, Pham H D and Lin M-F 2019 Rich magnetic quantization phenomena in AA bilayer silicene *Sci. Rep.* **9** 14799
[67] Slater J C and Koster G F 1954 Simplified LCAO method for the periodic potential problem *Phys. Rev.* **94** 1498
[68] Chiu Y-H, Ho J-H, Ho Y-H, Chuu D-S and Lin M-F 2009 Effects of a modulated electric field on the optical absorption spectra in a single-layer graphene *J. Nanosci. Nanotechnol.* **9** 6579–86
[69] Do T-N, Lin C-Y, Lin Y-P, Shih P-H and Lin M-F 2009 Configuration-enriched magneto-electronic spectra of AAB-stacked trilayer graphene *Carbon* **94** 619–32
[70] Nguyen D K, Tran N T T, Nguyen T T and Lin M-F 2018 Diverse electronic and magnetic properties of chlorination-related graphene nanoribbons *Sci. Rep.* **8** 17859
[71] Lin C-Y, Wu J-Y, Chang C-P and Lin M-F 2014 Magneto-optical selection rules of curved graphene nanoribbons and carbon nanotubes *Carbon* **69** 151–61
[72] Chiu C W, Lee S H, Chen S C and Lin M F 2014 Electronic excitations in doped monolayer graphenes *J. Appl. Phys.* **109** 113711
[73] Sun M, Tang W, Ren Q, Wang S, Yu J, Du Y and Zhang Y 2015 First-principles study of the alkali earth metal atoms adsorption on graphene *Appl. Surf. Sci.* **356** 668–73
[74] Chiu C-W, Huang Y-C, Shyu F-L and Lin M-F 2011 Influence of anisotropic dipole matrix element on optical response of AB-stacked graphene superlattice *Opt. Lett.* **36** 3136–8

[75] Chiu C-W, Huang Y-C, Shyu F-L and Lin M-F 2011 Excitation spectra of ABC-stacked graphene superlattice *Appl. Phys. Lett.* **98** 261920

[76] Ho Y-H, Chiu Y-H, Su W-P and Lin M-F 2011 Magneto-absorption spectra of Bernal graphite *Appl. Phys. Lett.* **99** 011914

[77] Wills J M and Cooper B R 1990 First-principles calculations for a model Hamiltonian treatment of hybridizing light actinide compounds *Phys. Rev.* B **42** 4682

[78] Tan X, Cabrera C R and Chen Z 2014 Metallic BSi_3 silicene: a promising high capacity anode material for lithium-ion batteries *J. Phys. Chem.* C **118** 25836–43

[79] Kong X, Chen Q and Sun Z 2013 Enhanced oxygen reduction reactions in fuel cells on H-decorated and B-substituted graphene *Chem. Phys. Chem.* **14** 514–9

[80] Esrafili M D and Behzadi H 2013 A comparative study on carbon, boron-nitride, boron-phosphide and silicon-carbide nanotubes based on surface electrostatic potentials and average local ionization energies *J. Mol. Model.* **19** 2375–82

[81] Reich S, Maultzsch J and Thomsen C 2002 Tight-binding description of graphene *Phys. Rev.* B **66** 035412

[82] Do T-N, Shih P-H, Gumbs G, Huang D, Chiu C-W and Lin M-F 2013 Diverse magnetic quantization in bilayer silicene *Phys. Rev.* B **97** 125416

[83] Chiu C-W, Huang Y-C, Chen S-C, Lin M-F and Shyu F-L 2011 Low-frequency electronic and optical properties of rhombohedral graphite *Phys. Chem. Chem. Phys.* **13** 6036–42

[84] Do T-N, Shih P-H, Chang C-P, Lin C-Y and Lin M-F 2016 Rich magneto-absorption spectra of AAB-stacked trilayer graphene *Phys. Chem. Chem. Phys.* **18** 17597

[85] Li T S and Lin M F 2006 Transport properties of finite carbon nanotubes under electric and magnetic fields *J. Phys. Condens. Matter.* **18** 10693–703

[86] Shih P-H, Do T-N, Huang B-L, Gumbs G, Huang D and Lin M-F 2019 Rich magneto-absorption spectra of AAB-stacked trilayer graphene *Carbon* **18** 608–14

[87] Chen S-C, Wu J-Y and Lin M-F 2018 Feature-rich magneto-electronic properties of bismuthene *New J. Phys.* **20** 062001

[88] Barraza S J and Denmark S E 2018 Synthesis, reactivity, functionalization, and ADMET properties of silicon-containing nitrogen heterocycles *J. Am. Chem. Soc.* **140** 6668–84

[89] Chiu Y H, Ho J H, Chang C P, Chuu D S and Lin M F 2018 Low-frequency magneto-optical excitations of a graphene monolayer: Peierls tight-binding model and gradient approximation calculation *Phys. Rev.* B **78** 245411

[90] Rakhmanov A L, Rozhkov A V, Sboychakov A O and Franco N 2012 Instabilities of the AA-stacked graphene bilayer *Phys. Rev. Lett.* **109** 206801

[91] Paul-Boncour V, Gupta M, Joubert J M, Percheron-Guegan A, Parentc P and Laffon C 2000 Investigation of the electronic properties of substituted $LaNi_5$ compounds used as material for batteries *J. Mater. Chem.* **10** 2741–7

[92] Ruegg A and Fiete G A 2011 Topological insulators from complex orbital order in transition-metal oxides heterostructures *Phys. Rev.* B **84** 201103

[93] Partoens B and Peeters F M 2006 From graphene to graphite: electronic structure around the K point *Phys. Rev.* B **74** 075404

[94] Lin S-Y, Chang S-L, Tran N T T, Yang P-H and Lin M-F 2015 H–Si bonding-induced unusual electronic properties of silicene: a method to identify hydrogen concentration *Phys. Chem. Chem. Phys.* **17** 26443

[95] Chang S-L, Lin S-Y, Lin S-K, Lee C-H and Lin M-F 2014 Geometric and electronic properties of edge-decorated graphene nanoribbons *Sci. Rep.* **4** 6038

[96] Lin S-Y, Chang S-L, Chen H-H, Su S-H, Huang J-C and Lin M-F 2016 Substrate-induced structures of bismuth adsorption on graphene: a first principles study *Phys. Chem. Chem. Phys.* **18** 18978
[97] Eremeev S V, Landolt G, Menshchikova T V, Slomski B, Koroteev Y M and Aliev Z S 2012 Atom-specific spin mapping and buried topological states in a homologous series of topological insulators *Nat. Commun.* **3** 635
[98] Chen L, Feng B and Wu K 2013 Observation of a possible superconducting gap in silicene on Ag (111) surface *Appl. Phys. Lett.* **102** 081602
[99] Wu J-Y, Chen S-C, Gumbs G and Lin M-F 2016 Feature-rich electronic excitations of silicene in external fields *Phys. Rev.* B **94** 205427
[100] Wu J-Y, Lin C-Y, Gumbsb G and Lin M-F 2015 The effect of perpendicular electric field on temperature-induced plasmon excitations for intrinsic silicene *RSC Adv.* **5** 51912
[101] Chen S-C, Wu C-L, Wu J-Y and Lin M-F 2016 Magnetic quantization of sp^3 bonding in monolayer gray tin *Phys. Rev.* B **94** 045410
[102] Giese T J and York D M 2005 Improvement of semiempirical response properties with charge-dependent response density *J. Chem. Phys.* **123** 164108
[103] Chen H H, Su S H, Chang S L, Cheng B Y, Chong C W, Huang J C A and Lin M F 2015 Long-range interactions of bismuth growth on monolayer epitaxial graphene at room temperature *Carbon* **93** 180–6
[104] Chuang Y-C, Wu J-Y and Lin M-F 2015 Electric field dependence of excitation spectra in AB-stacked bilayer graphene *Sci. Rep.* **3** 1368
[105] Dodaro J F, Kivelson S A, Schattner Y, Sun X Q and Wang C 2018 Phases of a phenomenological model of twisted bilayer graphene *Phys. Rev.* B **98** 075154
[106] Lin C-Y, Lee M-H and Lin M-F 2018 Coulomb excitations in ABC-stacked trilayer graphene *Phys. Rev.* B **98** 041408
[107] Lin C-Y, Huang B-L, Ho C-H, Gumbs G and Lin M-F 2018 Geometry-diversified Coulomb excitations in trilayer AAB stacking graphene *Phys. Rev.* B **98** 195442
[108] Chen C, Hayazawa N and Kawata S 2014 A 1.7 nm resolution chemical analysis of carbon nanotubes by tip-enhanced Raman imaging in the ambient *Nat. Commun.* **5** 3312
[109] Felner I, Asaf U, Levi Y and Millo O 2000 Tuning of the superconducting and ferromagnetic behavior by oxygen and hydrogen in $Eu_{1.5}Ce_{0.5}RuSr_2Cu_2O_{10-\delta}$ *Physica* C **334** 141–51
[110] Cui H-W, Li D-S and Fan Q 2013 Using nano hexagonal boron nitride particles and nano cubic silicon carbide particles to improve the thermal conductivity of electrically conductive adhesives *Electron. Mater. Lett.* **9** 1–5
[111] Wu J-Y, Chen S-C, Gumbs G and Lin M-F 2017 Field-induced diverse quantizations in monolayer and bilayer black phosphorus *Phys. Rev.* B **95** 115411
[112] Do T-N, Gumbs G, Shih P-H, Huang D and Lin M-F L 2017 Valley- and spin-dependent quantum Hall states in bilayer silicene *Phys. Rev.* B **100** 155403
[113] Gumbs G, Iurov A, Wu J-Y, Lin M F and Fekete P 2016 Plasmon excitations of multilayer graphene on a conducting substrate *Sci. Rep.* **6** 21063
[114] Axtell J C, Saleh L M A, Qian E A, Wixtrom A I and Spokoyny A M 2018 Synthesis and applications of perfunctionalized boron clusters *Inorg. Chem.* **57** 2333–50
[115] Liu C-C, Feng W and Yao Y 2011 Quantum spin Hall effect in silicene and two-dimensional germanium *Phys. Rev. Lett.* **107** 076802

[116] Wu J Y, Chen S C and Lin M F 2014 Temperature-dependent Coulomb excitations in silicone *New J. Phys.* **16** 125002

[117] Ravindran S, Chaudhary S, Colburn B, Ozkan M and Ozkan C S 2003 Covalent coupling of quantum dots to multiwalled carbon nanotubes for electronic device applications *Nano Lett.* **4** 447–53

[118] Tran N T T, Lin S-Y, Lin Y-T and Lin M-F 2016 Chemical bonding-induced rich electronic properties of oxygen adsorbed few-layer graphenes *Phys. Chem. Chem. Phys.* **18** 4000

[119] Lin S-Y, Lin Y-T, Tran N T T, Su W-P and Lin M-F 2016 Feature-rich electronic properties of aluminum-adsorbed graphenes *Carbon* **120** 209–18

[120] Lin Y-T, Lin S-Y, Chiu Y-H and Lin M-F 2015 Alkali-created rich properties in grapheme nanoribbons: chemical bondings *Sci. Rep.* **7** 1722

[121] Sharifi N, Ardjmand M, Ahangari M G and Ganji M D 2015 Si-decorated graphene: a superior media for lithium-ions storage *Struct. Chem.* **24** 1473–83

[122] Chen X and Chen B 2015 Macroscopic and spectroscopic investigations of the adsorption of nitroaromatic compounds on graphene oxide, reduced graphene oxide, and graphene nanosheets *Environ. Sci. Technol.* **49** 6181–9

[123] Li G, Luican A, Lopes dos Santos J M B, Castro Neto A H, Reina A, Kong J and Andrei E Y 2010 Observation of van Hove singularities in twisted graphene layers *Nat. Phys.* **6** 109–13

IOP Publishing

Shih-Yang Lin, Hsin-Yi Liu, Duy Khanh Nguyen, Ngoc Thanh Thuy Tran, Hai Duong Pham, Shen-Lin Chang, Chiun-Yan Lin and Ming-Fa Lin

Chapter 10

Silicon adsorption and substitution on graphene

10.1 Introduction

Carbon, one of the most versatile elements on our planet, can form many stable polymorphs in various dimensions owing to the diverse orbital hybridization in its unique electronic configuration. The carbon-based structures include 3D diamond [1], 3D graphites [2], 2D graphene [3], 1D graphene nanoribbons [4], cylindrical carbon nanotubes [5], carbon toroids, C_{60}-related fullerenes [6] and carbon onions [7]. These versatile morphologies demonstrate rich and unusual chemical bonds and dominant sp^2 bonding on the various symmetric surfaces, except for the sp^3 surfaces in diamond structure. Few- and multi-layer graphene systems have been synthesized successfully using different experimental methods [8, 9] since the first discovery of graphene in 2004 through mechanical exfoliation [10]. They exhibit a variety of novel and unusual properties as a consequence of their hexagonal symmetry, one-atom thickness and stacking configurations, e.g. semi-conducting or semi-metallic behaviors [11, 12], anomalous quantum Hall conductivities [13], diverse magnetic quantization phenomena [14], rich Coulomb excitations and decays [15], distinct magneto-optical selection rules [16], super-high mobility of charge carriers [17] and a very large Young's modulus [18]. To vary the fundamental properties and increase the potential for applications greatly, electronic states can be manipulated easily through the layer number [19], stacking configuration [20], deformation [21], sliding [22], twisting [23], gate voltage [24], magnetic field [25], and adatom adsorption [26, 27] and substitution [28–30].

Tuning fundamental properties is one of the critical topics in materials science, chemistry, physics and engineering. Chemical modification methods are the most effective. It is well known that pristine graphene possesses very strong σ bonding in a honeycomb lattice, leading to a quite active chemical environment with many $2p_z$-dangling bonds. The host carbon atoms appear to be able to couple with various

doi:10.1088/978-0-7503-3299-6ch10

adatoms [31], molecules [32] and even functional groups [33]. Many theoretical and experimental studies on surface reconstructions have shown adsorption-varied physical and chemical phenomena, such as the opening of observable band gaps [34], semiconductor–metal transitions, the absence or recovery of the Dirac-cone bands [35], spin-split energy bands under ferromagnetic configurations of specific adatoms [36], diverse van Hove singularities [37], and single- or multi-orbital hybridization in C–adatom bonds [38]. Interestingly, graphite intercalation compounds are the most efficient anode materials in Li^+-based batteries. Furthermore, battery efficiency can be enhanced by adding certain silicon materials, as identified from experimental measurements [39]. Si adsorption might play an important role in such potential applications. Another chemical modification is direct doping, the substitution of host carbon atoms by adatoms in the honeycomb lattice. To date, $B_xC_yN_z$ compounds have been produced successfully in the form of the stable structures of 1D cylindrical nanotubes [40] and 2D planar layers [41]. Their fundamental properties are very sensitive to changes in the components of different elements, e.g. the component-enriched electronic structures, optical absorption spectra [42] and transport properties [43]. The above-mentioned chemical modifications on graphene systems are worthy of detailed investigation. In particular, the complicated and strong Si–C bonds corresponding to, respectively, the adsorption and substitution cases, will induce important differences in orbital hybridization and thus varied fundamental properties [44].

Interestingly, according to the theoretical predictions of the first-principles method [45] and the tight-binding model [46], Si-substituted graphenes are highly suitable for exploring novel physical phenomena fully. These systems have demonstrated diverse electronic properties and optical excitations under magnetic quantization, which are absent in other condensed-matter systems. The low-lying valence and conduction bands are characterized well by the chemical bonding of C-2p_z and Si-3p_z orbitals. From the generalized tight-binding model and the Kubo formula, there are four kinds of highly degenerate Landau levels based on the spatial probability distributions and oscillation modes on the different substitution-enlarged sublattices and their complicated relations. Such magneto-electronic states lead to unusual magneto-optical selection rules. It should be noted that the generalized theoretical framework takes into account all the critical factors simultaneously, including non-uniform bond lengths, different site energies, various hopping integrals and external field effects without perturbation forms.

In this light, we focus on a theoretical framework which is suitable for atom adsorption and substitution simultaneously. This is demonstrated clearly by the silicon-doped graphene systems under first-principles calculations. After thorough analyses, diverse physical and chemical pictures, multi-orbital hybridization without spin configurations, will be proposed to explain the adsorption- and substitution-enriched honeycomb lattices, electronic energy spectra, spatial charge densities, and atom- and orbital-projected van Hove singularities. All the calculated results are consistent with one another under such mechanisms. Furthermore, they can account for the semiconductor–metal transitions during adatom adsorption and opening of the band gap in the substitution cases. The pure π and σ bonding, which are, respectively, created by the $2p_z$ and ($2s$, $2p_x$, $2p_y$) orbitals of carbon atoms, are changed dramatically by chemical modifications.

Whether the pure π and σ bonds become hybridized bonds or are absent completely is one focus of study. Various experimental measurements are available to verify the theoretical predictions such as, scanning tunneling microscopy (STM), transmission electron microscopy (TEM), low-energy electron diffraction (LEED) [47–49] for verifying structural properties, angle-resolved photoemission spectroscopy (ARPES) [50] for verifying electronic band structures and scanning tunneling spectroscopy (STS) experiments [51] for verifying density of states.

10.2 Silicon-adsorbed graphene systems

10.2.1 The geometric structure of Si-adsorbed graphene systems

The 2D layered carbon structure called graphene is made of a single-atom-thick hexagonal lattice of sp^2-bonded carbon atoms, in which the in-plane strong σ bonds of the carbon atoms creates a very flat 2D carbon layer that differs greatly from the buckled lattice in other 2D layered systems [52–54]. Clearly, the critical chemical bonds in graphene originate from the 2s, $2p_x$ and $2p_y$ orbitals and the $2p_z$ orbitals of the carbon atoms, where the former and the latter are responsible for the strong σ bonds and the weak π bonds, respectively. On the other hand, the competition between the sp^2 and sp^3 chemical bonds determines the geometric stability directly in the form of the buckled structures in the other group-IV layered systems. Furthermore, the planar C–C bond lengths of 1.42 Å are the shortest among all the group-IV layered systems. Under Si adsorption effects on the graphitic surface, the bridge site remains the most optimal adsorption position for the top and hollow sites, as illustrated in figure 10.1. This is very different in other adatom-adsorbed graphene systems owing mainly to the various orbital hybridization cases in the adatom–C bonds. A planar honeycomb structure can remain even in the presence of Si adatom adsorption, while the planar C–C bond lengths are extended to about 1.45–1.49 Å under the various adatom concentrations shown in table 10.1. This indicates clearly that the planar C–C bonds are low-powered since the multi-orbital hybridization in the Si–C bonds is shared by part of the carbon electrons.

10.2.2 The band structure of Si-adsorbed graphene systems

The first Brillouin zones for various unit cells are shown in figure 10.2, and are very useful for understanding the elongation effects. Such Brillouin zones schemes are used to represent in the energy–wave-vector space. The pristine monolayer graphene systems present band structures, as illustrated in figures 10.3(a) and (b), in which the strong σ bonds composed of C-2s, -$2p_x$ and -$2p_y$ orbitals result in asymmetry between the occupied valence bands and the unoccupied conduction bands. The occupied and unoccupied bands at low-lying energies cross linearly at the Dirac points so that isotropic Dirac-cone structures arise at the K and Γ points in the primitive and enlarged unit cells. Clearly, such pristine graphene systems belong to the zero-gap semiconductors because of the vanishing of the density of states at the Fermi level. Furthermore, the low-lying energy bands arise mainly from the free-

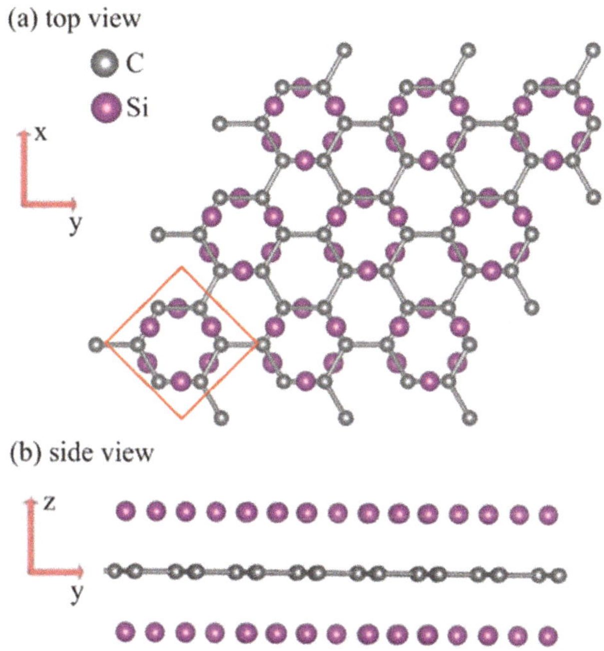

Figure 10.1. (a) Top view and (b) side view of the geometric structure of Si-adsorbed monolayer graphene [44]. The bridge site is the most optimal adsorption position for Si adatoms for the top and hollow sites.

Table 10.1. The energy gap E_g (eV)/semimetal (SM)/metal (M), C–C bond length (Å), Si–C bond length (Å) and Si height (Å) for Si-adsorbed graphene [44].

Configuration	Ratio of Si and C	Percentage	$E_g^{d(i)}$ (eV)/semimetal (SM)/metal (M)	C–C (Å)	Si–C (Å)	Si height (Å)
Pristine graphene	—	—	$E_g^d = 0$	1.42	—	—
Adsorption	Si:C = 6:6	100%	SM	1.492	2.535	2.422
	Si:C = 24:24	100%	SM	1.492	2.535	2.422
	Si:C = 3:6	50%	SM	1.490	2.514	2.400
	Si:C = 12:24	50%	SM	1.490	2.514	2.400
	Si:C = 1:6	16.6%	M	1.456	2.138	2.000
	Si:C = 1:8	12.5%	M	1.450	2.118	1.989
	Si:C = 1:24	4%	M	1.454	2.125	1.996
	Si:C = 1:32	3%	M	1.452	2.116	1.998

standing π bonds of the perpendicular C-2p$_z$ orbitals. The parabolic energy dispersions will replace the linear dispersions gradually when the valence and conduction state energies ($E^{c,v}$) increase. Moreover, there exist saddle points at the M point at −2.4 and 1.8 eV in the valence and conduction parabolic bands, respectively. The extreme point of parabolic energy dispersion initiated at the Γ

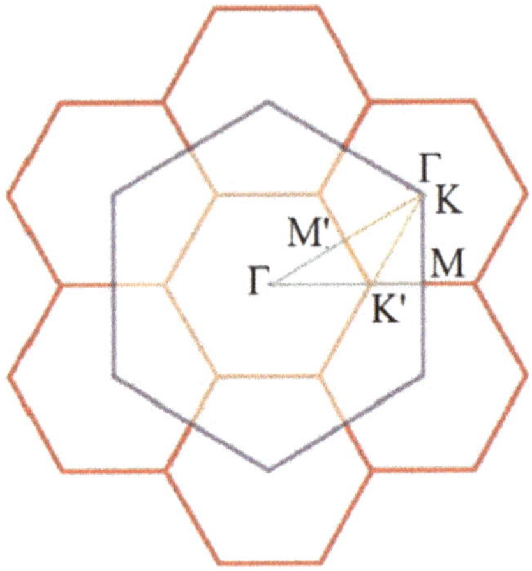

Figure 10.2. Brillouin zone scheme for various unit cells. ([55] Copyright 2019, Springer Nature. With permission of Springer.)

point in the valence band is situated at −3 eV, in which its electronic states is made up of C-2s, -2p_x and -2p_y orbitals.

Under Si adsorption effects, the electronic band structures present a major shift. The asymmetry in the occupied valence and unoccupied conduction bands becomes more obvious around the Fermi level. For the 100% Si double-sided adsorption case, the occupied and unoccupied bands intersect simultaneously with the Fermi level so that such a system can be regarded as a semimetal, as shown in figure 10.3(c). Obviously, the Dirac-cone structure is deformed at the band-edge state, in which the valence and conduction Dirac points are detached by about 0.5 eV. The highest occupied state between the bottom of the conduction band and the Fermi level can be regarded as free electrons. Meanwhile, the lowest unoccupied valence states between the top of the valence band and the Fermi level belong to free holes. Thus, such a system can be considered as a semimetal. In the 100% Si single-sided adsorption case (figure 10.3(d)), there exist free electrons and holes simultaneously, identical to the double-sided adsorption case (figure 10.3(c)). However, its free carriers decline since the number of energy bands that cross the Fermi level decrease under the reduction of adatom concentration. In addition, both C atoms and Si adatoms contribute strongly to most of the energy range in the band structures (figures 10.3(c) and (d)), where their dominance is illustrated by the red triangles and blue circles, respectively. In the single-adatom adsorption cases (figures 10.3(e) and (f)), only the lowest unoccupied valence states remain, thus these systems can be regarded as a metal within free holes or a p-type metal. From a detailed evaluation of the atom- and orbital-dominated energy bands, C atoms and Si adatoms co-dominate in most of the energy bands, where part of them can originate from either C atoms or Si adatoms under the 100% Si double-sided and single-sided adsorption cases. Altogether, the Si adatoms will dominate the low-lying and middle-lying energy bands strongly. The atom contribution in the sp^3–p bonds of Si–C is judged

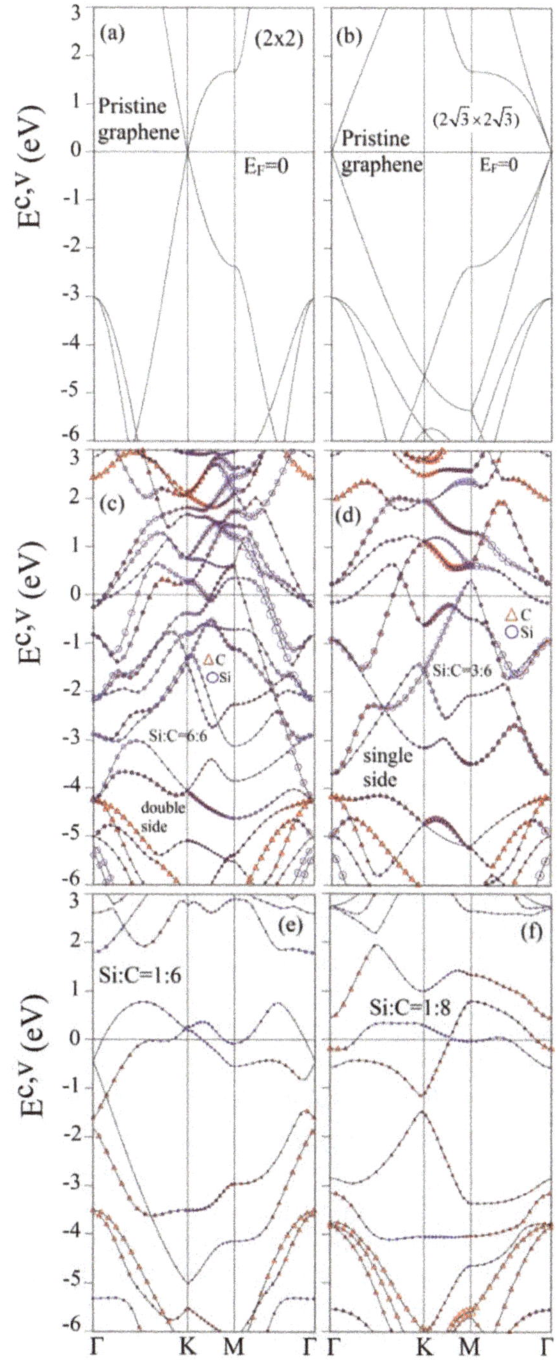

Figure 10.3. Band structures with dominance of C atoms and Si adatoms for Si-adsorbed graphene systems: (a) the pristine case, (b) pristine graphene under an enlarged unit cell identical to the 100% case, (c) 100% Si double-sided adsorption, (d) 100% Si single-sided adsorption, (e) 16.6% Si adsorption and (f) 12.5% Si adsorption. The red triangles and blue circles illustrate the dominance of the C atoms and Si adatoms, respectively [44].

to be about 4:1 in the Si 100% and 2:1 in Si 50% adsorption cases, as shown in figures 10.3(c) and (d), respectively. Such energy bands are contributed significantly by the four orbitals of Si-3s, -3p$_x$, -3p$_y$ and -3p$_z$ and the single orbital of C-2p$_z$. For the single-adatom cases (figures 10.3(e) and (f)), many energy bands are dominated solely by C atoms while only a few Si- and C-co-dominated bands remain at certain energies. Furthermore, it should be mentioned that the σ bands of C atoms are located at $E^v = -3.5$ and -3.2 eV, as shown in figures 10.3(e) and (f), respectively. This is evidence for shorter C–C bond lengths under single-adatom adsorption (figures 10.3(e) and (f)) compared with 100% adatom adsorption (figures 10.3(c) and (d)).

10.2.3 Spatial charge density of Si-adsorbed graphene

The spatial charge density (ρ) and the variation in the spatial charge density ($\Delta\rho$) are the critical physical quantities for understanding the orbital hybridization mechanism in chemical bonds. $\Delta\rho$ is constructed by subtracting the charge density of the pristine system and the isolated adatoms from composite systems. A pristine monolayer graphene system produces a very high charge distribution between C atoms, shown by the red region enclosed by a black rectangle in figure 10.4(a). This demonstrates the strong σ bond made of C-2s, -2p$_x$ and -2p$_y$ orbitals on the graphitic plane. The strong σ bonds are unbreakable under various Si adsorption cases, as shown in figures 10.4(b), (d) and (f). Away from the central line between the C atoms, the charge distribution is much lower, illustrated by the red rectangle in figure 10.4(a), in which it can be regarded as the π bonds of C-2p$_z$ orbitals. The weak π bonds are very sensitive to various Si adsorption cases, as indicated in figures 10.4(b), (d) and (f). The variation in the spatial charge distributions related to Si adatoms and C atoms is demonstrated clearly, as shown by $\Delta\rho$ in figures 10.4(c), (e), and (g). The charge density is strengthened close to the C atoms along the z-direction (red areas), demonstrating the transfer of some electrons from Si adatoms to C atoms. Moreover, there exists variation in the charge distribution along the x- and y-directions between the Si adatoms and C atoms/Si adatoms, confirming the (3s, 3p$_x$, 3p$_y$, 3p$_z$)–2p$_z$ and (3s, 3p$_x$, 3p$_y$, 3p$_z$)–(3s, 3p$_x$, 3p$_y$, 3p$_z$) orbital hybridization in the Si–C bonds and Si–Si bonds, respectively.

10.2.4 Density of states of Si-adsorbed graphene systems

The atom- and orbital-projected density of states (DOS) is a very useful physical quantity to provide a clear evidence of important features in the energy–wave-vector space. The critical points in the band structures create van Hove singularities in the DOS, where the saddle points, the local minima and the local maxima arise from the band-edged states. Generally, the DOS of a pristine graphene system exhibits a shoulder structure at −3 eV, a V-shaped structure in low-lying energies and divergent peaks located at −2.4 and 1.8 eV, which are symmetric in opposite directions centered at the Fermi level, as illustrated in figure 10.5(a). Such unique structures in the DOS result from the special features in the band structures, such as, respectively, the extreme points at the Γ point, the anisotropic Dirac-cone structure at the K point and the saddle points at the M points. In particular, the latter two

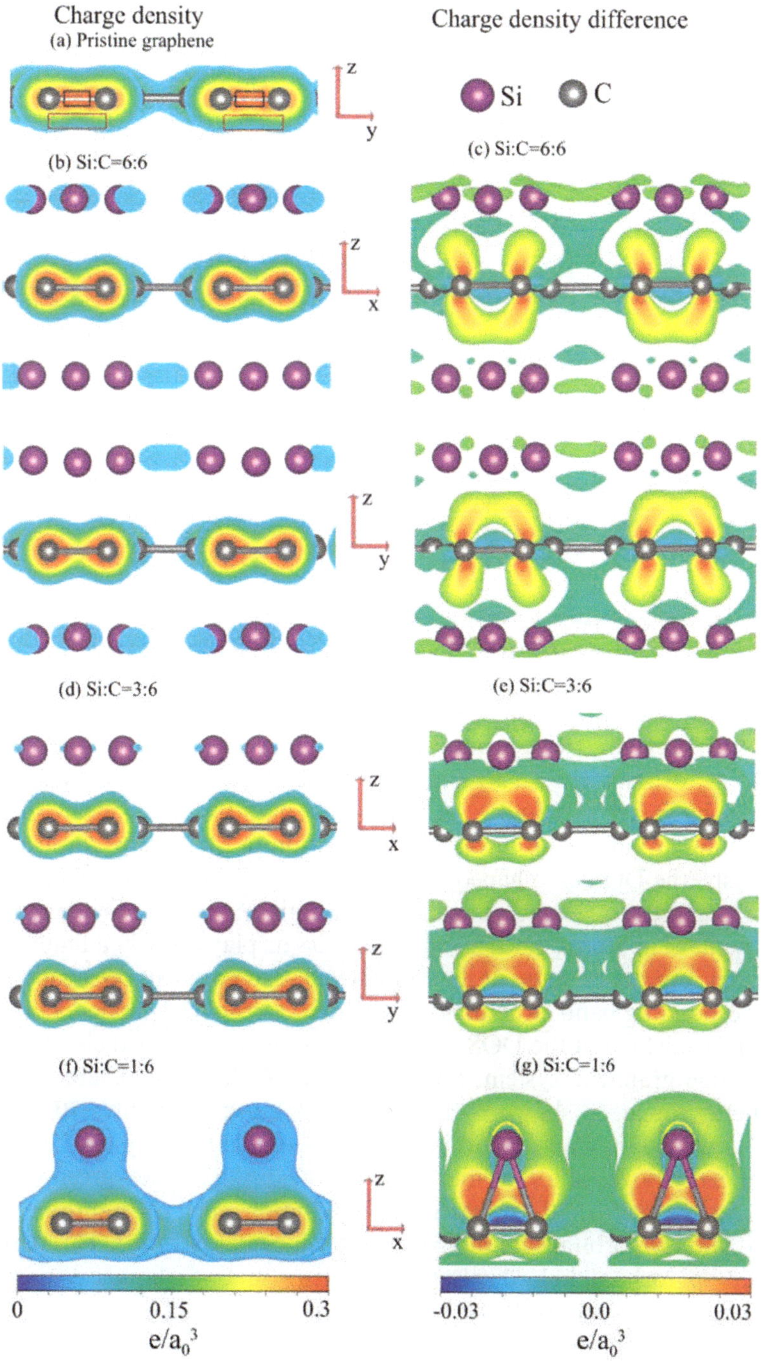

Figure 10.4. Spatial charge density for (a) pristine graphene; spatial charge density (left) and charge density difference (right) for (b) and (c) 100% Si double-sided adsorption, (d) and (e) 100% Si single-sided adsorption, and (f) and (g) 16.6% single-adatom adsorption [44].

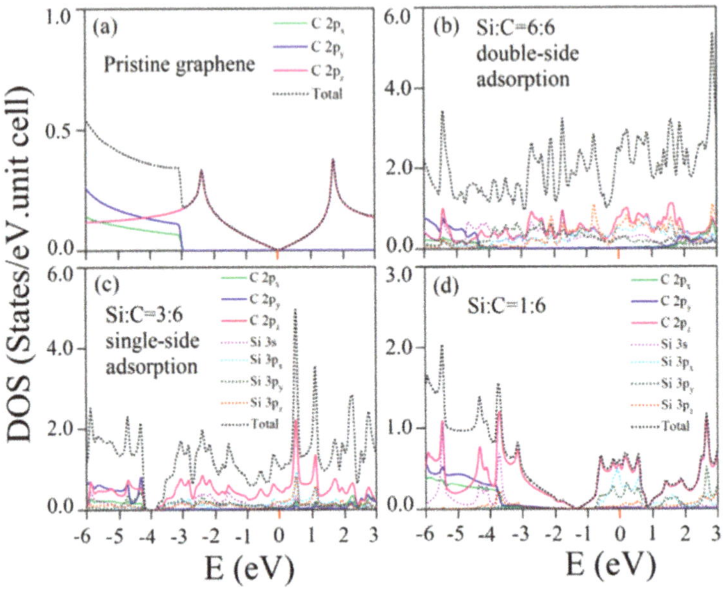

Figure 10.5. Atom- and orbital-projected DOS for (a) the pristine case, (b) 100% Si double-sided adsorption, (c) 100% Si single-sided adsorption and (d) 16.6% single-adatom adsorption [44].

structures arise from the π bands of C-2p$_z$ orbitals, as shown by the solid pink curve in figure 10.5(a). The extreme points at the Γ point are generated by the σ bands of C-2p$_x$ and -2p$_y$ orbitals shown by the solid green and blue curves, respectively. In addition, such π orbitals and σ orbitals are separated from each other clearly, demonstrating an unique sp^2 orbital hybridization in the planar graphene sheet.

Obviously, the DOS is reshaped fully under various Si adsorption configurations, as shown in figures 10.5(b)–(d), in which most of the extra van Hove singularities are created by the Si adatoms. Such a reshaped DOS displays clearly a finite existence of the electronic density of states at the Fermi level, that confirms the conducting states in its corresponding band structures. In contrast, zero-gap semiconducting behavior leads to the vanishing of the DOS at the Fermi level, as observed clearly in the DOS of the pristine graphene system. Specifically, the C-2p$_z$ orbitals are present in the whole energy range, as shown by the solid pink curves in figures 10.5(b)–(d). The four Si-3s, -3p$_x$, -3p$_y$ and 3p$_z$ orbitals contribute significantly to the wide-energy range initiated above −4 eV, as illustrated by the dashed violet, green, blue and red curves in figures 10.5(b)–(d), respectively. In addition, such C-2p$_z$ orbitals and Si-3s, -3p$_x$, -3p$_y$ and -3p$_z$ orbitals are merged in the wide-energy range. This demonstrates that the p–sp^3 orbital hybridization mechanism in C–Si bonds is present in Si-adsorbed graphene systems. It should be mentioned that the weak π bonds of C-2p$_z$ orbitals are very sensitive to the multi-orbital hybridization in the adatom–C bonds. On the other hand, the Si adsorption effects barely modify the strong σ bonds of the C-2p$_x$ and -2p$_y$ orbitals, where such pristine σ orbitals are identified by their isolated shoulder structure at −4.2, −4.1 and −3.6 eV in the 100% double-sided adsorption (figure 10.5(b)), the 100% single-sided adsorption (figure 10.5(c)) and the single-adatom adsorption (figure 10.5(d)), respectively. Clearly, the pristine σ orbital induced isolated shoulder structure of the single-adatom adsorption case is located at a shallower energy than in the former two cases. This implies that the C–C chemical bonds in the single-adatom adsorption case are stronger than those in the 100% Si double-sided and single-sided adsorption cases.

10.3 Silicon-substituted graphene systems

10.3.1 The geometric structure of Si-substituted graphene systems

The 2D planar structure can be present in silicon–carbon honeycomb lattices even under various adatom substitution cases. This indicates that the strong quasi-σ chemical bonds remain in the Si–C bonds, originating from the in-plane sp^2–sp^2 orbital hybridization. Under various substitution cases, the C–C and Si–C bond lengths vary between about 1.35–1.47 Å and 1.62–1.83 Å, respectively, as shown in table 10.2. The fully Si-substituted graphene system presented in figure 10.6(a), the pristine silicon–carbon compound, is stabilized with the optimal Si–C bond length of 1.78 Å, which is much longer than that in graphene. This indicates that the strength in the quasi-σ Si–C bonds declines greatly, which can be figured out by a significant decrease in the spatial charge density in Si–C bonds. Moreover, the various adatom distribution configurations under specific adatom concentrations include the ortho-, para- and meta-substituted cases shown in figures 10.6(b) and (c) and the single-adatom substituted case shown in figure 10.6(d), in which the ortho- and para-substituted cases are degenerate only at 50% concentration, illustrated in figure 10.6(b).

10.3.2 Electronic band structure of Si-substituted graphene systems

The electronic band structures are reshaped completely in the presence of various Si substitution configurations, as shown in figures 10.7(a)–(f). The valence and conduction bands are obviously asymmetric about the Fermi level under various substitution cases, e.g. the 100% substitution configuration in figure 10.7(a), the 50% ortho and meta substitution configurations in figures 10.7(b) and 10.7(c), and the

Table 10.2. Energy gap E_g (eV)/metal (M)/semimetal (SM), C–C bond length (Å) and Si–C bond length (Å) for Si-substituted graphene systems [44].

Configuration	Ratio of Si and C	Percentage	$E_g^{d(i)}$ (eV)/semimetal (SM)/metal (M)	C–C (Å)	Si–C (Å)
Substitution	Si:C = 3:3	100%	$E_g^i = 2.56$	—	1.78
	Si:C = 2:4 ortho	50%	$E_g^d = 0.04$	1.478	1.833
	Si:C = 2:4 meta	50%	$E_g^d = 0.56$	1.465	1.829
	Si:C = 1:5 single-adatom	20%	$E_g = 0$	1.351	1.621
	Si:C = 1:17 single-adatom	5.8%	$E_g = 0$	1.372	1.631
	Si:C = 1:23 single-adatom	4.3%	$E_g = 0$	1.379	1.649

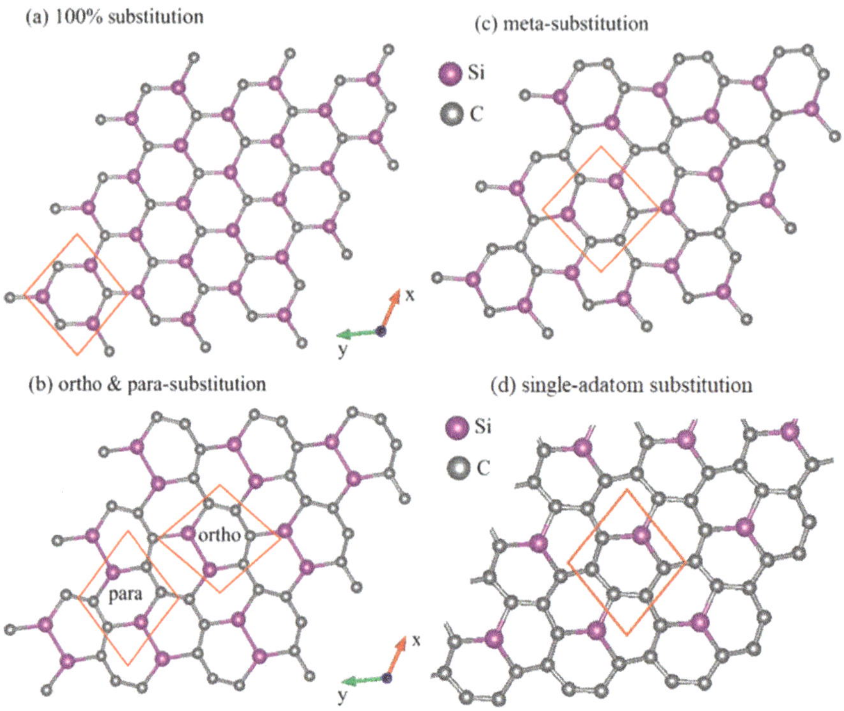

Figure 10.6. Top-view structures of Si-substituted graphene systems for (a) 100% Si substitution, (b) 50% Si ortho/para substitution, (c) 50% Si meta substitution and (d) single-adatom substitution [44].

20%, 5.8% and 4.3% single-adatom substitution configurations in figures 10.7(d), (e) and (f), respectively. The band structures under various substitution and adsorption cases differ completely from each other, as shown in figures 10.7 and 10.3, respectively. Most of the various substitution configurations are semiconductors within the finite or vanished energy gaps. The full Si substitution configuration presents semiconducting behavior with a direct band gap of 2.56 eV. Direct or indirect band gaps are determined by the highest valence band and the lowest conduction band at the band-edge state. The magnitude of the band gaps decreases when the Si substitution concentrations decrease and they are varied sharply by the various adatom distribution configurations, e.g. the zero energy gap in the 50% ortho substitution configuration presented in figure 10.7(b) and the direct energy gap of 0.56 eV in the 50% meta substitution configuration shown in figure 10.7(c). In addition, the anisotropic Dirac-cone structure occurs at certain k-points between Γ and K, as shown in figure 10.7(b), in which only one Dirac point intersects with the Fermi level, so that its electronic DOS becomes invisible. Under the single-adatom substitution cases, semiconducting behavior within a vanished band gap arises, where only one Dirac point in the conduction band intersects with the Fermi level near the Γ point, as illustrated in figures 10.7(d)–(f). Furthermore, the linear Dirac-cone structure at the Γ point in pristine systems (figures 10.3(a) and (b)) is separated and distorted significantly (figure 10.7(b)) or even destroyed completely

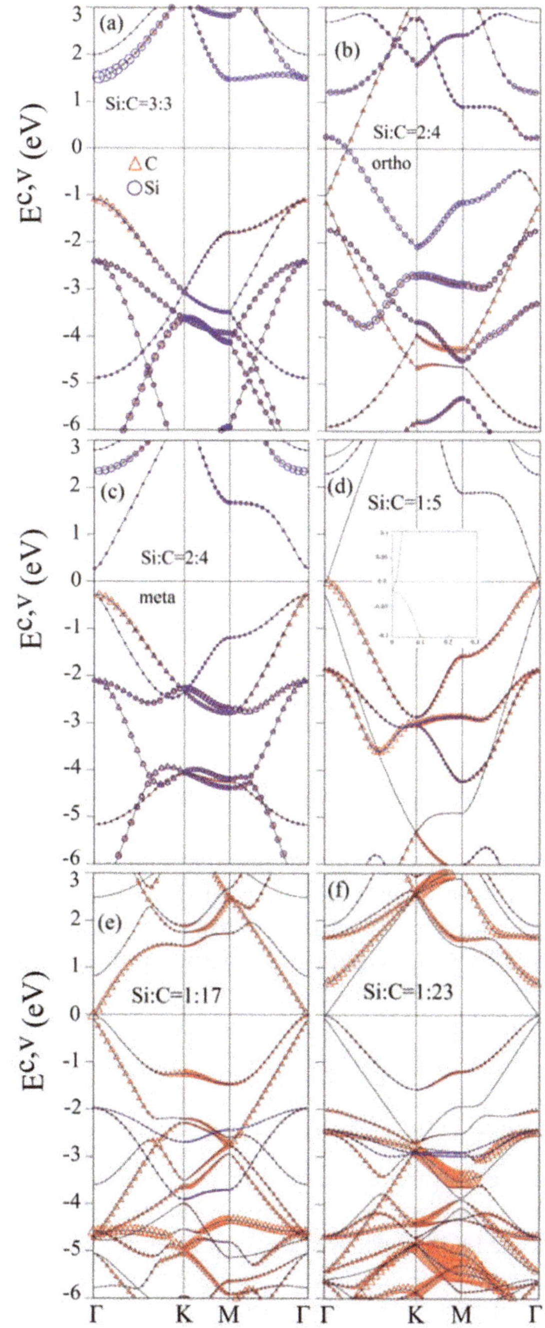

Figure 10.7. The Si adatom- and C atom-dominated electronic band structures of Si-substituted graphene systems for (a) 100% Si substitution, (b) 50% Si ortho substitution, (c) 50% Si meta substitution, (d) 20% single-adatom substitution, (e) 5.8% single-adatom substitution and (f) 4.3% single-adatom substitution. The blue circles and red triangles represent the dominance of the Si adatoms and C atoms, respectively [44].

(figures 10.7(a) and (c)). Also, this Dirac-cone structure is disfigured significantly without separation under the single-adatom substitution cases shown in figures 10.7(d)–(f). The energy bands near the Fermi level are initiated from the Γ point, but are almost independent of the M and K points. All the energy bands are co-dominated mainly by the Si adatoms and C atoms under high substitution concentrations, as illustrated by the blue circles and red triangles in figures 10.7(a)–(c), respectively. However, the Si adatom and C atom co-dominated energy bands remain only at certain energies under the single-adatom substitution shown in figures 10.7(d)–(f). Moreover, only the σ bands due to C-2p_x and 2p_y orbitals disappear completely. This testifies clearly to the existence of quasi-σ and quasi-π bonds arising from (3s, 3p_x, 3p_y)–(2s, 2p_x, 2p_y) and 3p_z–2p_z orbital hybridization, respectively.

10.3.3 Spatial charge density of Si-substituted graphene

The spatial charge density is varied greatly by the different Si substitution configurations. The Si–C bonds in the full Si substitution configuration exhibit a high charge density distribution between two neighboring atoms, illustrated in figure 10.8(a). Planar silicon–carbon honeycomb structures are created by the quasi-σ bonds, where their spatial charge strength drops significantly compared to the pristine σ bonds in graphene systems shown in figure 10.4(a). This reduced charge strength matches the longer Si–C bonds of 1.78 Å, compared to the shorter C–C bonds of 1.42 Å. This reduced charge strength is also revealed in other substitution cases, e.g. 50% substitution under the ortho and meta configurations in figures 10.8(c) and (e) and 20% substitution under the single-adatom configuration in figure 10.8(g), respectively. In addition, quasi-π bonds of the perpendicular orbitals also arise near the boundary. They present non-well-behaved charge distributions compared to the pristine graphene system in figure 10.4(a). As a result, the charge strength in the quasi-π bonds is much weaker than that in the quasi-σ bonds. The obvious variation of charge density on the (z, y)- and (z, x)-planes, shown in figures 10.8(b), (d), (f) and (h), suggests significant (3s, 3p_x, 3p_y)–(2s, 2p_x, 2p_y) and 3p_z–2p_z orbital hybridization in the Si–C bonds. The coexistence of multi- and single-orbital interactions are further supported by the atom- and orbital-projected DOSs.

10.3.4 Density of states of Si-substituted graphene systems

All the electronic band features and the critical orbital hybridization mechanism in the chemical bonds of Si-substituted graphene systems can be verified directly by the DOS. Vanishing of the DOS appears in the vicinity of the Fermi level or at the Fermi level, as shown in figure 10.9, resulting from the finite- or zero-gap semiconducting behaviors in the electronic band structures, respectively. In addition, the special characteristics in the DOS are produced by the van Hove singularities, including the distorted V-shaped structure across the Fermi level and prominent asymmetric peaks. Clearly, the DOS shows that the Si-3p_z and C-2p_z orbitals emerge simultaneously at low-lying energies, as illustrated by the dashed red curves and solid pink curves in figures 10.9(a)–(d). Moreover, the Si-3s, -3p_x and -3p_y and C-2s, -2p_x and

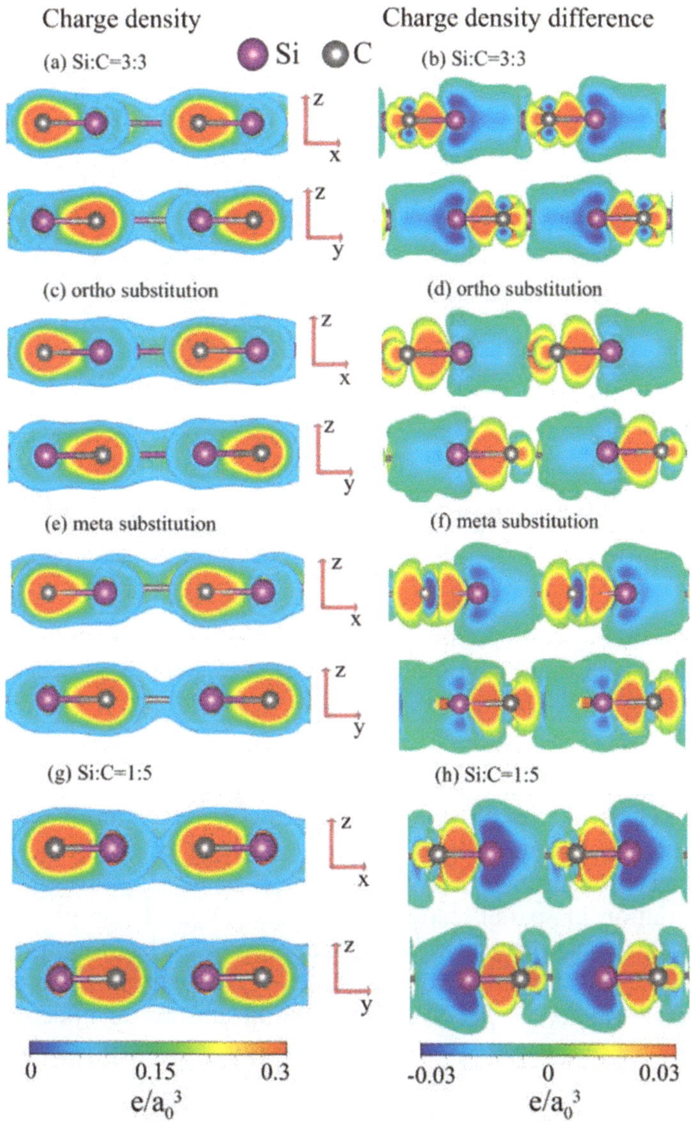

Figure 10.8. Total charge density (left) and its variation (right) for the (a) and (b) 100% substitution configuration, (c) and (d) 50% ortho substitution configuration, (e) and (f) 50% meta substitution configuration, and (g) and (h) 20% single-adatom substitution configuration, respectively [44].

-2p_y orbitals also present a distinct combination at the wide-energy range in the valence band. This shows clearly that the unique orbital hybridization mechanism of p–p and sp^2–sp^2 is formed in the Si–C bonds. As a result, this p–p and sp^2–sp^2 hybridization mechanism confirms that the Si-substituted graphene systems are governed by quasi-π and quasi-σ chemical bonds.

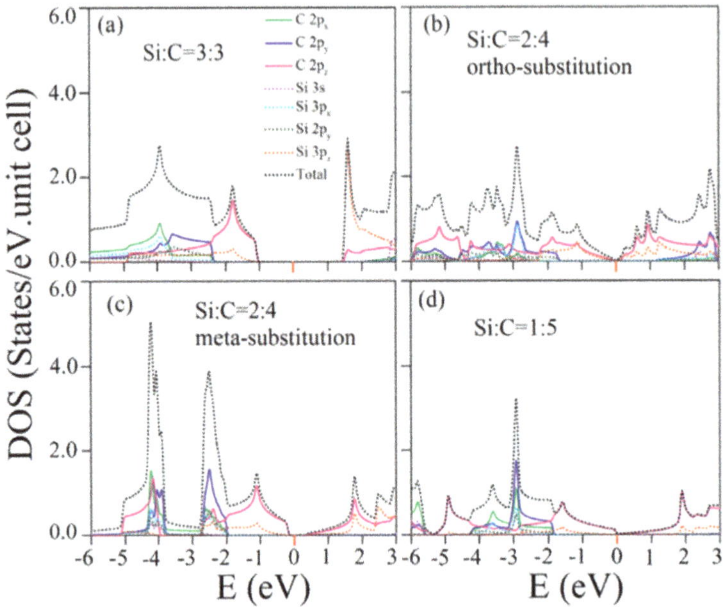

Figure 10.9. Atom- and orbital-projected DOS for Si-substituted graphene systems: (a) 100% substitution configuration, (b) 50% ortho substitution configuration, (c) 50% meta substitution configuration and (d) 20% substitution configuration [44].

10.4 Conclusion

The feature-rich properties of Si-doped graphene systems are investigated using a complete theoretical framework under first-principles calculations, including the adatom-varied geometric structures, electronic band structures, spatial charge densities, variation in spatial charge density and DOS. These physical quantities undergo a dramatic change under various doping cases. The Si-induced feature-rich properties are revealed by the proposed multi-orbital hybridization mechanism in the critical chemical bonds. The Si adsorption and substitution effects in monolayer graphene systems result in free carriers and energy gaps, respectively. In addition, semiconducting, metallic and semi-metallic states are also revealed in Si-doped graphene systems. On the geometric side, the Si-adsorbed graphene systems present a planar graphitic structure with the Si-adsorbed position optimized at the bridge site for the hollow and top sites. Such a planar structure shows minor change in the σ bonds of the C-2s, -2p_x and -2p_y orbitals, in which the C-2s, -2p_x and -2p_y orbitals do not participate in Si–C bonding. The Si–C bond lengths vary from 2.1 to 2.5 Å, which can be well-suited to form significant multi-orbital hybridization. The C–C bond lengths exhibit very little variation. Through detailed analyses and studies of the spatial charge density, charge density difference and DOS, the sp^3–p orbital hybridization mechanisms are assumed to govern Si-adsorbed graphene systems. For the Si-substituted monolayer graphene systems, the planar geometric structure continues to exist. As a result, strong Si–C bonds are formed in the Si-substituted structures, where the Si–C bond lengths are in the range of 1.62–1.83 Å, which is much shorter than those in the Si-adsorbed structures. This shows that there are more C orbitals hybridized with the Si orbitals. Therefore, the sp^2–sp^2 and p–p orbital hybridization mechanism in Si–C bonds is deduced to dominate in Si-substituted graphene systems. The Si-induced rich fundamental properties have great potential for nanoscale applications. The complete theoretical framework developed using first-principles calculations can be generalized thoroughly for the doping effects in emergent materials.

References

[1] Williams O A 2011 Nanocrystalline diamond *Diam. Relat. Mater.* **20** 621–40

[2] Wallace P R 1947 The band theory of graphite *Phys. Rev.* **71** 622

[3] Novoselov K S, Fal'ko V I, Colombo L, Gellert P R, Schwab M G and Kim K 2012 A roadmap for graphene *Nature* **490** 192

[4] Han M Y, Özyilmaz B, Zhang Y and Kim P 2007 Energy band-gap engineering of graphene nanoribbons *Phys. Rev. Lett.* **98** 206805

[5] Wilder J W, Venema L C, Rinzler A G, Smalley R E and Dekker C 1998 Electronic structure of atomically resolved carbon nanotubes *Nature* **391** 59

[6] Dresselhaus M S, Dresselhaus G, Rao A M and Eklund P C 1996 Optical properties of C60 and related materials *Synth. Met.* **78** 313–25

[7] Sano N, Wang H, Chhowalla M, Alexandrou I and Amaratunga G A J 2001 Synthesis of carbon 'onions' in water *Nature* **414** 506–7

[8] Faugeras C *et al* 2008 Few-layer graphene on SiC, pyrolitic graphite, and graphene: a Raman scattering study *Appl. Phys. Lett.* **92** 011914

[9] Jo G *et al* 2010 Large-scale patterned multi-layer graphene films as transparent conducting electrodes for GaN light-emitting diodes *Nanotechnology* **21** 175201

[10] Li D and Kaner R B 2008 Graphene-based materials *Science* **320** 1170–1

[11] Koshino M 2010 Interlayer screening effect in graphene multilayers with ABA and ABC stacking *Phys. Rev.* B **81** 125304

[12] Decker R *et al* 2011 Local electronic properties of graphene on a BN substrate via scanning tunneling microscopy *Nano Lett.* **11** 2291–5

[13] Ostrovsky P M, Gornyi I V and Mirlin A D 2008 Theory of anomalous quantum Hall effects in graphene *Phys. Rev.* B **77** 195430

[14] Huang Y K, Chen S C, Ho Y H, Lin C Y and Lin M F 2014 Feature-rich magnetic quantization in sliding bilayer graphenes *Sci. Rep.* **4** 7509

[15] Shin S Y, Hwang C G, Sung S J, Kim N D, Kim H S and Chung J W 2011 Observation of intrinsic intraband π-plasmon excitation of a single-layer graphene *Phys. Rev.* B **83** 161403

[16] Ho Y H, Chiu Y H, Lin D H, Chang C P and Lin M F 2010 Magneto-optical selection rules in bilayer Bernal graphene *ACS Nano* **4** 1465–72

[17] Wei D and Kivioja J 2013 Graphene for energy solutions and its industrialization *Nanoscale* **5** 10108–26

[18] Jiang J W, Wang J S and Li B 2009 Young's modulus of graphene: a molecular dynamics study *Phys. Rev.* B **80** 113405

[19] Lee J, Novoselov K S and Shin H S 2010 Interaction between metal and graphene: dependence on the layer number of graphene *ACS Nano* **5** 608–12

[20] Zhang Z, Huang H, Yang X and Zang L 2011 Tailoring electronic properties of graphene by π–π stacking with aromatic molecules *J. Phys. Chem. Lett.* **2** 2897–905

[21] Min K and Aluru N R 2011 Mechanical properties of graphene under shear deformation *Appl. Phys. Lett.* **98** 013113

[22] Feng X, Kwon S, Park J Y and Salmeron M 2013 Superlubric sliding of graphene nanoflakes on graphene *ACS Nano* **7** 1718–24

[23] Brown L, Hovden R, Huang P, Wojcik M, Muller D A and Park J 2012 Twinning and twisting of tri- and bilayer graphene *Nano Lett.* **12** 1609–15

[24] Oostinga J B, Heersche H B, Liu X, Morpurgo A F and Vandersypen L M 2008 Gate-induced insulating state in bilayer graphene devices *Nat. Mater.* **7** 151

[25] Zhang Y *et al* 2006 Landau-level splitting in graphene in high magnetic fields *Phys. Rev. Lett.* **96** 136806

[26] Chan K T, Neaton J B and Cohen M L 2008 First-principles study of metal adatom adsorption on graphene *Phys. Rev.* B **77** 235430

[27] Xu C, Brown P A, Lu J and Shuford K L 2015 Electronic properties of halogen-adsorbed graphene *J. Phys. Chem.* C **119** 17271–7

[28] Krasheninnikov A V, Lehtinen P O, Foster A S, Pyykkö P and Nieminen R M 2009 Embedding transition-metal atoms in graphene: structure, bonding, and magnetism *Phys. Rev. Lett.* **102** 126807

[29] Sun M, Ren Q, Zhao Y, Chou J P, Yu J and Tang W 2017 Electronic and magnetic properties of 4D series transition metal substituted graphene: a first-principles study *Carbon* **120** 265–73

[30] Lv R *et al* 2012 Nitrogen-doped graphene: beyond single substitution and enhanced molecular sensing *Sci. Rep.* **2** 586

[31] Miura Y, Kasai H, Dino W A, Nakanishi H and Sugimoto T 2003 Effective pathway for hydrogen atom adsorption on graphene *J. Phys. Soc. Jpn* **72** 995–7

[32] Dai J, Yuan J and Giannozzi P 2009 Gas adsorption on graphene doped with B, N, Al, and S: a theoretical study *Appl. Phys. Lett.* **95** 232105

[33] Lee G, Lee B, Kim J and Cho K 2009 Ozone adsorption on graphene: *ab initio* study and experimental validation *J. Phys. Chem.* C **113** 14225–9

[34] Gao H, Wang L, Zhao J, Ding F and Lu J 2011 Band gap tuning of hydrogenated graphene: H coverage and configuration dependence *J. Phys. Chem.* C **115** 3236–42

[35] Sahin H and Ciraci S 2012 Chlorine adsorption on graphene: chlorographene *J. Phys. Chem.* C **116** 24075–83

[36] Tran N T T, Nguyen D K, Lin S Y, Gumbs G and Fa-Lin M 2019 Fundamental properties of transition-metals-adsorbed graphene *ChemPhysChem.* **10** 1905

[37] Tran N T T, Lin S Y, Glukhova O E and Lin M F 2016 π-bonding-dominated energy gaps in graphene oxide *RSC Adv.* **6** 24458–63

[38] Lin S Y, Lin Y T, Tran N T T, Su W P and Lin M F 2017 Feature-rich electronic properties of aluminum-adsorbed graphenes *Carbon* **120** 209–18

[39] Shahrokhi M and Leonard C 2017 Tuning the band gap and optical spectra of silicon-doped graphene: many-body effects and excitonic states *J. Alloys Compd.* **693** 1185–96

[40] Yoshioka T, Suzuura H and Ando T 2003 Electronic states of BCN alloy nanotubes in a simple tight-binding model *J. Phys. Soc. Jpn* **72** 2656–64

[41] Raidongia K, Nag A, Hembram K P S S, Waghmare U V, Datta R and Rao C 2010 BCN: a graphene analogue with remarkable adsorptive properties *Chemistry–Eur. J.* **16** 149–57

[42] Dawlaty J M *et al* 2008 Measurement of the optical absorption spectra of epitaxial graphene from terahertz to visible *Appl. Phys. Lett.* **93** 131905

[43] Barreiro A, Lazzeri M, Moser J, Mauri F and Bachtold A 2009 Transport properties of graphene in the high-current limit *Phys. Rev. Lett.* **103** 076601

[44] Nguyen D K, Tran N T T, Gumbs G and Lin M F 2018 Rich essential properties of Si-doped graphene (arXiv:1810.09635)

[45] Yan W J, Xie Q, Qin X M, Zhang C H, Zhang Z Z and Zhou S Y 2017 First-principle analysis of photoelectric properties of silicon–carbon materials with graphene-like honeycomb structure *Comput. Mater. Sci.* **126** 336–43

[46] Liu M H *et al* 2015 Scalable tight-binding model for graphene *Phys. Rev. Lett.* **114** 036601

[47] Xue J *et al* 2011 Scanning tunnelling microscopy and spectroscopy of ultra-flat graphene on hexagonal boron nitride *Nat. Mater.* **10** 282

[48] Borysiuk J *et al* 2009 Transmission electron microscopy and scanning tunneling microscopy investigations of graphene on 4 H-SiC (0001) *J. Appl. Phys.* **105** 023503

[49] Man K L and Altman M S 2012 Low energy electron microscopy and photoemission electron microscopy investigation of graphene *J. Phys. Condens. Matter* **24** 314209

[50] Mo S K 2017 Angle-resolved photoemission spectroscopy for the study of two-dimensional materials *Nano Converg.* **4** 6

[51] Li G, Luican A and Andrei E Y 2009 Scanning tunneling spectroscopy of graphene on graphite *Phys. Rev. Lett.* **102** 176804

[52] Lalmi B *et al* 2010 Epitaxial growth of a silicene sheet *Appl. Phys. Lett.* **97** 223109

[53] Li L *et al* 2014 Buckled germanene formation on Pt(111) *Adv. Mater.* **26** 4820–4

[54] Chen R B, Chen S C, Chiu C W and Lin M F 2017 Optical properties of monolayer tinene in electric fields *Sci. Rep.* **7** 1849

[55] Nguyen D K, Tran N T T, Chiu Y H and Lin M F 2019 Concentration-diversified magnetic and electronic properties of halogen-adsorbed silicene *Sci. Rep.* **9** 13746

Chapter 11

Silicene on Ag(111)

11.1 Si-adlayer–Ag(111)

Monolayer silicene on Ag(111) presents rich geometric and electronic properties, owing mainly to the extremely non-uniform chemical environments in the different moiré superlattices, greatly enhanced buckling and significant Si–Ag multi-orbital hybridization. Substrate effects are taken into consideration fully by using a sufficient width of silver. Four-layer Ag atoms, as shown in figure 11.1, are utilized in the first-principles calculations. The optimal geometry shows clearly that there exist 72 atoms enclosed under a greatly enlarged unit cell due to the (3×3) and (4×4) structures in the Si and Ag layers, respectively. The lattice constant on the (x, y)-plane is ~11.48 Å (table 11.1). The various bond lengths, the height differences between two interlayer sublattices and the interlayer distances for two neighboring layers reflect directly the above-mentioned mechanisms. The intralayer Si–Si, Si–Ag and Ag–Ag bond lengths are different within an enlarged unit cell, in which they lie in the ranges of 2.304–2.339, 2.597–2.942 and 2.828–3.254 Å, respectively. The height differences of the first-layer silicene and silver are, respectively, ~0.46 and 0.473 Å, where three Si atoms in the former even reach ~1.10–1.26 Å. This indicates clearly the strengthened non-orthogonality of the π and σ bonding in silicene (increased sp^3 bonding). Also note that the buckling degree declines quickly for the second-layer silver and the other layers. Moreover, the interlayer distances, which correspond to the Si–Ag layers and the first, second and third Ag–Ag layers, are about 2.038–3.403, 2.313–2.646, 2.218–2.511 and 2.359–2.578 Å. Generally, they are larger than or comparable to those in bilayer silicene systems (details in section 5.1). A very large fluctuation is revealed in the heterojunction, such that the Si–Ag chemical bonds are expected to provide significant multi-orbital hybridization (discussed later). Various types of chemical bonding (non-uniform hopping integrals) appear to arise in the intralayer and interlayer atomic interactions associated with specific moiré superlattices; therefore, such results will generate high barriers

doi:10.1088/978-0-7503-3299-6ch11

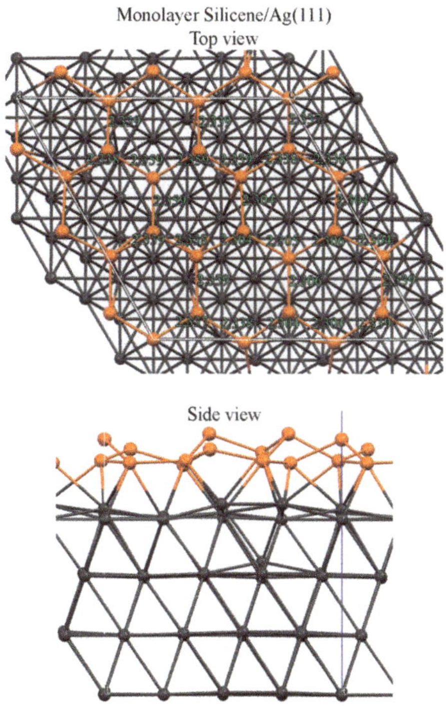

Figure 11.1. The optimal geometric structure of silicene–Ag(111) with a four-layer substrate (top and side views). The area enclosed by the white line corresponds to a moiré superlattice (3 × 3, 4 × 4). The red and gray spheres, respectively, correspond to the silicon and silver atoms.

for simulating this compound through the tight-binding model. The theoretical predictions for the main geometric features could be examined using high-resolution STM, TEM and LEED measurements (details in sections 3.1 and 3.2).

Up-to-date experimental measurements conducted on optimal geometric structures for Si-adlayer–Ag(111) compounds have shown moiré superlattices for the Si and Ag layers of $3 \times 3/4 \times 4$, 3×3, $2\sqrt{3} \times 2\sqrt{3}$, $\sqrt{3} \times \sqrt{3}$, $\sqrt{7} \times \sqrt{7}$ and $\sqrt{13} \times \sqrt{13}$ [1–6]. Rich stacking configurations in the Si–Ag heterojunctions have been synthesized successfully in experimental laboratories. Such phenomena are similar to the twisted structures of few-layer graphene systems [7], illustrating clearly the geometry-varied fundamental properties. For example, the different zone-folding effects, in cooperation with the very strong chemical bonding with a specific substrate, are expected to create diverse physical, chemical and material properties.

The Si-adlayer–Ag(111) compound, as shown in figures 11.2(a) and (b), presents unusual band structures. Within the pure zone-folding effect of the specific 3×3 moiré superlattice, pristine silicene (figure 11.2(a)) possesses more energy subbands for a greatly reduced first Brillouin zone. It can identify slightly separated Dirac-cone structures due purely to the $3p_z$ orbitals across the Fermi level at the Γ point, but not the K point; furthermore, they appear in a doubly degenerate form. The deeper σ valence subbands are initiated from $E_v \leqslant -1.20$ eV under the Γ point. With

Table 11.1. The geometric, electronic and magnetic properties of Si-adlayer–Ag(111), AB-bt Si-bilayer–Ag(111), and monolayer silicene in terms of the Si–Si, Si–Ag and Ag–Ag bond lengths, the height differences between the A and B sublattices in silicene, the net magnetic moments, the lattice constants under a specific moiré superlattice and the ground state energies per unit cell. The interlayer distances for two silicene layers, silicene and substrate, and two silver layers.

	Si–Si bond length (Å)	Si–Ag bond length (Å)	Ag–Ag bond length (Å)	Height difference of first Si layer	Height difference of second Si layer	Height difference of first Ag layer	Magnetic moment	Lattice constant	Ground state energy (eV)
Monolayer Si–Ag(111)	2.304–2.339	2.597–2.942	2.828–3.254	1.192	—	0.473	NM	11.48	−257.233
AB bilayer Si–Ag(111)	2.313–2.359	2.495–2.873	2.856–2.988	0.865	1.177	0.315	1.01 μ_B/FM	11.48	−350.328
Pristine silicene	2.279	—	—	0.451	—	—	NM	11.61	−86.086

	Si–Si layer distance (Å)	Si–Ag layer distance (Å)	Ag–Ag layer distance 1	Ag–Ag layer distance 2	Ag–Ag layer distance 3
Monolayer Si–Ag(111)	—	2.038–3.403	2.313–2.646	2.218–2.511	2.359–2.578
AB bilayer Si–Ag(111)	2.426–2.449	2.119–3.299	2.215–2.669	2.311–2.521	2.338–2.498

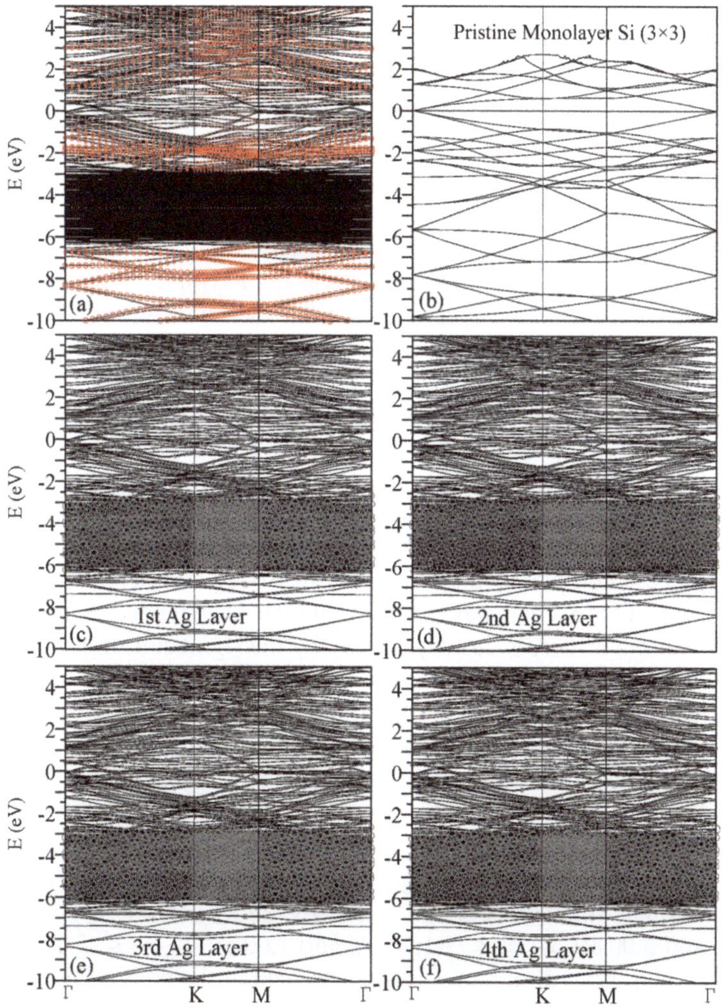

Figure 11.2. Band structures for (a) the Si-adlayer–Ag(111) compound and (b) monolayer silicene, where the atom dominance is shown in the former. Also shown in (c)–(f) are those for the contributions of the four Ag layers.

a decreasing state energy, the $3p_z$-π and ($3p_x$, $3p_y$)-σ valence bands might be mixed together through anti-crossing behaviors (details in section 4.1). Apparently, 1/3 and 2/3 of the electronic valence states are dominated by, respectively, the former and the latter. Specifically, the −3.20 eV valence band near the Γ point should originate from the π-band bottom state, compared to a non-folding band structure (figure 4.2). In short, there are thirteen valence states at the Γ point within the energy range of $-3.20\ \mathrm{eV} \leqslant E^v \leqslant 0\ \mathrm{eV}$.

The pristine zone-folding silicene states are changed dramatically through the quite strong substrate effects, as shown in figure 11.2(b). Many band-edge states, e.g. the extreme and saddle points in energy–wave-vector space, are created in the presence of heterojunctions. The asymmetric spectra of the electron and hole energy subbands around the Fermi level are enhanced greatly by the interlayer atomic interactions between the neighboring Si–Ag and Ag–Ag layers. Compared to monolayer silicene (figure 11.2(a)), this compound presents more valence and conduction energy subbands because of the many silver atoms in the specific moiré superlattice. That is to say, there exists a very complicated electronic structure. For the low-lying energy spectrum, four energy subbands, which exhibit the non-uniform energy dispersions along the high-symmetry directions, cross E_F and thus have occupied and unoccupied states. Such an unusual result indicates a obvious semi-metallic property in the presence of low 2D free carrier density. However, monolayer silicene belongs only to the narrow-gap semiconductors. The low-energy valence and conduction bands, with $|E^{c,v}| < 1.0$ eV, are almost dominated by silver atoms; therefore, the substrate is expected to make critical contributions to the other fundamental properties. This phenomenon illustrates the importance of the Ag–Ag chemical bonding in the entire band structure of a Si-adlayer–Ag(111) compound. Apparently, the original Dirac-cone structures across the Fermi level are completely absent under the Si–Ag interlayer interactions at the heterojunctions. The π chemical bonding, which is extended continuously on a silicene honeycomb lattice, is destroyed fully by the significant multi-orbital hybridization in the Si–Ag chemical bonds. Furthermore, the initial σ valence bands behaves thus. As a result, it is impossible to define the π- and σ-band widths, or their physical and chemical pictures have become meaningless. Also, we cannot simulate the first-principles band structure using the tight-binding model.

The whole band structure could be classified into certain energy ranges, according to the silicon and silver dominance (the red and blue circles in figure 11.2(b)). Electronic energy spectra in the ranges of -10.0 eV $\leqslant E^v \leqslant -6.50$ eV are dominated by Si atoms, -6.50 eV $\leqslant E^v \leqslant -3.00$ eV and -1.00 eV $\leqslant E^v \leqslant 1.00$ eV by Ag atoms, and 3.00 eV $\leqslant E^v \leqslant -1.00$ eV and 1.00 eV $\leqslant E^v \leqslant 5.00$ eV by Si and Ag atoms. Similar atom dominance is also revealed in the bilayer case except for the low-lying valence and conduction bands. The Ag layer dependence is very strong in the above-mentioned ranges, as shown in figures 11.2(c)–(f). That is to say, each silver layer makes a comparable contribution to the electronic states, and only the first Ag layer is not sufficient to account for the fundamental electronic properties. The atom-dominated band structure in Si-adlayer–Ag(111) is related closely to the intralayer Si–Si and Ag–Ag hopping integrals and the interlayer Si–Ag and Ag–Ag ones. The atom- and orbital-decomposed electronic states are further examined using the important van Hove singularities (see figure 11.4). In addition, similar atom dominance can be found in the bilayer case (discussed later) except for the low-energy bands across the Fermi level.

There exist close relationships between the complicated band structures and moiré superlattices. Several stable geometries of Si-adlayer–Ag(111) compounds, identified from experimental studies [1–6], possess enlarged unit cells on the silicon and silver layers. The significant zone-folding and strong substrate effects appear to be combined together directly under any conditions, thus leading to geometry-varied electronic energy spectra. For example, according to first-principles calculations, in figure 11.2(b) under the 3×3 and 4×4 superlattices, it is impossible to observe the modified Dirac-cone structures, the simple energy dispersions near the Fermi level, the specific Fermi momenta and the finite band gap. ARPES measurements on the whole valence bands (details in section 3.3) might become very difficult as a result of certain valence subbands with non-uniform k-dependence across E_F simultaneously, in particular for those due to the larger silver atom weights.

The spatial charge density presents drastic changes after the formation of the heterojunction in figures 11.3(a)–(c), being combined directly with the atom- and orbital-decomposed van Hove singularities (see figure 11.4), and might be able to examine the critical multi-orbital hybridization in the Si–Si, Si–Ag and Ag–Ag chemical bonds effectively. In addition, the intralayer and interlayer atomic interactions survive simultaneously in the Ag–Ag bonds. The Si–Ag chemical bonds, which enhance the buckling of the Si honeycomb lattice and create the large deformation on the first Ag layer (figure 11.1), play a critical role in the unusual carrier density. From the (x, z)-plane projection, the four silicon atoms (the blue spheres in the unit cell) of the A and B sublattices have significant orbital hybridization with three silver atoms (white spheres) at a specific heterojunction, as shown in figure 11.3(a). Interestingly, the first pair in the A and B sublattices exhibits only identical chemical bonding with the first Ag atom; furthermore, the third and fourth Si atoms, respectively, correspond to the second and third Ag atoms. As a result, there are two kinds of orbital hybridization in the Si–Ag bonds, i.e. the charge distributions are different around them. Sufficiently high charge densities are revealed between the Si and Ag atoms, in which they are deduced to be transferred from the original (3s, $3p_x$, $3p_y$, $3p_z$) and (4d, 5p, 5s) orbitals (four and nine orbitals). This mechanism, the very complicated multi-orbital hybridization of four–nine bonding, is further supported by the Si- and Ag-co-dominated band structure under certain energy ranges (e.g. -3.0 eV $\leqslant E^v \leqslant -1.0$ eV and 1.0 eV $\leqslant E^v \leqslant 5.0$ eV in figures 11.2(a)–(f)), and the non-negligible contributions of the atom- and orbital-decomposed densities of states within the specific range of -1.0 eV $\leqslant E^v \leqslant 1.0$ eV (see figures 11.4(a) and (b)). Apparently, both the π and σ chemical bonding in the Si honeycomb lattice are affected strongly and/or distorted by the substrate effects, such that their well-defined behavior is completely absent in the fundamental properties. Such viewpoints would become meaningless in accounting for Ag-substrate-related silicene systems. Specifically, the charge density, which is localized about the Ag atom center, is very high in terms of the wide and dark red region. This phenomenon originates mainly from the eleven valence electrons in each silver atom. The observable carrier densities are revealed in the nearest- and next-nearest Ag–Ag atomic interactions (the green and light blue regions). They might be the weakest among the Ag–Ag, Ag–Si and Si–Si bonds.

The main features of the very complicated band structure in figure 11.2 are reflected directly in the 2D van Hove singularities, as illustrated in figure 11.4(a) and (b). Generally, the band-edge states, which correspond to the extreme points, saddle-point structures, constant-energy loops and partially flat dispersions, are capable of generating shoulders, logarithmically symmetric peaks, square-root asymmetric

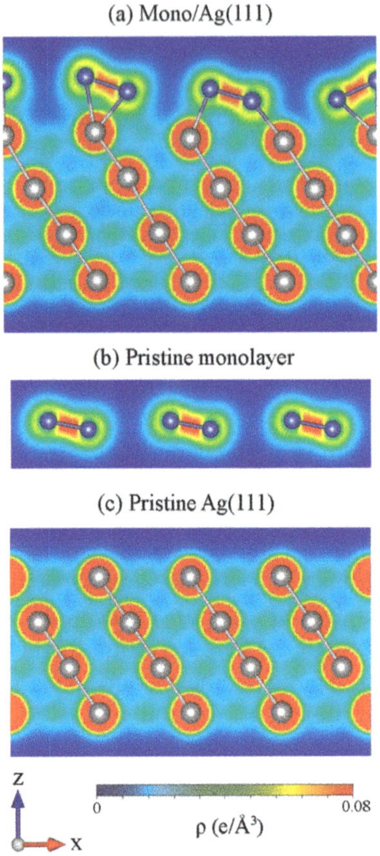

Figure 11.3. (a) The spatial charge density in silicene–Ag(111), (b) monolayer silicene and (c) four-layer silver substrate in the presence of the (x, z)-plane.

peaks and delta-function-like peaks. There are too many van Hove singularities to distinguish the neighboring special structures well; therefore, only the merged and broadening structures can be observed in the calculated results, such as the Si and Ag atom contributions shown by the black curves in figures 11.4(a) and (b), respectively. The above-mentioned four kinds of van Hove singularities might become ambiguous because of the strong zone-folding effects on the strong enhancement of the critical points. The density of states appears to be finite and small at the Fermi level, indicating the semi-metallic property. This behavior is similar to those in few-layer graphenes [7] and AB-stacked graphite [8]. However, the V-shaped structure across or near the Fermi level, which corresponds to the modified Dirac-cone structures, disappears completely. Furthermore, the distribution range of Si-3p_z is more than 10 eV (the green curve), which is much larger than that (~6 eV) in monolayer silicene (discussed in section 4.1). These properties are consistent with the absence of the well-defined π-electronic energy bands and chemical bonding (figures 11.2 and 11.3). A similar phenomenon is revealed in the σ-electronic orbitals, where the 3s and (3p_x + 3p_y) orbitals are represented by the red

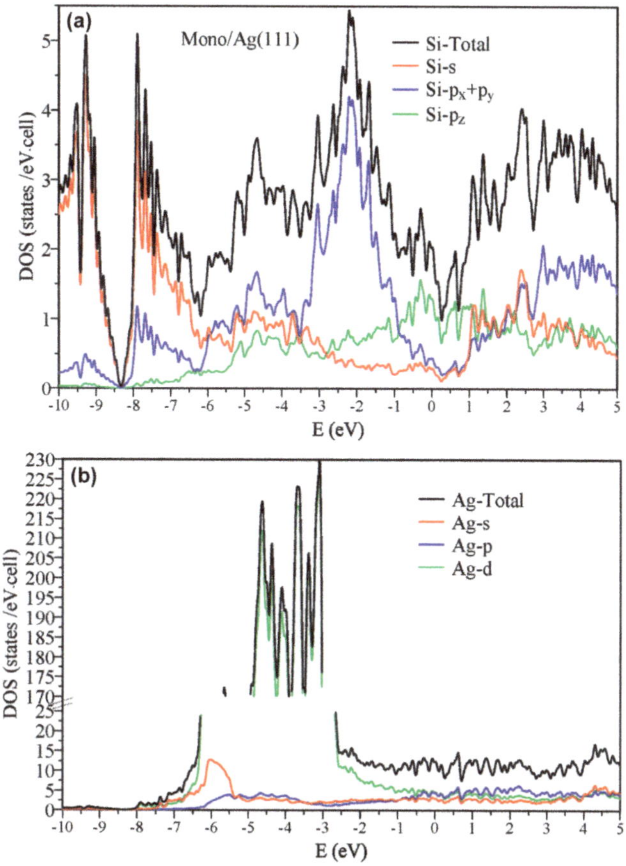

Figure 11.4. The atom- and orbital-projected densities of states in the Si-adlayer/Ag(111) system: the contributions arising from (a) the total, 3s-, ($3p_x$ + $3p_y$)- and $3p_z$-Si orbitals (the black, red, blue and green curves, respectively) and (b) the total, 5s-, 5p- and 4d-Ag orbitals (the black, red, blue, green curves, respectively), in which $5p_x$, $5p_y$ and $5p_z$ make comparable contributions.

and blue curves. That is to say, the 3s-, $3p_x$-, $3p_y$- and $3p_z$-orbitals of silicon atoms, which make non-negligible contributions at the same energies of the special structures except for $E < -7.0$ eV, need to be investigated simultaneously, e.g. the significant chemical bonding in the Si–Si and Si–Ag bonds. Specifically, at very deep energies, the 3s orbitals dominate the energy spectrum of the Si-adlayer–Ag(111) compound; therefore, the third σ-electronic bonding could survive under strong silicene–substrate interactions.

Compared to the monolayer silicon atom contribution (figure 11.4(a)), the four-layer atoms of the silver substrate present dominant or co-dominant features in most of the energy ranges except for the very low range ($E < -7.0$ eV). This behavior arises mainly from the nine kinds of active orbitals in terms of (4d, 5p, 5s) in each silver atom and many Ag atoms under a specific moiré superlattice. Moreover, the nine 4d valence electrons make the most important contributions in the occupied energy spectrum among the 5s, 5p and 4d orbitals (the red, blue and green curves), in particular within the range of $-6.0\text{ eV} \leqslant E^v \leqslant -1.0\text{ eV}$. It should be noted that $5p_x$, $5p_y$ and $5p_z$ have almost identical contributions, being quite different in the silicene case.

Such a result might reflect directly the significant characteristics of transition-metal atoms, i.e. the hopping integrals and/or the chemical bonding of 4d orbitals could create a significant valence-electronic energy spectrum due to the Ag–Ag bonds. Furthermore, all the 5s, 5p and 4d orbitals possess comparable densities of states within the range of $-1.0\text{ eV} \leqslant E^v \leqslant 1.0\text{ eV}$ across the Fermi level, accompanied by the dominant Si-3p_z orbitals. This illustrates the very complicated multi-orbital hybridization at the heterojunction hybridization associated with the Ag–Si bonds.

11.2 The AB-bt bilayer silicene–Ag(111) compound

A bilayer silicene subsystem on Ag(111) could present plenty of quasi-stable stacking, where the most stable AB-bt configuration under the pristine case (details in section 5.1) is chosen for a model investigation. Its moiré superlattice and lattice constant are almost identical to those in the monolayer–substrate compound, as shown in figures 11.5 and table 11.1. The number of nearest-neighbor chemical

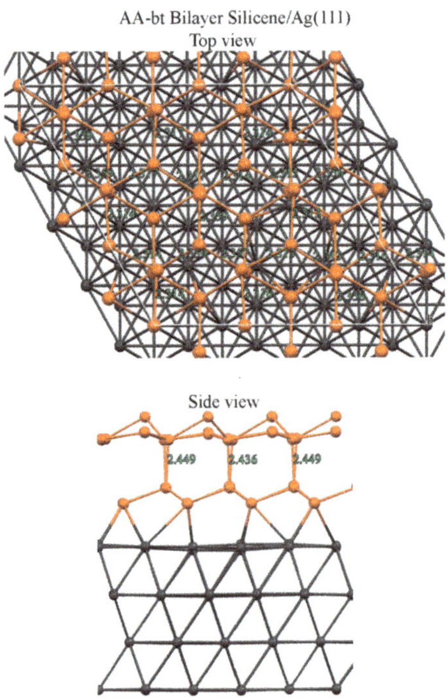

Figure 11.5. Top and side views for an AB-bt Si-bilayer on a four-layer substrate Ag(111), where the area enclosed by the white lines indicates a moiré superlattice (3 × 3, 4 × 4). The red and gray spheres belong to the silicon and silver atoms, respectively.

bonds are three for intralayer silicene, one for bilayer AB-bt stacking, three for the Si–Ag heterojunction, six for Ag–Ag along the [111] direction and six for intralayer silver.

Such a result reflects directly the fact that the silicon and silver atoms are very active elements in generating quite strong multi-orbital hybridization. There exist non-uniform Si–Si, Si–Ag and Ag–Ag bond lengths within the ranges of 2.313–2.359, 2.495–2.873 and 2.856–2.988 Å. The height differences, which are revealed in the first Si layer, the second Si layers and the initial Ag layer (measured from the heterojunction) are ~0.865, ~0.419 and ~0.319 Å, respectively. Compared to a Si-adlayer compound (figure 11.1), the buckling degree is reduced in the first Si layer and decreases slightly in the Ag layer. The interlayer distances corresponding to the neighboring silicenes, the Si–Ag bilayer, the first Ag–Ag layer, the second Ag–Ag layer and the third Ag–Ag layer lie in the ranges of, respectively, 2.426–2.449, 2.119–3.299, 2.215–2.669, 2.311–2.521 and 2.338–2.498 Å. The sufficiently short distances and very obvious fluctuation in the heterojunction appear to indicate that the interlayer atomic interactions belong to strong multi-orbital hybridization, but not the weak van der Waals interactions. In addition to the optimal geometric structure, the layer number, stacking configuration and substrate play important roles in the other fundamental properties.

Bilayer AB-bt silicene presents an unusual band structure, as discussed in detail in section 5.2. The main features, which correspond to an electronic energy spectrum within the 3 × 3 superlattice in figure 11.6(b), include the dominance of spin–orbit coupling on the low-lying valence and conduction bands, spin-up and spin-down degenerate states (a vanishing magnetic moment), an indirect band gap of $E_g \sim 0.50$ eV, the absence of a modified Dirac-cone structure, non-uniform and oscillatory energy dispersions, and undefined π and σ valence bands due to non-van der Waals interactions (significant interlayer multi-orbital hybridization). However, when this pristine bilayer silicene has very strong interlayer and intralayer interactions with or within the silver substrate, the original electronic properties are transformed dramatically.

The bilayer AB-bt silicene–Ag(111) compound, as shown in figures 11.6(a) and (c), presents spin-split valence and conduction bands. This is completely different to the spin-degenerate properties of the Si-adlayer–substrate system (figure 11.2(b)) and pristine bilayer silicene (figure 11.6(b)). Spin-splitting can be observed in the whole energy range, in which the largest energy spacing might reach ~0.10 eV. The low-lying valence and conduction bands within the range of -2.0 eV $\leqslant E^{c,v} \leqslant 2.0$ eV, which cross the Fermi levels (figure 11.6(c)), are capable of creating spin-dominated occupied and unoccupied electronic states with direction-dependent Fermi momenta. It is well known that the magnitudes of the magnetic moments are determined mainly by these. However, it is almost impossible to evaluate the strongly anisotropic Fermi momenta from non-uniform energy spectra and thus estimate the 2D free carrier density and the net magnetic moment. According to numerical VASP calculations, the ferromagnetic moment within a moiré unit cell, which is competed by the occupied spin-up and spin-down states, is ~1.01 μ_B. The

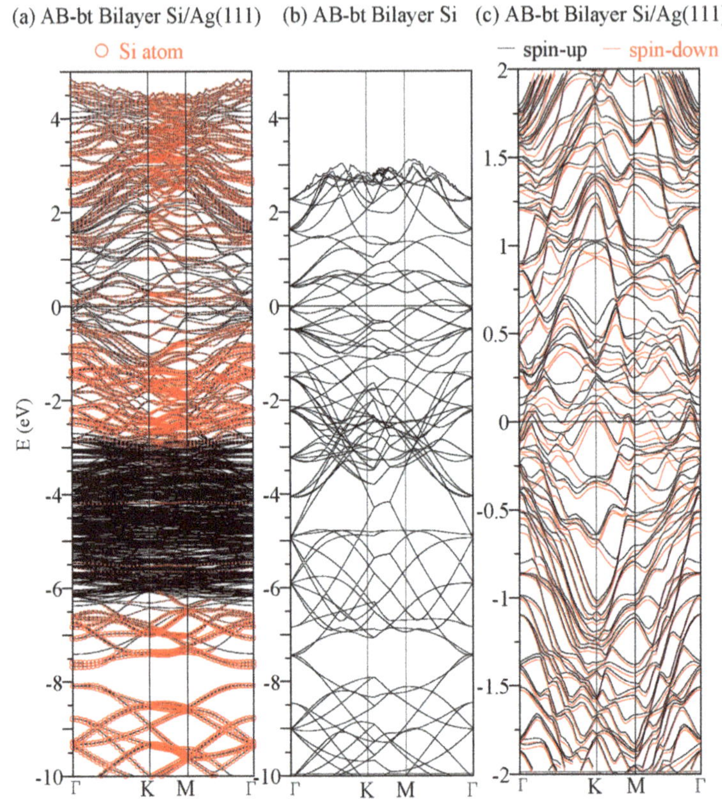

Figure 11.6. Electronic energy spectra for (a) AB-bt bilayer silicene on Ag(111) and (b) a pristine system without a substrate. The atom dominance is shown only in the former. Also shown in (c) is the enlarged band structure within $|E| \leqslant 2.0$ eV with the spin-split electronic states.

calculated value should be sufficiently large for experimental SQUID verification (details in section 3.4).

The spin-dependent magnetic configuration is worthy of closer examination. The low-lying electronic states originate mainly from the silver atoms, since the weights due to the silicon atoms are almost negligible (the very small red circles in figure 11.6(b)). Furthermore, they are independent of the spin–orbit coupling, which is in sharp contrast to that of pristine AB-bt bilayer or monolayer silicene (figure 11.6(a)). The transition-metal atoms appear to be capable of creating spin-related many-particle interactions, e.g. Hubbard-like on-site spin-up and spin-down Coulomb potentials [9]. This unusual Hamiltonian includes, at least, the multi-orbital hopping integrals with complicated intralayer and interlayer terms, the site energies due to the different orbitals and the spin-dependent electron–electron Coulomb interactions [10]. Interestingly, the bilayer AB-bt and monolayer silicene–Ag(111) belong to, respectively the ferromagnetic and non-magnetic condensed-matter systems. The concise physical pictures which are responsible for this important difference between the two compounds cannot be found as of yet.

However, the ferromagnetic configuration is revealed clearly in the spatial spin distribution (discussed later).

High-resolution ARPES and SP-ARPES measurements (details in section 3.3) are available for examining the theoretical predictions for the significant substrate effects on electronic energy spectra. These methods have verified successfully the diverse occupied bands for the emergent 2D systems, e.g. the varied band structures in the few-layer C-, Si-, Ge-, Sn-, Pb-, Sb- and Bi-related materials [11–15] in the presence of a heterojunction. For example, a modified valence Dirac cone near the K/K′ point, with a band gap of ~0.6 eV, is revealed in the Si-adlayer–Ag(111) under 3×3 and 4×4 moiré superlattices for the Si and Ag layers, respectively [1]. The existence of the π valence band close to the Fermi level indicates significant Si–Ag atomic interactions and Si atom dominance in the enlarged unit cell. On the other hand, the quasi-graphene band structures would disappear completely because of the strong orbital hybridization at the boundary (the large substrate-atom weight), e.g. the occupied energy spectra for silicene–ZrB_2 (0001) [16], bilayer tinene–Bi_2Te_3 [111] [17] and Pb-adlayer–Au(111) [18, 19]. The various optimal geometries appear to be able to create varied band structures. The current 3×3 and 4×4 cells of the Si-adlayer and bilayer AB-bt silicene on a Ag(111) substrate present rich and unique features, a vanishing Dirac-cone structure, non-uniform energy dispersions, multi-subbands across the Fermi level, semi-metallic behavior, silver-dominated low-lying valence bands, undefined π and σ bands, and spin-degenerate and spin-split states corresponding to, respectively, the monolayer and bilayer cases. Experimental studies could provide some information about the intralayer and interlayer Si–Si, interlayer Si–Ag, and intralayer and interlayer Ag–Ag chemical bonding.

To explore the substrate effects on the bilayer AB-bt silicene fully, the significant orbital hybridization in various chemical bonds is examined directly using the spatial charge distributions, as shown in figures 11.7(a) and (b). Under the (x, z)-projection, critical Si–Ag bonds at the heterojunction arise in an unusual form, in which six silicon atoms on the A and B sublattices exhibit very strong multi-orbital hybridization with four silver atoms. The Si B-sublattice atoms of the first and second pairs are, respectively, able to bond with the first and second Ag atoms; furthermore, the fifth and sixth single atoms of the former correspond to the third and fourth atoms of the Ag. The chemical bonding for bilayer and monolayer silicene systems appears to contrast sharply in the presence or absence of the silver substrate. There are higher carrier densities in the Si–Ag bonds of the bilayer AB-bt silicene–Ag(111), such as the yellow or even the orange regions between the silicon and silver atoms in figure 11.8(a). This greatly enhances the Si atom dominance of the low-lying energy bands with -1.0 eV $\leqslant E^{c,v} \leqslant$ 10.0 eV (figure 11.6(a) and (c)). Part of the π- and σ-electronic orbitals appear to be transferred from the first Si layer to the hetero-junction, thus leading to obvious destruction of the well-behaved chemical bonding. Seriously distorted π and σ bonding is also induced by the interlayer Si–Si atomic interactions (the red regions between the first and second Si layers). That is to say, significant interlayer Si–Ag and Si–Si orbital hybridization is responsible for the undefined π and σ states, e.g. the thorough absence of a modified Dirac cone in the band structure and a V-shaped van Hove singularity (figure 11.6(a) and 11.9(a)). Their

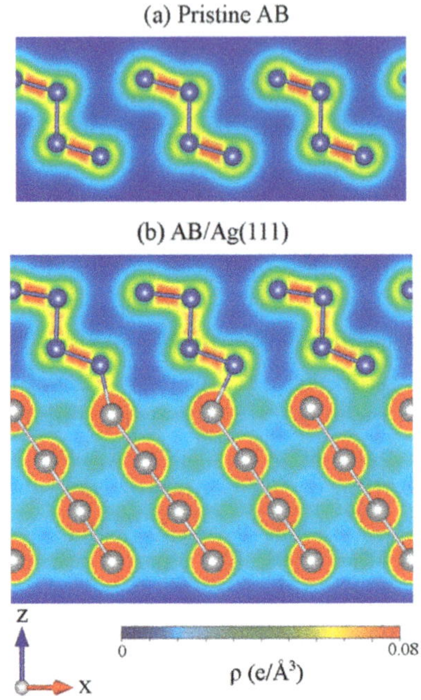

Figure 11.7. Charge distribution of (a) pristine AB-bt bilayer silicene and (b) AB-bt bilayer silicene on Ag(111).

critical multi-orbital hybridization is identified with care from the atom- and orbital-decomposed densities of states and the atom-dominated band structure within the specific energy ranges of -1.0 eV $\leqslant E^{c,\,v} \leqslant$ 1.0 eV and $-$ 3.0 eV $\leqslant E^{v} \leqslant -1.0$ eV, respectively (figures 11.9(a) and (b), and figures 11.6(a) and (c)). As a result, the Si–Ag and Si–Si bonds, respectively, belong to the (3s, $3p_x$ + $3p_y$, $3p_z$)–(4d, 5p, 5s) (four–nine-orbital) and sp^3 bonding. For the Ag–Ag bonds, both the bilayer- and monolayer-related systems (figures 11.7(a) and 11.3(a)) behave similarly, including a wide and high carrier density centered about each Ag atom (the large-radius red sphere), a non-negligible charge distribution in the nearest- and next-nearest atomic interactions (the green and light blue regions), and the weakest of the three kinds of chemical bonds. Moreover, the very complicated chemical bonding is deduced to arise mainly from the multi-orbital hybridization of nine 4d orbitals, which is supported by the density of states and the band structure within the energy range of -6.0 eV $\leqslant E^{c,v} \leqslant$ 1.0 eV (figures 11.9(b) and 11.6(a) and (c)).

The ferromagnetic configuration in the bilayer AB-bt silicene–Ag(111) compound can be further analysed using the spatial spin density distribution, as shown by figures 11.8(a) and (b) and table 11.2. There appears to exist an extremely non-uniform spin arrangement, since its configuration is very sensitive to the silicon and silver atoms, their positions and available orbitals. The spin-up electronic states originate mainly from the six silicon atoms away from the heterojunction or in the B sublattice of the upper layer, while the other Si and Ag atoms produce only small

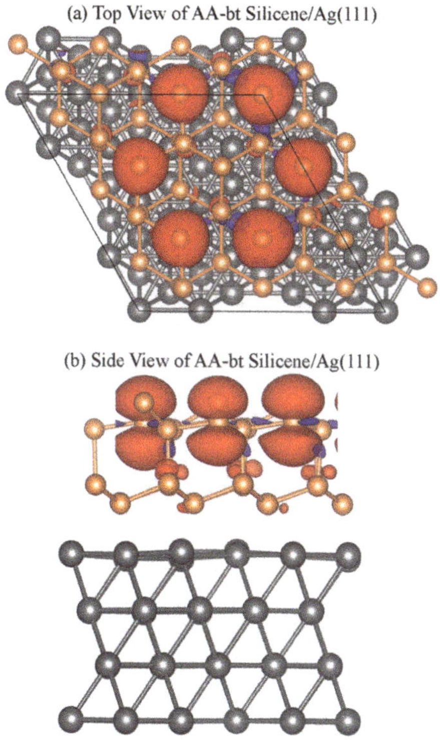

Figure 11.8. The spatial spin density distribution of bilayer AB-bt silicene–Ag(111) with a four-layer substrate: (a) top view and (b) side view (the red and blue regions represent the spin-up and spin-down states, respectively).

distributions. The spin-up magnetic moment around the six silicon atoms (the red regions) lies in the range of $0.53 \leqslant \mathbf{M} \leqslant 0.69$ in units of μ_B, in which the ($3p_x$, $3p_y$, $3p_z$) and 3s orbitals, respectively, have dominant and minor roles (a ratio of ~10:1). The other three Si atoms in the B sublattice and the nine atoms in the A-sublattice of the second Si layer exhibit spin-down configurations with the low enclosed volumes (the blue regions). As a result, the weakly anti-ferromagnetic behaviors are suppressed strongly by the ferromagnetic behaviors. For the magnetic property near the heterojunction, all the silicon atoms in the first layer display similar strengths of magnetic moments (~0.001–0.003 μ_B), where the contributions due to the 3s and 3p orbitals are comparable to each other. Interestingly, most of the silver atoms have a minor spin-up configuration, but a few of them present weak or vanishing spin-down configurations. In addition, their nine 4d orbitals do not have prominent contributions; therefore, they are unable to create unusual magnetic phenomena. It should be noted that the total summation of magnetic moments from the first-layer silicene and four-layer silver might affect the net value, i.e. they are non-negligible in spin-dependent phenomena. The above-mentioned spin arrangement is consistent with atom- and spin-dominated low-lying occupied and unoccupied states across the Fermi level (the spin-split energy bands in figure 11.6(c)), and the atom- and spin-

Table 11.2. The magnetic moments, with the projected contributions arising from the different atoms and orbitals.

Atom		Magnetic moment (μ_B)				Atom		Magnetic moment (μ_B)				Atom		Magnetic moment (μ_B)			
		s	p	d	total			s	p	d	total			s	p	d	total
Ag	1	−0.001	−0.001	0.001	−0.001	Ag	33	0	−0.001	0	−0.001	Si	65	−0.002	0	0	−0.001
Ag	2	0	−0.001	0.001	0	Ag	34	0	−0.001	0	−0.001	Si	66	−0.001	0	0	−0.001
Ag	3	0	−0.001	0.001	−0.001	Ag	35	0	−0.002	0	−0.001	Si	67	−0.002	0.001	0	−0.001
Ag	4	0	−0.001	0.001	0	Ag	36	0	−0.001	0	−0.001	Si	68	−0.001	0	0	−0.001
Ag	5	0	−0.001	0.001	0	Ag	37	0	−0.001	0.001	0	Si	69	−0.002	0	0	−0.001
Ag	6	−0.001	−0.001	0.001	0	Ag	38	0	−0.001	0.001	0	Si	70	−0.001	0	0	−0.001
Ag	7	0	−0.001	0.001	−0.001	Ag	39	0	−0.002	0	−0.001	Si	71	−0.002	−0.001	0	−0.002
Ag	8	0	−0.001	0.001	−0.001	Ag	40	0	−0.002	0	−0.001	Si	72	−0.001	−0.001	0	−0.002
Ag	9	0	−0.001	0.001	−0.001	Ag	41	0	−0.001	0.001	0	Si	73	−0.002	0.001	0	−0.001
Ag	10	0	−0.001	0.001	−0.001	Ag	42	0	0	0	0	Si	74	−0.001	0	0	−0.001
Ag	11	0	−0.001	0.001	−0.001	Ag	43	0	−0.001	0.001	0	Si	75	−0.002	0.001	0	−0.001
Ag	12	0	−0.001	0.001	−0.001	Ag	44	0	−0.001	0	−0.001	Si	76	−0.001	−0.001	0	−0.002
Ag	13	−0.001	−0.001	0.001	0	Ag	45	0	−0.001	0	−0.001	Si	77	−0.002	−0.001	0	−0.002
Ag	14	0	−0.001	0.001	−0.001	Ag	46	0	−0.001	0.001	0	Si	78	−0.001	−0.001	0	−0.003
Ag	15	0	−0.001	0.001	−0.001	Ag	47	0	−0.001	0.001	0	Si	79	−0.002	−0.001	0	−0.002
Ag	16	0	−0.001	0.001	0	Ag	48	0	−0.001	0	−0.001	Si	80	−0.001	0	0	−0.001
Ag	17	0.001	−0.001	0.001	0.001	Ag	49	0	−0.001	0	−0.001	Si	81	−0.002	0	0	−0.001
Ag	18	0	0.001	0	0	Ag	50	0	−0.002	0.001	−0.001	Si	82	−0.001	0	0	−0.001
Ag	19	0	−0.001	0.001	0	Ag	51	0	−0.002	0	−0.001	Si	83	−0.005	−0.052	0	−0.057
Ag	20	−0.001	0	0	0	Ag	52	0	−0.001	0	−0.001	Si	84	0.001	0.006	0	0.006
Ag	21	0	0.001	0	0	Ag	53	0	−0.001	0	−0.001	Si	85	−0.006	−0.061	0	−0.067
Ag	22	−0.001	0	0	0	Ag	54	0	−0.002	0	−0.001	Si	86	0.001	0.006	0	0.006
Ag	23	−0.001	−0.001	0.001	−0.001	Ag	55	−0.001	−0.001	0	−0.002	Si	87	0	0.002	0	0.002
Ag	24	−0.001	−0.001	0.001	−0.001	Ag	56	0	−0.002	0.001	−0.001	Si	88	0	0.004	0	0.004
Ag	25	0	−0.001	0.001	0	Ag	57	0	−0.001	0.001	0	Si	89	−0.005	−0.049	0	−0.053
Ag	26	−0.001	0.002	0	0.001	Ag	58	0	−0.001	0.001	0	Si	90	0	0.004	0	0.004

(*Continued*)

Table 11.2. (*Continued*)

Atom		Magnetic moment (μ_B)				Atom		Magnetic moment (μ_B)				Atom		Magnetic moment (μ_B)			
		s	p	d	total			s	p	d	total			s	p	d	total
Ag	27	0	−0.001	0.001	0	Ag	59	0	−0.002	0.001	−0.001	Si	91	0	0.002	0	0.002
Ag	28	−0.001	−0.001	0.001	−0.001	Ag	60	0	−0.002	0	−0.001	Si	92	0.001	0.005	0	0.006
Ag	29	−0.001	0	0	0	Ag	61	0	−0.001	0	−0.001	Si	93	−0.005	−0.052	0	−0.057
Ag	30	−0.001	0.002	0	0.001	Ag	62	0	−0.001	0.001	0	Si	94	0	0.004	0	0.004
Ag	31	−0.001	0.002	0	0.001	Ag	63	0	−0.001	0	−0.001	Si	95	0	0.002	0	0.002
Ag	32	0	0.001	0	0	Ag	64	0	−0.001	0.001	−0.001	Si	96	0.001	0.006	0	0.006
												Si	97	−0.005	−0.051	0	−0.056
												Si	98	0.001	0.005	0	0.006
												Si	99	−0.007	−0.062	0	−0.069
												Si	100	0.001	0.005	0	0.006
													total	−0.065	−0.326	0.033	−0.358

dependent densities of states near E_F (the spin-up and spin-down states in figures 11.9(a) and 11.9(b)).

The atom-, orbital- and spin-projected densities of states (figures 11.9(a) and 11.9(b)), combined with the atom-dominated and spin-split band structure (figures 11.6(a) and (c)), is available for detailed determination of the critical orbital hybridization in the Si–Si, Si–Ag and Ag–Ag bonds, and the ferromagnetic spin configuration. The (3s, $3p_x + 3p_y$, $3p_z$) and (4d, 5p, 5s) orbitals, which arise from, respectively, the silicon and silver atoms, make significant contributions to many van Hove singularities. Generally, the most important differences among the orbital-decomposed special structures lie only in their intensities, since all of them are merged together. The modified V-shape across the Fermi level, the π-band width and the initial σ-electronic structure are absent, as found for in Si-adlayer–Ag(111) (figure 11.6(a) and (b)). Similar results also include the 3s-σ-electronic energy

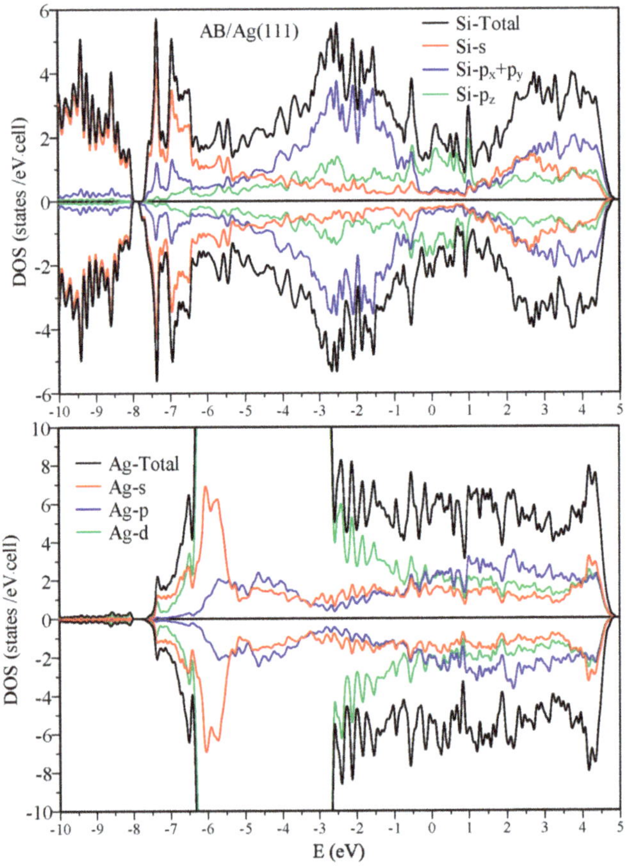

Figure 11.9. (Upper panel) The Si-atom-decomposed van Hove singularities for the AB-bt Si-bilayer–Ag(111). (Lower panel) The Ag-atom-decomposed van Hove singularities for the AB-bt Si-bilayer–Ag(111).

spectrum below $E \sim -7.0$ eV, the 4d-dominated valence states within -6.0 eV $< E <$ -1.0 eV and the ($3p_z$, 5s, 5p, 4d)-co-dominated low-lying region of $|E| < 1.0$ eV for the significant chemical bonding in, respectively, the Si–Si, Ag–Ag and Si–Ag bonds.

The Si-, Ag- and spin-dominated densities of states near the Fermi level, as illustrated in figures 11.9, can determine the magnetic behavior. Most of the deeper valence energy spectra, which lie in the range $E < -1.50$ eV, make negligible contributions to the net magnetic moment, since the spin-up and spin-down states, with almost the same density of states, arise simultaneously, e.g. the red and blue curves for Si and the black and green curves for Ag. On the other hand, within the critical low-energy range for Si dominance, the spin-up density of states is much higher than that of the spin-down configuration, while for Ag it is comparable only in value. Consequently, the dominant ferromagnetic configurations originate from the Si atoms, and the weak or significant ferromagnetic configurations originate from the Ag atoms. After direct combination with the atom-, layer- and orbital-decomposed magnetic moment (table 11.2), the spatial spin distribution (figures 11.8(a) and (b)), the atom- and orbital-projected densities of states (figures 11.9), and the $3p_z$ orbitals of the specific second-layer six silicon atoms are predicted to play the most important roles in the varied magnetic phenomenon. In short, the bilayer AB-bt silicene–Ag(111) and pristine system exhibit ferromagnetic and non-magnetic behaviors, respectively, in which the spin-dependent many- and single-particle interactions (the spin–orbit coupling and the on-site Coulomb potentials) need to be included in reliable simulations of the phenomenological models (e.g. the determination of various hopping integrals in the tight-binding model).

It is well known that STS measurements are very useful in directly identifying the main features of band structures due to the band-edge states (unusual van Hove singularities; details in section 3.2). Up-to-date experimental results are able to verify the critical band properties across the Fermi level in graphene-related sp^2-bonding materials, such as the metallic and semiconducting behaviors in the radius- and chirality-enriched carbon nanotubes [20], the magnitude of the band gap being inversely proportional to the graphene nanoribbon width [21], the dimension-varied special structures in 1D, 2D and 3D carbon-related condensed-matter systems [20–23], and the stacking- and layer-number-enriched singularities in few-layer twisted graphene systems [24, 25]. However, no similar STS experiments have been conducted on Si-, Ge-, Sn- and Pb-substrate systems, which is related closely to their initial synthesis methods.

References

[1] Feng B *et al* 2012 Evidence of silicene in honeycomb structures of silicon on Ag(111) *Nano Lett.* **12** 3507–11

[2] Chiappe D, Grazianetti C, Tallarida G, Fanciulli M and Molle A 2012 Local electronic properties of corrugated silicene phases *Adv. Mater.* **24** 5088–93

[3] Jamgotchian H, Colignon Y, Hamzaoui N, Ealet B, Hoarau J Y, Aufray B and Bibérian J P 2012 Growth of silicene layers on Ag(111): unexpected effect of the substrate temperature *J. Phys. Condens. Matter* **24** 172001
[4] Lin C L *et al* 2012 Structure of silicene grown on Ag(111) *Appl. Phys. Express* **5** 045802
[5] Sone J, Yamagami T, Aoki Y, Nakatsuji K and Hirayama H 2014 Epitaxial growth of silicene on ultra-thin Ag(111) films *New J. Phys.* **16** 095004
[6] Arafune R *et al* 2013 Structural transition of silicene on Ag(111) *Surf. Sci.* **60** 297–300
[7] Bistritzer R and MacDonald A H 2011 Moiré bands in twisted double-layer graphene *Proc. Natl Acad. Sci.* **108** 12233–7
[8] Rong Z Y and Kuiper P 1993 Electronic effects in scanning tunneling microscopy: moiré pattern on a graphite surface *Phys. Rev.* B **48** 17427
[9] Wei P and Qi Z Q 1994 Insulating gap in the transition-metal oxides: a calculation using the local-spin-density approximation with the on-site Coulomb U correlation correction *Phys. Rev.* B **49** 10864
[10] Kane C L and Mele E J 2005 Quantum spin Hall effect in graphene *Phys. Rev. Lett.* **95** 226801
[11] Reina A *et al* 2008 Large area, few-layer graphene films on arbitrary substrates by chemical vapor deposition *Nano Lett.* **9** 30–5
[12] Dávila M E and Le Lay G 2016 Few layer epitaxial germanene: a novel two-dimensional Dirac material *Sci. Rep.* **6** 20714
[13] Matthes L, Pulci O and Bechstedt F 2013 Massive Dirac quasiparticles in the optical absorbance of graphene, silicene, germanene, and tinene *J. Phys. Condens. Matter* **25** 395305
[14] Kordyuk A A 2014 ARPES experiment in fermiology of quasi-2D metals *Low Temp. Phys.* **40** 286–96
[15] Arakane T *et al* 2012 Tunable Dirac cone in the topological insulator $Bi_{2-x}Sb_xTe_{3-y}Se_y$ *Nat. Commun.* **3** 636
[16] Gill T G *et al* 2017 Metallic atomically-thin layered silicon epitaxially grown on silicene/ZrB_2 *2D Mater.* **4** 021015
[17] Wang W, Zhang G and Li X 2008 Manipulating growth of thermoelectric Bi_2Te_3/Sb multilayered nanowire arrays *J. Phys. Chem.* C **112** 15190–4
[18] Green M P, Hanson K J, Carr R and Lindau I 1990 STM observations of the underpotential deposition and stripping of Pb on Au (111) under potential sweep conditions *J. Electrochem. Soc.* **137** 3493–8
[19] Toney M F, Gordon J G, Samant M G, Borges G L, Melroy O R, Yee D and Sorensen L B 1995 *In-situ* atomic structure of underpotentially deposited monolayers of Pb and Ti on Au (111) and Ag(111): a surface x-ray scattering study *J. Phys. Chem.* **99** 4733–44
[20] Olk C H and Heremans J P 1994 Scanning tunneling spectroscopy of carbon nanotubes *J. Mater. Res.* **9** 259–62
[21] Söde H *et al* 2015 Electronic band dispersion of graphene nanoribbons via Fourier-transformed scanning tunneling spectroscopy *Phys. Rev.* B **91** 045429
[22] Li G, Luican A and Andrei E Y 2009 Scanning tunneling spectroscopy of graphene on graphite *Phys. Rev. Lett.* **102** 176804
[23] Kobayashi Y, Fukui K I, Enoki T, Kusakabe K and Kaburagi Y 2005 Observation of zigzag and armchair edges of graphite using scanning tunneling microscopy and spectroscopy *Phys. Rev.* B **71** 193406

[24] Luican A *et al* 2011 Single-layer behavior and its breakdown in twisted graphene layers *Phys. Rev. Lett.* **106** 126802

[25] Brihuega I *et al* 2012 Unraveling the intrinsic and robust nature of van Hove singularities in twisted bilayer graphene by scanning tunneling microscopy and theoretical analysis *Phys. Rev. Lett.* **109** 196802

IOP Publishing

Silicene-Based Layered Materials
Essential properties
Shih-Yang Lin, Hsin-Yi Liu, Duy Khanh Nguyen, Ngoc Thanh Thuy Tran, Hai Duong Pham, Shen-Lin Chang, Chiun-Yan Lin and Ming-Fa Lin

Chapter 12

Silicene–, germanene– and stanene–graphene composites

Composite bilayer materials, which consist of two different monolayer subsystems, are highly suitable for studying unique fundamental properties and well-known phenomena without up-to-date experimental verification, such as the two monolayers of boron-nitride and graphene [1–4] and the silicene–graphene van der Waals heterostructure [5]. For example, butterfly energy spectra are predicted to be present in the magnetic-field-dependent Landau-level (LL) energies for a monolayer carbon honeycomb lattice [6], in which the periodic oscillatory behavior is the so-called Aharonov–Bohm (AB) effect with a period of a flux quantum ($\varphi_0 hc/e$). The traditional quantum AB effect has been identified through electron-beam scattering [7], although it is impossible to verify from a pure monolayer material, since the experimental magnetic-field strength cannot reach the ~10 T corresponding to the magnetic flux through a very small hexagon. Recently, a bilayer BN–C composite system was synthesized successfully and this unusual 2D material was utilized to examine the periodic magneto-electronic energy spectrum in detail [8]. Its main features include a moiré superlattice, a very weak van der Waals interlayer interaction, and a barely modified Dirac-cone structure due to the two zero- and large-gap subsystems. The low-lying Landau π-electronic states are able to generate a lot of valence and conduction subgroups under the enlarged unit cell; that is, there are more LL subgroups with periodic magneto-electronic oscillatory energy spectra. Such phenomena have been confirmed in experimental laboratories, since the flux quantum is achieved by many hexagons by a observable magnetic-field strength. Whether the unique AB effect can survive in bilayer twisted graphene systems, with a very small rotation angle relative to each other [9], is worthy of theoretical study using the generalized tight-binding model.

Clearly, the hybridized bilayer systems, which are based on the group-IV elements, will become mainstream materials. Recently, graphene–silicene–Ru[0001] with an

doi:10.1088/978-0-7503-3299-6ch12

unusual van der Waals heterostructure has been fabricated successfully using chemical deposition [10, 11]. A monolayer graphene layer is first synthesized on a Ru[0001] substrate. Then, silicene is built-up below the C layer through silicon atom intercalation, in which the Si nanostructures are very sensitive to their quantity. Under higher dosages, a silicene layer is revealed clearly in STM images (details of STM in section 3.2). The experimental results are combined directly with first-principles calculations to identify a unique moiré superlattice, the (8×8)–$(\sqrt{21} \times \sqrt{21})$–$(7 \times 7)$ enlarged configuration. The interlayer atomic interactions, which arise from the graphene and silicene, and the silicene and Ru substrate, play a critical role in determining the most stable configuration. For VASP studies on bilayer graphene–silicene composite systems in the absence of a substrate [12], the main results include distinct moiré superlattices, anisotropic low-energy valence and conduction bands, and a modified V-shaped density of states with a vanishing value at the Fermi level. In addition, there are also some density functional theory calculations for graphene-based 2D van der Waals heterostructures, e.g. theoretical predictions for the varied electronic properties in graphene–MoS_2 [13, 14], graphene–ZnO [15, 16] and graphene–phosphorene [17, 18], The above-mentioned 2D heterojunctions might be able to provide promising future applications in electronic, electrochemical, optical and memory devices [5, 19, 20].

This chapter is focused on graphene-based bilayer composites using the other group-IV materials as a specific adlayer, in particular, the various moiré superlattices, and the orbital-bonding-varied fundamental properties. The spin-configuration-induced intrinsic interactions and the van der Waals interlayer forces are included in the detailed first-principles calculations. The Si and Ge adlayers and the Sn and Pb adlayers are classified into two categories, since their low-energy physical properties depend on, respectively, π bonding and sp^3 bonding. The various ground state energies, which correspond to the different enlarged unit cells or moiré superlattices for two subsystems, are analyzed with care to achieve optimal geometric structures. Most importantly, the π and σ bonding, which are well defined in the two independent subsystems, are expected to exhibit significant modification and hybridization. Their characteristics will be examined thoroughly using the modified Dirac-cone structures with simultaneous free electrons and holes, their initial states at the stable K and/or Γ valleys, the increase in valence and conduction energy subbands due to the zone-folding effect, the frequent subband crossing and anticrossing, the normal or irregular π and σ bonding, the strong atom and orbital dependence, the unusual charge distributions in non-uniform buckled moiré superlattices and the bilayer-coupling-enriched van Hove singularities. Interestingly, the composite bilayer system should be in sharp contrast to the two independent subsystems in terms of the fundamental properties. Full discussion is provided on the important differences between the silicene–graphene and bilayer silicenes and graphenes in terms of the geometric, electronic and magnetic properties. Similar comparisons between the graphene–group-IV bilayer composites are very useful in illustrating fully the interlayer atomic interactions (chemical bonding) and the element-dependent moiré superlattices.

12.1 Si–graphene bilayer systems

Bilayer silicene–graphene composite systems (figures 12.1(a) and (b), and 12.2(a) and (b)), present unusual coupling of moiré superlattices which are quite different to the normal stacking configurations of pure bilayer silicenes (figure 5.1) [21, 22] and graphenes (figure 5.5) [23], but similar to those of twisted bilayer graphene systems [24]. Generally, there are many metastable geometric structures. Based on first-principles calculations, two kinds of composite configurations, $(\sqrt{3} \times \sqrt{3})$–$(\sqrt{7} \times \sqrt{7})$ (R = 10.9°) and $(\sqrt{7} \times \sqrt{7})$–$(\sqrt{4} \times \sqrt{4})$ (R = 19.1°) in silicon–carbon honeycomb lattices, are chosen for the model study. The former shows a lattice constant of 6.36 Å, an interlayer distance of 3.35 Å and a buckled height of 0.55 Å, as revealed in figures 12.1(a) and (b), while the values for the latter are 9.88, 3.33 and 0.59 Å, respectively (figures 12.2(a) and (b)). A specific moiré superlattice includes six Si and 14 C atoms, therefore an enlarged unit cell leads to the zone-folding effect within the greatly reduced first Brillouin zone. However, the Si–Si and C–C bond lengths in honeycomb lattices are almost identical to those in the two independent subsystems (tables 12.1 and 12.2). This represents only negligible effects on the σ-electronic bonding. The interlayer distance is just equal or very close to that (3.35 Å) of AB-stacked bilayer graphene [25]. This result suggests that similar van der Waals interactions play a critical role in the bilayer coupling. They are weak, but become very important under combined effects with moiré superlattices. That is to say, the single-orbital hybridization of $3p_z$–$2p_z$ will be the most important type of interlayer chemical bonding. Such atomic interactions are not very complicated; therefore, the interlayer hopping integrals in the Hamiltonian of the tight-binding model can be obtained under good numerical fitting [25]. Moreover, the buckled structure (0.45 Å) [26] of silicene is enhanced greatly by the significant interlayer atomic interactions with graphene. The non-orthogonality of the π and σ bonding in a silicene subsystem might become more obvious. In addition, the spin-dependent interactions are

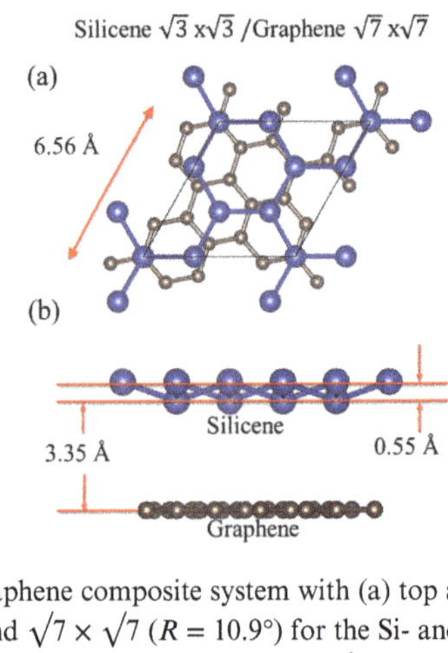

Figure 12.1. A bilayer silicene–graphene composite system with (a) top and (b) side views, with enlarged unit cells characterized by $\sqrt{3} \times \sqrt{3}$ and $\sqrt{7} \times \sqrt{7}$ (R = 10.9°) for the Si- and C-honeycomb lattices (the blue and brown spheres), respectively. The lattice constant in (a) is 6.36 Å, and the interlayer distance and the buckled height in (b) are 3.35 Å and 0.55 Å.

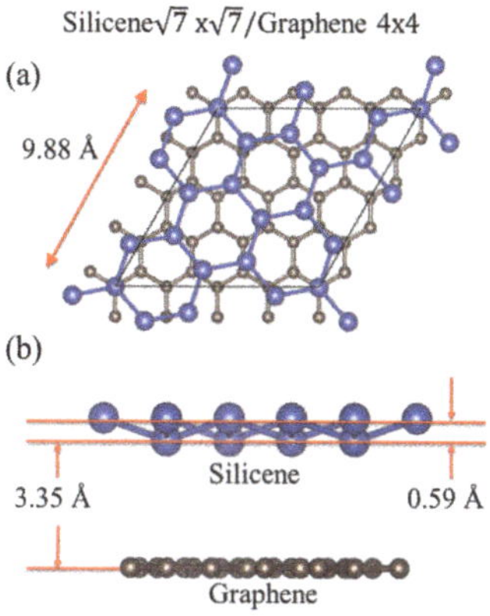

Figure 12.2. The optimal geometry for a specific metastable bilayer silicene–graphene composite with (a) top and (b) side views, corresponding to a moiré superlattice of ($\sqrt{7} \times \sqrt{7}$)–($\sqrt{4} \times \sqrt{4}$) (R = 19.1°) in the silicon and carbon honeycomb lattices, respectively. Also shown in (a) is the lattice constant of 9.88 Å and in (b) the interlayer distance and the buckled height of 3.35 and 0.59 Å, respectively.

negligible, compared to the interlayer orbital hybridization. The spin-induced splitting phenomenon and magnetic configuration are completely absent.

Bilayer silicene–graphene composite systems present rich and unique electronic energy spectra, as shown in figures 12.3(a)–(d). The critical mechanisms appear to originate from a significant combined effect, the obvious superpositioning of zone folding and interlayer coupling. They affect the main features of the band structures and Bloch wave functions strongly, compared to the two independent subsystems (monolayer silicene and graphene in figures 12.3(c) and (d), respectively). There are more carbon-dominated electronic states (red spheres) and fewer silicon-dominated electronic states (blue spheres), owing mainly to, respectively, the higher and lower atom numbers in an enlarged unit cell. Many valence and conduction energy subbands are highly asymmetric to each other about the Fermi level. The low-lying valence and conduction energy subbands are very sensitive to the moiré superlattices (the enlarged unit cells in the metal-stable configurations). For the stacking configuration of ($\sqrt{3} \times \sqrt{3}$)–($\sqrt{7} \times \sqrt{7}$) (R = 10.9°) (figure 12.3(a)), the low-energy band property is determined mainly by two distinct stable valleys which are initiated from, respectively, the K/K′ and Γ points. For the K/K′ and Γ points, their modified Dirac-cone structures, with and without an energy spacing of ~0.055 eV, are dominated by the carbon and silicon atoms, respectively. That is, they arise mainly from the graphene and silicene honeycomb lattices. Most importantly, their Fermi levels are situated above and below the modified conduction and valence Dirac point, since the $2p_z$ orbital energy is deeper than the $3p_z$ one. As a result, there exist few free conduction electrons or valence holes simultaneously with the same free carrier density (identical contributions to the transport property). This silicene–graphene composite should belong to the semimetallic systems, as it is generated purely by the weak overlap of the valence and conduction subbands. Such phenomenon will be further confirmed from the

Table 12.1. The Si–Si and C–C bond lengths and the total ground state energy per unit cell for a bilayer silicene–graphene composite system with different moiré superlattices of $\sqrt{3} \times \sqrt{3}$ and $\sqrt{7} \times \sqrt{7}$ (R = 10.9°).

Bond	Length (Å)
Si1–Si2	2.248 83
Si1–Si4	2.248 47
Si1–Si5	2.248 72
Si2–Si3	2.246 15
Si2–Si6	2.247 26
Si3–Si4	2.2465
Si3–Si5	2.246 44
Si4–Si6	2.247 25
Si5–Si6	2.247 09
C1–C14	1.426 83
C1–C12	1.426 83
C1–C4	1.426 82
C2–C3	1.426 56
C2–C13	1.425 52
C2–C5	1.425 93
C3–C6	1.4259
C3–C14	1.425 82
C4–C5	1.425 78
C4–C7	1.425 15
C5–C8	1.426 55
C6–C7	1.425 53
C6–C9	1.426 56
C7–C10	1.426 47
C8–C9	1.425 93
C8–C11	1.425 53
C9–C12	1.425 75
C10–C11	1.426 46
C10–C13	1.426 44
C11–C14	1.425 16
C12–C13	1.425 19

density of states at E_F (discussed later in figure 12.5(a)). A similar energy spectrum is revealed in the stacking configuration of ($\sqrt{7} \times \sqrt{7}$)–(4 × 4) (R = 19.1°) (figure 12.3(d)). Furthermore, both free conduction electrons and valence holes survive at the same K/K′ valley. However, the Fermi velocity (~2.60 eV Å^{-1}) in the Si-induced Dirac-cone structure is lower than that (~5.60 eV Å^{-1}) in the C-created structure, reflecting directly the 3p$_z$- and 2p$_z$-π bonding strengths. Whether the Si- and C-dominated Dirac cones arise simultaneously can be verified using high-resolution ARPES measurements [27].

Table 12.2. Similar to table 12.1, but for moiré superlattices of $\sqrt{7} \times \sqrt{7}$ and 4×4 (R = 19.1°).

Bond	Length (Å)	Bond	Length (Å)	Bond	Length (Å)
Si1–Si4	2.237 39	C1–C2	1.426 79	C17–C18	1.425 03
Si1–Si12	2.237 37	C2–C3	1.426 05	C18–C19	1.4262
Si1–Si14	2.237 39	C3–C4	1.425 99	C19–C20	1.426 82
Si2–Si3	2.235 85	C4–C5	1.425 09	C20–C21	1.426 47
Si2–Si5	2.235 53	C5–C6	1.426 32	C21–C22	1.425 93
Si2–Si13	2.235 52	C6–C7	1.426 24	C22–C23	1.4257
Si3–Si6	2.2355	C7–C8	1.424 94	C23–C24	1.426 38
Si3–Si14	2.235 86	C1–C8	1.426 81	C24–C17	1.425 92
Si4–Si5	2.235 82	C2–C9	1.424 97	C18–C25	1.425 96
Si4–Si7	2.235 15	C4–C11	1.426 05	C20–C27	1.426 18
Si5–Si8	2.235 87	C6–C13	1.426 92	C22–C29	1.426 46
Si6–Si7	2.235 48	C8–C15	1.426 05	C24–C31	1.425 68
Si6–Si9	2.235 86	C9–C10	1.426 17	C25–C26	1.426 12
Si7–Si10	2.237 06	C10–C11	1.426 87	C26–C27	1.424 98
Si8–Si9	2.235 55	C11–C12	1.425 87	C27–C28	1.425 46
Si8–Si11	2.235 59	C12–C13	1.425 83	C28–C29	1.426 83
Si9–Si12	2.2358	C13–C14	1.426 12	C29–C30	1.4257
Si10–Si11	2.237 07	C14–C15	1.426 03	C30–C31	1.426 41
Si10–Si13	2.237 09	C15–C16	1.425 88	C31–C32	1.426 75
Si11–Si14	2.235 15	C16–C9	1.425 54	C32–C25	1.425 98
Si12–Si13	2.235 08	C10–C17	1.426 44	C26–C1	1.426 72
		C12–C19	1.425 79	C28–C3	1.425 95
		C14–C21	1.425 03	C30–C5	1.4259
		C16–C23	1.426 86	C32–C7	1.4255

Before further examination of the π- and σ-electronic properties in a bilayer composite, one needs to understand fully those of the two independent subsystems. Monolayer silicene and graphene exhibit a full π valence band along the KMΓ path with a band width of 3.15 and 7.60 eV, respectively. The linear energy bands near the K valley possess Fermi velocities of 5.10×10^5 ms^{-1} for silicene and 8.10×10^5 m s^{-1} for graphene, which are somewhat modified by the interlayer coupling. The three σ bands within the ΓMK structure, which are dominated by the ($3p_x + 3p_y$), ($3p_x - 3p_y$) and 3s orbitals for silicene and the ($2p_x + 2p_y$), ($2p_x - 2p_y$) and 2s orbitals for graphene, possess band widths of 2.7, 5.7 and 3.6 eV, and 8.5, 10.5 and 6.5 eV, respectively. The π and σ states of the former exhibit obvious anticrossing behavior due to their non-orthogonalities. The well-behaved π- and σ-electronic phenomena present dramatic changes after bilayer coupling through van der Waals interactions.

The π- and σ-electronic valence subbands, which are determined mainly by the carbon and/or silicon atoms (the red and blue spheres), can be studied roughly using the zone-folding band structures, owing mainly to the moderately or slightly distorted chemical bonding. Interestingly, for the stacking configuration of

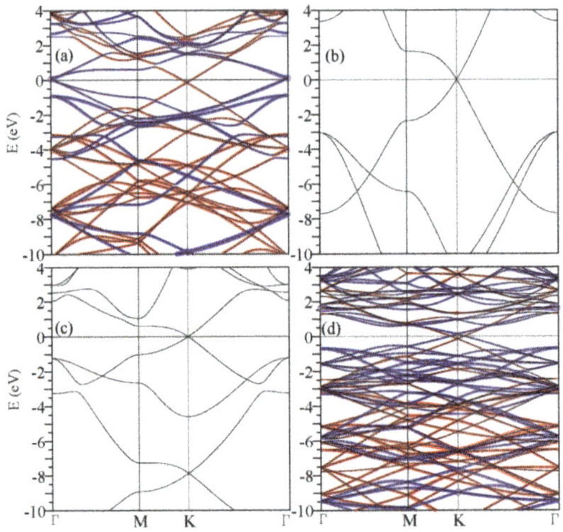

Figure 12.3. Valence- and conduction-state energy spectra along the 2D high-symmetry points in the reduced first Brillouin zone for bilayer silicene–graphene composite systems within the specific moiré superlattices of (a) $(\sqrt{3} \times \sqrt{3})$–$(\sqrt{7} \times \sqrt{7})$ ($R = 10.9°$) in the silicon honeycomb lattice and (d) $(\sqrt{7} \times \sqrt{7})$–$(\sqrt{4} \times \sqrt{4})$ ($R = 19.1°$) in the carbon honeycomb lattice, respectively. The band structure in the absence of interlayer coupling for monolayer (b) graphene and (c) silicene.

$(\sqrt{3} \times \sqrt{3})$–$(\sqrt{7} \times \sqrt{7})$ ($R = 10.9°$) (figure 12.3(a)) the C- and Si-dominated π subband and the first two σ subbands are initiated, respectively, from the K/Γ, Γ/Γ and Γ/Γ valleys with state energies of −0.20/0.16, −3.15/−0.85 and −3.15/0.85 eV. The Si-dominated ones remain concave-downward parabolic energy dispersions; furthermore, their double degeneracy is affected little by the interlayer van der Waals interactions. Specifically, the Fermi velocity of the Si- and C-dominated π bands is about 5.24×10^5 and 8.39×10^5 m s^{-1}, respectively.

Similar band structures are revealed in the other quasi-stable silicene–graphene composite systems, and the only difference lies in whether the Si-related Dirac cones could survive near the Γ or K/K′ point. For example, the bilayer systems of $\sqrt{7} \times \sqrt{7}$ and $\sqrt{4} \times \sqrt{4}$ ($R = 19.1°$) (figure 12.3(d)) possess initial π and σ valence subbands at the K and Γ valleys, respectively. Their Fermi velocities, corresponding to the Si- and C-related π bands are, respectively, $\sim 4.83 \times 10^5$ and $\sim 8.02 \times 10^5$ m s^{-1}. The initial formation of the π- and σ-band structure illustrates clearly their existence in the C and Si honeycomb lattices using modified bonding. However, frequent anticrossing and crossing of the valence and conduction subbands appear at $E^v <$ −1.0 eV, such that it would be very difficult to characterize the entire π- and σ-band widths well at deeper energies. In general, an avoided crossing occurs when two energy subbands possess, at least, a comparable common component (or same-order multi-independent modes). Here the atom- and orbital-decomposed contributions (the densities of states in figures 12.5(a), (b) and (c)) represent the critical factors. Through careful calculations and analyses, the anticrossing behaviors are attributed to the non-orthogonal π and σ chemical bonding in the silicene layer, the $3p_z$–$2p_z$

single-orbital interlayer hybridization with non-uniform hopping integrals, and the different ionization potentials of the 2s, 2p, 3s and 3p orbitals.

The phenomenological model, i.e. the tight-binding mode, might be reliable and suitable for bilayer silicene–graphene composite systems in obtaining a thorough understanding of the fundamental properties, that is, it would be very useful in providing full information about the coupling-varied physical, chemical and material phenomena. From the first-principles calculations, the main features of the band structures include a greater number of valence and conduction subbands, an asymmetric electron and hole energy spectrum about the Fermi level, slightly modified Dirac-cone structures, and identified π- and σ-band widths and their anticrossing behavior. These are related closely to, respectively, the moiré superlattice, the interlayer single-orbital $3p_z$–$2p_z$ atomic interactions, the different orbital energies of $3p_z$ and $2p_z$, the weak but significant distortion of chemical bonding, and the non-orthogonalities of ($3s$, $3p_x$, $3p_y$)–$3p_z$ and $3p_z$–$2p_z$. These critical mechanisms would need to be included in an important Hamiltonian under the tight-binding model, so that the fitted low-lying energy bands are close to those in figures 12.3(a) and (d). Most importantly, the intrinsic atomic interactions possess different on-site Coulomb energies due to the ionization potentials, and intralayer and interlayer hopping integrals associated with four Si and C orbitals (sp^3 and sp^2 in the silicon and carbon honeycomb lattices, respectively, and various $3p_z$–$2p_z$ orbital hybridization cases within a moiré superlattice). How to solve the position-dependent coupling hopping integrals becomes a critical issue, for which an empirical formula has been successful in studying the low-energy band structures of bilayer graphenes and coaxial carbon nanotubes [28, 29]. It is then able to deal with the complicated magnetic quantization phenomena through the generalized tight-binding model [30], as developed previously for various magneto-electronic properties, rich magneto-optical absorption spectra and their selection rules, unusual quantum Hall conductivities, and varied magnetoplasmon modes and inter-LL transitions [30].

The theoretical predictions conducted on wave-vector-dependent band structures below the Fermi level could be examined using high-resolution ARPES (details in section 3.3). Previous measurements have identified the low-energy valence bands of layered graphene systems, which are initiated only from the K/K′ valleys, such as the linear Dirac-cone structure in twisted bilayer graphenes and monolayer graphene, the parabolic and linear dispersions in bilayer and trilayer AB stacking, respectively, and the linear, partially flat and sombrero-shaped bands in trilayer ABC stacking [31, 32]. These varied electronic energy spectra are verified to be associated only with the pure and unique interlayer $2p_z$–$2p_z$ orbital hybridization in normal and enlarged honeycomb lattices, according to the good agreement with the first-principles method and the tight-binding model. Moreover, the significant chemical bonding between the 2D group-IV materials and substrates is confirmed directly from the largely separated Dirac points (semiconducting behavior), their complete absence and new energy bands across the Fermi level (the metallic phenomenon). Further experimental measurements are required to examine the occupied band structures in bilayer silicene–graphene composite systems. The main features include the coexistence and separation of two distinct modified Dirac-cone structures with or without energy spacing at the K/K′ and Γ valleys associated with zone folding, the free conduction electrons and valence holes due to the weak band overlaps, the lower/higher Fermi velocity in a silicon/carbon honeycomb lattice, the normal formation of C- and Si-dominated σ-electronic valence subbands, and the frequent crossing and anticrossing between the high-symmetry points for the $E^v < -1.0$ eV valence bands. Careful analyses of ARPES measurements are capable of providing sufficient information about the unusual moiré superlattice, the stacking configuration, the intralayer orbital hybridization in silicene and graphene (the significant sp^3 and pure sp^2 bonding, respectively), and their interlayer coupling through the $3p_z$–$2p_z$ orbital hybridization.

The spatial charge distributions and their variation (ρ and $\Delta\rho$, details section 4.1) after silicene–graphene coupling (figures 12.4(a)–(j)) directly show weak, but significant, interlayer atomic interactions, in particular for $\Delta\rho$. For ($\sqrt{3} \times \sqrt{3}$)–($\sqrt{7} \times \sqrt{7}$) ($R = 10.9°$), the σ-electronic charge densities between, respectively, two silicon and two carbon atoms present redistribution behaviors (figures 12.4(a) and (d)), compared to the monolayer cases (figures 12.4(b) and (e)). There are more charges around atoms, but the opposite is true for the middle region. This illustrates clearly the reduced strength of σ bonding on the silicene and graphene layers. In the buckled silicene honeycomb lattice, the non-orthogonality of the σ and π bonding becomes more obvious in the presence of interlayer atomic interactions, or their

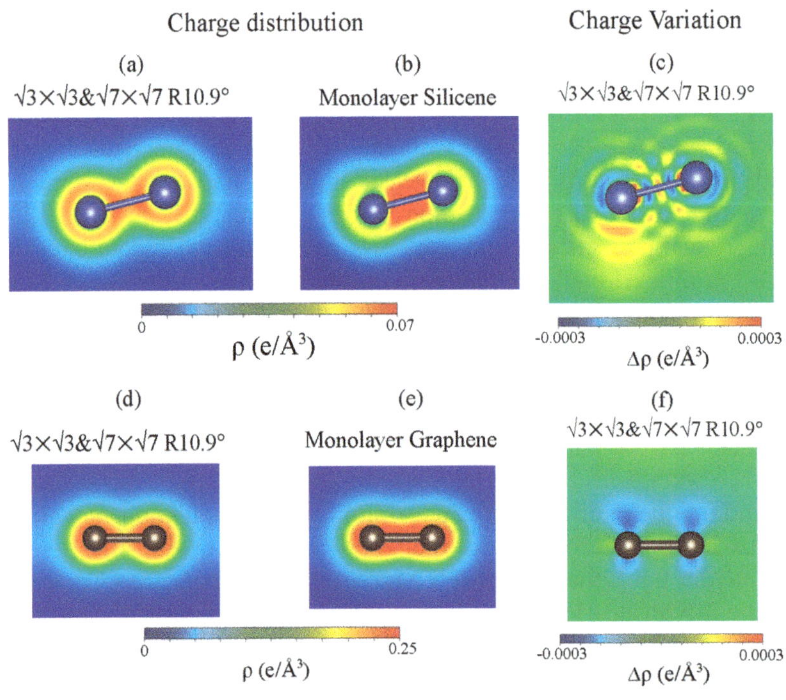

Figure 12.4. The spatial charge density distributions ((a), (b), (d) and (e)) and the variation of charge densities ((c) and (f)) after the coupling of composite silicene–graphene bilayer systems under various conditions: (a) and (c) $\sqrt{3} \times \sqrt{3}$, (d) and (f) $\sqrt{7} \times \sqrt{7}$ ($R = 10.9°$), (b) monolayer silicene and (e) single-layer graphene.

hybridization is becoming stronger under the interlayer $3p_z$–$2p_z$ bonding. These features are consistent with the charge density variation in the Si–Si and C–C bonds (figures 12.4(c) and (f)). The $\Delta\rho$ of silicene presents certain observable regions related to the σ- and π-electronic orbitals, while that in graphene exhibits only changes along the perpendicular direction through the $2p_z$ orbitals. Similar phenomena for charge density distributions and their variation are revealed in other silicene–graphene composite systems, e.g. those of the $(\sqrt{7} \times \sqrt{7})$–$(\sqrt{4} \times \sqrt{4})$ ($R = 19.1°$) configuration (not shown). Most importantly, the interlayer $3p_z$–$2p_z$ single-orbital hybridization is responsible for the two modified Dirac-cone structures across the Fermi level (figures 12.3(a)–(d)).

The atom- and orbital-decomposed van Hove singularities, as shown in figures 12.5(a)–(c), are very useful in determining the critical chemical bonding (the important orbital hybridization). The 2D band-edge states, which correspond to the Dirac cones, local extrema, saddle points, constant-energy loops and partial flat

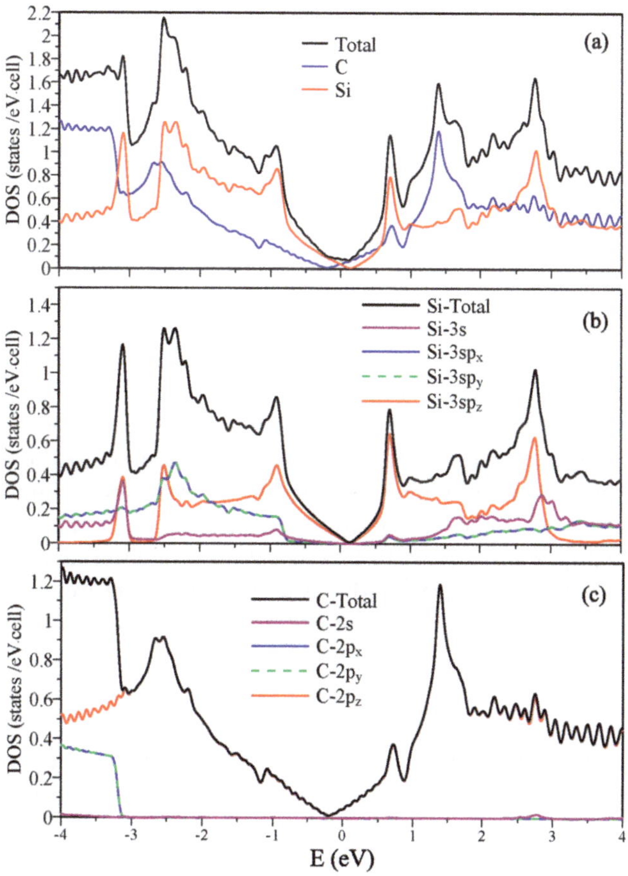

Figure 12.5. The silicon-, carbon- and (s, p_x, p_y, p_z)-orbital-projected density of states for bilayer silicene–graphene composites under the different cases: $(\sqrt{3} \times \sqrt{3})$–$(\sqrt{7} \times \sqrt{7})$ ($R = 10.9°$) in terms of (a) Si and C atom contributions, (b) Si orbital contributions and (c) C orbital contributions.

energy dispersions present, respectively, V-shapes, shoulders, symmetric peaks in the logarithmically divergent form, square-root divergent peaks and delta-function-like peaks. V-shaped structures appear to be revealed above or below the Fermi level. The silicene–graphene composite of ($\sqrt{3} \times \sqrt{3}$)–($\sqrt{7} \times \sqrt{7}$) (R = 10.9°) shows a modified low-lying plateau across E_F, which is the linear superposition of two V-shaped structures due to the Si-3p$_z$ and C-2p$_z$ orbitals (the blue and red curves in figure 12.5(a)). It should be noted that the slope of each V-shape is inversely proportional to the Fermi velocity. The two V-shaped structures due to the Si-3p$_z$ and C-2p$_z$ orbitals occur at ∼15 and −20 meV, since they arise from the Γ- and K-valley Dirac cones (figure 12.3(a)). The existence of few free conduction electrons and valence holes reflects directly the Coulomb site-energy difference of the Si-3p$_z$ and C-2p$_z$ orbitals (or shows their hybridization as significant interlayer atomic interactions). Furthermore, a finite density of states at E_F clearly indicates semi-metallic behavior.

Interestingly, the strongest evidence for single-orbital interlayer orbital hybridization can be identified from the π and π^* peaks in the logarithmic structure. Clearly they originate from the saddle M points of the π valence and π^* conduction bands (figures 12.3(a)–(c)). Such peaks represent one of the most important characteristics of sp^2-bonding-dominated materials, such as graphene, silicene and germanene. Two honeycomb lattices in a silicene–graphene composite are coupled together under van der Waals interactions, so the Si- and C-induced π and π^* peaks arise simultaneously. That is to say, there exist two strong π peaks at $E \sim -1.02$ and −2.65 eV, and two strong π^* peaks at $E \sim 0.65$ and 1.50 eV (the black solid curve in figure 12.5(a)). Specifically, the first and second π peaks are merged with the special structures, respectively, due to the initial and saddle-point electronic states (the Γ and M points in figure 12.3(a)) of the Si-(3p$_x$, 3p$_y$) σ-electronic orbitals (the deep blue curve and dashed green curve in figure 12.5(b)). Moreover, the initial C-(2p$_x$, 2p$_y$)-orbital states produce an obvious shoulder structure at $E \sim -3.20$ eV (figure 12.5(c)). In addition, more extra van Hove singularities, compared to those of monolayer systems (not shown), should be associated with the zone-folding effects of the extremely non-uniform chemical environment (figure 12.1). The above-mentioned results show the dominant behavior of sp^2 bonding, while the significant sp^3 bonding in a buckled silicene lattice leads to merged structures of four orbitals (figure 12.5(b)). They are also revealed in the density of states of the ($\sqrt{7} \times \sqrt{7}$)–($\sqrt{4} \times \sqrt{4}$) (R = 19.1°) configuration (not shown).

High-resolution STS measurements are sufficiently sensitive to examine the merged, split and modified van Hove singularities in a silicene–graphene composite material due to the weak, but important, van der Waals interactions. Such experiments have been successful in the identification of significant coupling effects in few-layer graphene systems, e.g. an almost symmetric V-shape vanishing at the Fermi level for a monolayer system (a zero-gap semiconductor), logarithmically symmetric peaks in twisted bilayers [32], a gate-voltage-induced energy gap under bilayer AB and trilayer ABC stacking, a delta-function-like peak centered about the Fermi level for trilayer and penta-layer ABC stacking, and a sharp dip structure near E_F

combined with a pair of square-root peaks in trilayer AAB stacking (a narrow-gap semiconductor with constant-energy loops). Further experimental studies conducted on the modified plateau structure across E_F (the linear superposition of the two V-shapes centered at positive and negative energies), two logarithmic π and π^* peaks, and certain obvious shoulders at deeper energies (the initiation and termination of the π and σ bands), could verify the intralayer π and σ bonding and the interlayer $3p_z$–$2p_z$ single-orbital hybridization. This is very useful in establishing a reliable tight-binding model in the presence of a moiré superlattice (the non-uniform interlayer hopping integrals in an enlarged unit cell).

Interestingly, the bilayer silicene–graphene composite is completely different to monolayer silicene or graphene and any bilayer silicene or graphene system in terms of the fundamental physical properties (details in chapters 4 and 5, and section 12.1). The significant differences appear to include the planar or buckled honeycomb lattices, the optimal geometric parameters, the primitive unit cells, the atom- and orbital-dominated energy subbands, the asymmetry of electron and hole energy spectra about the Fermi level, the semimetallic or semiconducting behaviors, the original, modified, distorted or destroyed Dirac-cone band structures, the varied chemical bonding in the spatial charge distributions, the intralayer sp^2 or sp^3 orbital hybridization and the interlayer $2p_z$–$2p_z$, $2p_z$–$3p_z$ or ($3s$, $3p_x$, $3p_y$, $3p_z$)–($3s$, $3p_x$, $3p_y$, $3p_z$) hybridization, and the zone-folding- and hopping-integral-enriched van Hove singularities in the density of states. The strong competition and/or cooperation between the sp^2 and sp^3 bonding leads to different heights in the two silicene-based A and B sublattices. The moiré superlattices could be enlarged greatly by the twisted effects, stacking modulations, the coupling of two distinct monolayer subsystems, and uniform or non-uniform external electric and magnetic fields. van der Waals interactions could survive in most bilayer systems, except for the strongly coupled silicene–silicene systems. The asymmetry degree of the electron and hole bands about the Fermi level is estimated to follow the sequence silicene–silicene > silicene–graphene > graphene–graphene > silicene > graphene. The valence and conduction band or subband overlaps, which are found only in bilayer silicene–graphene composites and graphene systems, induce low-density conduction electrons and valence holes simultaneously and thus belong to the semimetals. Furthermore, the highest free carrier density arises in AA-stacked bilayer graphene with the highest stacking symmetry. The two monolayer systems and bilayer silicenes exhibit only almost negligible- or finite-gap behaviors. Generally, the Dirac-cone-related energy spectra are completely absent in bilayer silicenes under the very important intralayer and interlayer sp^3 chemical bonding. The pristine or slightly modified Dirac cones survive only in monolayer silicene and graphene, AA- and AA′-stacked graphenes, and silicene–graphene composites. The low-lying energy bands are initiated from the K/K′ valleys for monolayer silicene and graphene, all the bilayer graphenes, and specific bilayer silicene–graphene composites, while they can appear at both the K and Γ valleys simultaneously for the other bilayer silicene–graphene systems. The pure or hybridized π bonding and the defined or undefined sp^2 and sp^2 chemical bonding display varied charge distributions in the five kinds of silicene- and graphene-related systems, including in terms of their positions, strengths and profiles. The multi- or single-orbital hybridization in various chemical bonds, as well as the uniform or extremely non-uniform chemical environments in the honeycomb lattices and the moiré superlattices. will dominate the main features of the electronic energy spectra and van Hove singularities in the density of states and thus the other fundamental properties. The band-edge states, which correspond to the Dirac, extreme, saddle and dispersionless points, and quasi-1D loops in the energy–wave-vector space, are able to create the V-shape, shoulder, symmetric peak in the logarithmic form, delta-function-like peak and square-root divergent peak. The energy, number, intensity and form of these van Hove singularities could be regarded other important characteristics of silicene- and graphene-based materials.

In addition to the geometric and electronic properties, the above-mentioned monolayer and bilayer systems present rich and unique magnetic quantization phenomena, according to up-to-date theoretical calculations [30] and magnetic measurements [33]. The diverse magneto-electronic, magneto-optical and transport properties include the well-behaved, perturbed and undefined LLs, the sublattice- and spin-dominated LLs, the square-root, linear and abnormal magnetic-field dependence, the almost symmetric or the highly asymmetric electron and hole spectra about the Fermi level, the crossing, non-crossing and anticrossing behaviors, the single- and multi-valley initial structures, the gate-voltage-enriched LLs, the specific and extra magneto-selection rules, the single, double and twin delta-function-like absorption peaks, and the normal or irregular quantum Hall conductivities and their strong dependence on the stacking configurations and layer numbers [30, 34]. Clearly there exist certain important differences among these five kinds of emergent materials. For example, monolayer graphene and silicene exhibit Dirac-cone-quantized magneto-electronic states with high degeneracy. However, the latter is changed dramatically under an external gate voltage through the spin- or valley-split LLs, owing mainly to the coupling of electric fields and spin–orbit interactions. Stacking-varied LLs are observed easily in the sliding bilayer graphenes [30]. Their low-energy features are characterized by carbon-$2p_z$ orbitals, while four-orbital hybridization is responsible for the magnetic quantization of bilayer silicene [35]. A bilayer silicene–graphene composite system is expected to present unusual double-valley LLs, as quantized magnetically from the low-lying modified Dirac cones at the K and Γ points (figure 12.2(a)). Such topics are worthy of further investigation.

12.2 Ge–graphene bilayer composites

A specific bilayer germanene–graphene composite system, with the stacking configurations of 3×3 and 5×5 under a zero twisted angle, is shown in figures 12.6(a) and (b). The two subsystems are represented by, respectively, the blue and red spheres. Based on detailed first-principles calculations and careful analyses, this material possesses the smallest moiré superlattice among all the metastable bilayer composites, corresponding to 18 germanium and 50 carbon atoms in the enlarged unit cells of the honeycomb lattices. The Ge–Ge and C–C bond lengths are in, respectively, the ranges of ~2.463–2.467 Å and ~1.420–1.423 Å (table 12.3). The modulations of

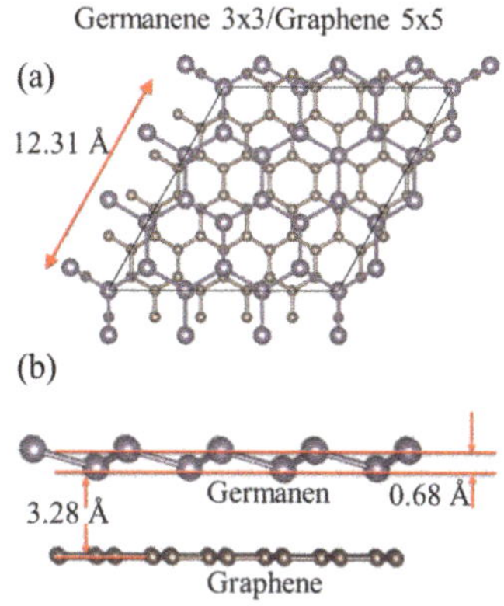

Figure 12.6. A bilayer germanene–graphene composite observed on the (a) (x, y)- and (b) (x, z)-plane projections, with moiré superlattices of 3 × 3 and 5 × 5 (R = 0°) for the germanium and carbon honeycomb lattices (the blue and red spheres), respectively. (a) shows the lattice constant of 12.31 Å and (b) the interlayer distance of 3.28 Å and the buckled height of 0.68 Å.

the chemical bonds are very small. Most importantly, the enhanced lattice constant on the (x, y)-plane is 12.31 Å (figure 12.6(a)), the interlayer distance is 3.28 Å and the buckled height on germanene is 0.68 Å (figure 12.6(b)). Specifically, the carbon honeycomb lattice remains a planar structure, as observed in the silicene–graphene cases. Moreover, its interlayer distance is slightly smaller than or comparable to those of bilayer graphenes and silicene–graphene systems [5]. This suggests weak, but very important, interlayer van der Waals interactions. A germanene–graphene composite is determined to possess critical $2p_z$–$4p_z$ interlayer orbital hybridization, which is further confirmed by the atom-dominated energy spectra, the spatial charge distributions, and the atom- and orbital-projected van Hove singularities (see figures 12.7–12.9). Apparently, quite complicated interlayer hopping integrals, which are created by the various effective interlayer atomic interactions, arise in the specific moiré superlattice with an extremely non-uniform chemical, physical and material environment.

The band structures of germanene–graphene composites are changed dramatically after significant interlayer atomic interactions. The pristine monolayer graphene and germanene ($E_g \sim 24$ meV) belong to, respectively, the zero- and finite-gap semiconductors at the K/K′ points, as demonstrated clearly in the reduced first Brillouin zones of the 5 × 5 and 3 × 3 enlarged unit cells (figures 12.7(a) and (b)). However, the composite system presents very complicated electronic energy spectra in figures 12.7(c) and (d) with carbon- and germanium-atom dominance, respectively. It is an indirect-gap semiconductor of $E_g^i \sim 3.3$ meV (the inset in figure 12.7(c)), in which the highest valence-band state and the lowest conduction band state are at, respectively, the K and Γ valleys. Interestingly, these two energy bands come mainly from graphene and germanene. Furthermore, the low-lying electronic states near the K and Γ valleys are contributed mostly by the carbon and germanium honeycomb lattices, respectively. The former exhibits a modified Dirac-cone structure with an energy spacing of Dirac points of about 70 meV (the red

Table 12.3. The geometric bond lengths and the total ground state per unit cell for a bilayer germanene–graphene composite with moiré superlattices of 3 × 3 and 5 × 5 without a relative rotation angle.

Bond	Length (Å)	Bond	Length (Å)	Bond	Length (Å)	Bond	Length (Å)
C1–C2	1.422 08	C12–C21	1.422 63	C36–C37	1.420 41	Ge1–Ge2	2.465 75
C2–C3	1.420 56	C14–C23	1.422 02	C37–C38	1.422 26	Ge2–Ge3	2.464 42
C3–C4	1.421 84	C16–C25	1.422 01	C38–C39	1.422 27	Ge3–Ge4	2.465 74
C4–C5	1.422 61	C18–C27	1.422 64	C39–C40	1.420 42	Ge4–Ge5	2.465 72
C5–C6	1.420 71	C20–C29	1.422	C40–C31	1.421 66	Ge5–Ge6	2.464 43
C6–C7	1.420 69	C21–C22	1.420 71	C32–C41	1.4217	Ge6–Ge1	2.465 73
C7–C8	1.422 72	C22–C23	1.420 62	C34–C43	1.421 74	Ge2–Ge7	2.464 46
C8–C9	1.421 68	C23–C24	1.421 96	C36–C45	1.421 49	Ge7–Ge8	2.465 72
C9–C10	1.420 62	C24–C25	1.421 99	C38–C47	1.422 29	Ge8–Ge9	2.463 16
C10–C1	1.422 02	C25–C26	1.420 62	C40–C49	1.421 46	Ge4–Ge9	2.4632
C2–C11	1.420 58	C26–C27	1.420 66	C41–C42	1.420 62	Ge9–Ge10	2.463 16
C4–C13	1.421 11	C27–C28	1.421 76	C42–C43	1.420 59	Ge10–Ge11	2.465 73
C6–C15	1.420 64	C28–C29	1.421 48	C43–C44	1.421 43	Ge6–Ge11	2.464 47
C8–C17	1.421 17	C29–C30	1.421 45	C44–C45	1.421 98	Ge11–Ge12	2.464 81
C10–C19	1.420 62	C30–C21	1.421 68	C45–C46	1.421 45	Ge12–Ge7	2.464 82
C11–C12	1.421 84	C22–C31	1.420 72	C46–C47	1.420 42	Ge8–Ge13	2.465 78
C12–C13	1.421 09	C24–C33	1.420 74	C47–C48	1.420 42	Ge13–Ge14	2.464 41
C13–C14	1.420 76	C26–C35	1.420 68	C48–C49	1.421 44	Ge14–Ge15	2.464 41
C14–C15	1.421 99	C28–C37	1.420 42	C49–C50	1.422 06	Ge10–Ge15	2.465 78
C15–C16	1.421 97	C30–C39	1.420 43	C50–C41	1.421 41	Ge15–Ge16	2.464 82
C16–C17	1.420 72	C31–C32	1.422 72	C42–C1	1.422 05	Ge16–Ge17	2.466 77
C17–C18	1.421 16	C32–C33	1.421 15	C44–C3	1.421 42	Ge12–Ge17	2.466 82
C18–C19	1.421 72	C33–C34	1.421 16	C46–C5	1.4217	Ge17–Ge18	2.466 78
C19–C20	1.421 43	C34–C35	1.422 61	C48–C7	1.421 68	Ge18–Ge13	2.464 83
C20–C11	1.4214	C35–C36	1.421 76	C50–C9	1.421 43	Ge14–Ge1	2.465 77
						Ge16–Ge3	2.464 85
						Ge18–Ge5	2.464 85

sphere near the Fermi level and K valley). This indicates clearly the slight, but significant, modification of the π-electronic bonding through the interlayer van der Waals interactions. A similar structure is revealed in the germanium honeycomb lattice near the Γ valley (the blue spheres in figure 12.7(d)). Within the low-energy range of -1.0 eV $\leqslant E^{c,v} \leqslant 1.0$ eV, there is also one conduction and two valence energy subbands that are initiated from the Γ valley. They are identified from the coupling effects of the sp^2 bonding and the spin–orbit interactions (details in chapter 4), which is attributed to the intrinsic properties of 2D germanene systems. However, such mechanisms are deduced to induce barely observable modifications of the graphene Dirac cone. For the deeper valence and high conduction energy subbands, frequent crossing and some anticrossing arise. This illustrates the critical

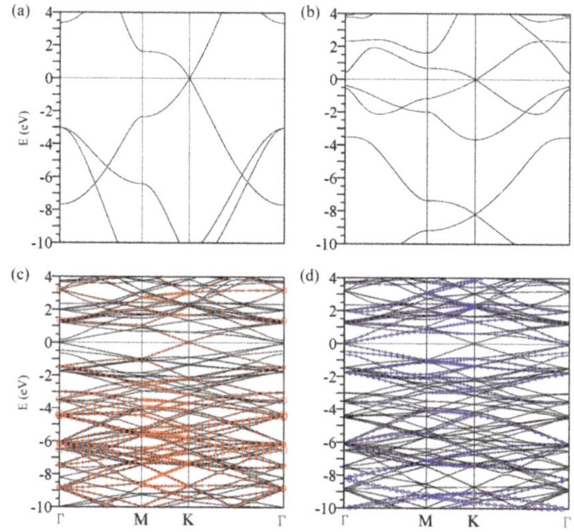

Figure 12.7. Electronic energy spectra with strong wave-vector dependence along the 2D high-symmetry points in the reduced first Brillouin zone for bilayer germanene–graphene composites with the specific moiré superlattices of 3 × 3 and 5 × 5 (R = 0°): (a) pristine graphene and (b) germanene, and composite bilayer systems with (c) carbon- (red spheres) and (d) germanium-atom dominance (blue spheres).

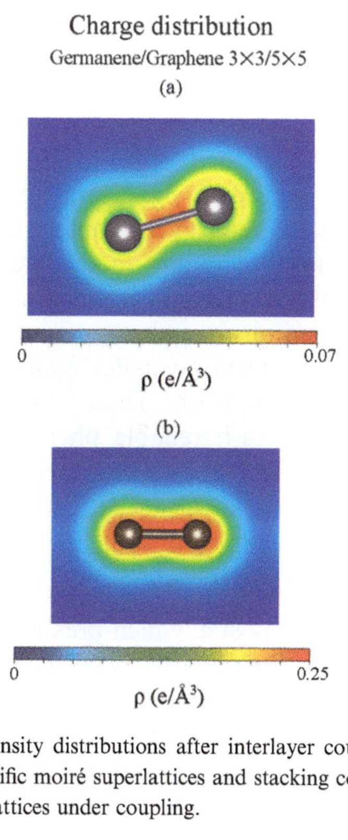

Figure 12.8. The spatial charge density distributions after interlayer coupling in a composite germanene–graphene bilayer system under specific moiré superlattices and stacking configurations: (a) 3 × 3 germanium and (b) 5 × 5 carbon honeycomb lattices under coupling.

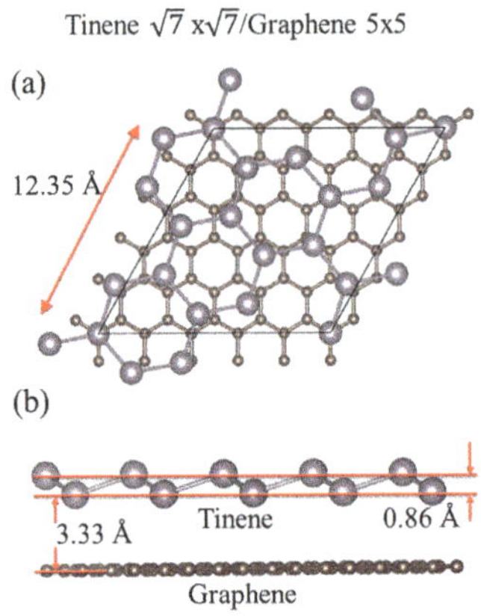

Figure 12.9. A bilayer tinene–graphene composite system on the (a) (x, y)- and (b) (x, z)-planes, with quasi-stable moiré superlattice of $\sqrt{7} \times \sqrt{7}$ and 5×5 ($R = 19.1°$) for the germanium and carbon honeycomb lattices (the blue and red spheres), respectively. Also shown in (a) is the lattice constant of 12.35 Å and in (b) an interlayer distance of 3.33 Å and a buckled height of 0.86 Å.

mechanisms, the zone folding, and the orbital- and spin-dependent interactions. It appears to be difficult to identify the π- and σ-band widths from the complex band structure.

The total charge densities and their variation after coupling effects, as shown for a germanene–graphene composite in figures 12.8(a) and (b), are very useful in understanding the critical orbital hybridization in the intralayer Ge–Ge and C–C bonds and interlayer Ge–C interactions. The π and σ bonding (figure 12.8(a)) could survive in the germanium honeycomb lattice, indicating the dominant feature of the latter. Their non-orthogonality, which is demonstrated clearly in the deformed charge density distributions, indicates significant sp^3 bonding. This behavior is somewhat enhanced through the interlayer coupling (figure 12.8(c)). On the other hand, the two types of chemical bonding in the graphene layer (figure 12.8(b)) remain orthogonal to each other in the presence of bilayer superposition, and its $2p_z$-orbital charge density shows an obvious change along the z-direction, as shown in figure 12.8(d). As a result, the important van der Waals interactions are identified to arise from mixed $2p_z$–$4p_z$ hybridization. These concise pictures of the chemical bonds could be utilized to establish reliable phenomenological models, e.g. the intralayer and interlayer hopping integrals due to the multi- and/or single-orbital hybridization of the tight-binding model.

12.3 Sn–graphene bilayer materials

A bilayer stanene–graphene composite, which presents the specific stacking configuration of $(\sqrt{7} \times \sqrt{7})$–$(5 \times 5)$ with a relative rotational angle of $R = 19.1°$, is chosen for the model study (figures 12.10(a) and (b)). The tin and carbon honeycomb lattice subsystems are illustrated by the blue and red spheres, respectively. According to accurate first-principles calculations and detailed analyses, such a material has 14 Sn and 50 C atoms in the enlarged unit cells, corresponding to the smallest moiré

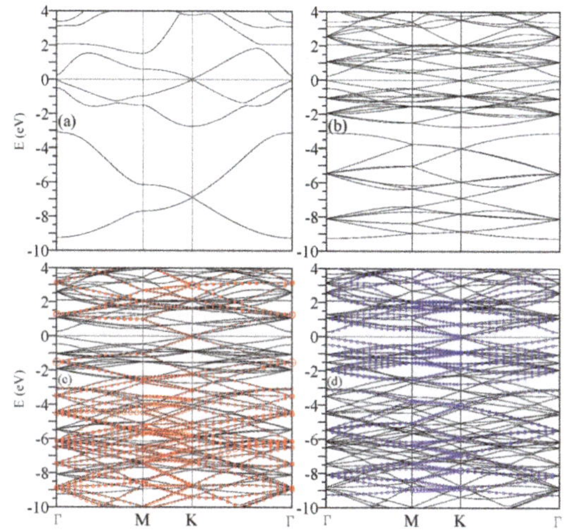

Figure 12.10. Band structure along the 2D high-symmetry points in the reduced first Brillouin zone for monolayer stanine under the (a) 1 × 1 and (b) $\sqrt{7} \times \sqrt{7}$ structures, and for a bilayer stanene–graphene composite system with the specific stacking configuration of ($\sqrt{7} \times \sqrt{7}$)–(5 × 5) ($R = 19.10°$) with the (c) the C dominance (blue spheres) and (d) Sn dominance (red spheres).

superlattice among all the metastable bilayer composite systems. The Sn–Sn and C–C bond lengths lie in, respectively, the ranges of ~2.828–2.832 Å and ~1.425–1.428 Å (table 12.4). Interestingly, the greatly enhanced lattice constant on the (x, y)-plane is 12.35 Å (figure 12.10(a)), the interlayer distance is 3.33 Å and the buckled height on germanene is 0.86 Å (figure 12.10(b)). However, graphene retains a planar structure, as revealed in the silicene– and germanene–graphene bilayer composites (figures 12.1 and 12.6). Moreover, bilayer group-IV-graphene systems possess comparable interlayer distances. Apparently, this indicates weak, but important, interlayer van der Waals interactions. A bilayer tinene–graphene composite will be studied thoroughly and will be found to have very important $2p_z$–$5p_z$ interlayer orbital hybridization through the atom-dominated energy subbands, spatial charge distributions and variation, and atom- and orbital-projected densities of states (see figures 12.11). Furthermore, the effective interlayer atomic interactions should be quite complicated under a specific moiré superlattice as a result of the extremely non-uniform chemical and physical environment. It is also noted that both sp^3 bonding and spin–orbit coupling are non-negligible for the greatly buckled Sn honeycomb lattices.

The bilayer tinene–graphene composite systems exhibit unusual band structures, as shown clearly by the stable ($\sqrt{7} \times \sqrt{7}$)–(5 × 5) structure in figures 12.11(a) and (b). Their main features are directly those (figures 12.7(a) and 12.11(b)) of two independent subsystems under minor, but observable, modifications, in particular for the low-lying energy subbands. Obviously, two pairs of modified Dirac-cone structures, which are associated with C and Sn honeycomb lattices (the red and blue spheres), arise at the K valley. They are doubly degenerate near the Fermi level in

Table 12.4. The geometric bond lengths and the total ground state per unit cell for a bilayer stanene–graphene composite system with a moiré superlattice of $(\sqrt{7} \times \sqrt{7})$–$(5 \times 5)$ under a relative rotation angle of $R = 19.1°$.

Bond	Length (Å)	Bond	Length (Å)	Bond	Length (Å)
C1–C2	1.427 35	C28–C29	1.425 83	Sn1–Sn4	2.827 74
C2–C3	1.425 52	C29–C30	1.427 93	Sn1–Sn12	2.827 62
C3–C4	1.425 62	C30–C21	1.426 83	Sn1–Sn14	2.829 08
C4–C5	1.426 36	C22–C31	1.426 26	Sn2–Sn3	2.830 58
C5–C6	1.426 26	C24–C33	1.426 75	Sn2–Sn5	2.832 11
C6–C7	1.425 23	C26–C35	1.425 27	Sn2–Sn13	2.831 38
C7–C8	1.425 32	C28–C37	1.425 51	Sn3–Sn6	2.832 12
C8–C9	1.425 85	C30–C39	1.4263	Sn3–Sn14	2.8302
C9–C10	1.425 53	C31–C32	1.426 35	Sn4–Sn5	2.8306
C10–Sn1	3.611 92	C32–C33	1.427 48	Sn4–Sn7	2.829 42
C2–C11	1.425 51	C33–C34	1.425 84	Sn5–Sn8	2.831 02
C4–C13	1.427 51	C34–C35	1.425 32	Sn6–Sn7	2.830 94
C6–C15	1.4257	C35–C36	1.4268	Sn6–Sn9	2.831 19
C8–C17	1.425 85	C36–C37	1.426 27	Sn7–Sn10	2.828 54
C10–C19	1.425 56	C37–C38	1.425 63	Sn8–Sn9	2.832
C11–C12	1.4259	C38–C39	1.4256	Sn8–Sn11	2.831 13
C12–C13	1.425 83	C39–C40	1.4255	Sn9–Sn12	2.830 35
C13–C14	1.4267	C40–C31	1.425 48	Sn10–Sn11	2.828 29
C14–C15	1.426 05	C32–C41	1.4256	Sn10–Sn13	2.8285
C15–C16	1.425 31	C34–C43	1.425 85	Sn11–Sn14	2.829 21
C16–C17	1.426 74	C36–C45	1.427 88	Sn12–Sn13	2.829 48
C17–C18	1.427 51	C38–C47	1.425 61		
C18–C19	1.425 55	C40–C49	1.425 84		
C19–C20	1.4251	C41–C42	1.425 57		
C20–C11	1.425 91	C42–C43	1.4255		

(Continued)

Table 12.4. (*Continued*)

Bond	Length (Å)	Bond	Length (Å)	Bond	Length (Å)
C12–C21	1.425 36	C43–C44	1.425 89		
C14–C23	1.425 29	C44–C45	1.426 15		
C16–C25	1.4261	C45–C46	1.425 91		
C18–C27	1.426 37	C46–C47	1.425 52		
C20–C29	1.426 17	C47–C48	1.426 26		
C21–C22	1.425 24	C48–C49	1.427 83		
C22–C23	1.425 67	C49–C50	1.426 17		
C23–C24	1.426 07	C42–C1	1.427 37		
C24–C25	1.425 31	C44–C3	1.425 13		
C25–C26	1.425 74	C46–C5	1.425 53		
C26–C27	1.426 27	C48–C7	1.426 81		
C27–C28	1.425 47	C50–C9	1.425 94		

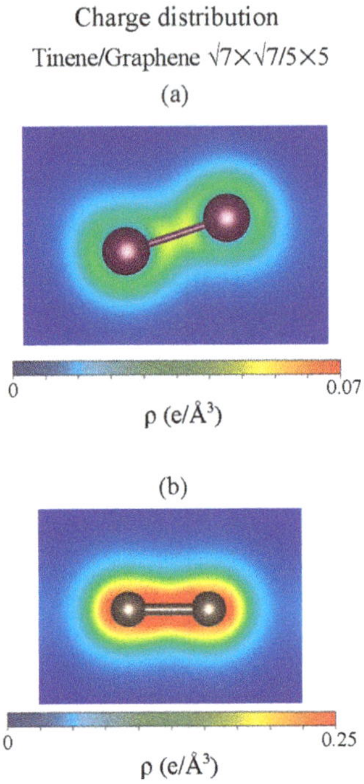

Figure 12.11. The spatial charge densities after the formation of a composite tinene–graphene bilayer system under the specific moiré superlattice of ($\sqrt{7} \times \sqrt{7}$)–(5×5) (R = 19.1°) for (a) tin and (b) carbon honeycomb lattices.

the absence of spin degrees of freedom. Furthermore, the energy spacing due to the separated valence and conduction Dirac points is about 60.0 meV. It should be noted that an indirect gap of $E^i_g \sim 30.0$ meV is determined by the lowest unoccupied conduction band states and the highest occupied valence-band state at, respectively, the K and Γ valleys. The former and the latter correspond, respectively, to the π- and σ-electronic energy subbands (details in chapter 4). Interestingly, the low-energy subbands of $|E^{c,v}| < 1.0$ eV, which are initiated from the Γ valley, are dominated by the Sn atoms (the blue spheres). In short, a quite complicated band structure, with a lot of crossing and some anticrossing, is related strongly to the coupled effects, the zone folding of the bilayer composite, the dominant sp^2 bonding in the honeycomb lattices, the significant sp^3 bonding of tinene, the interlayer $2p_z$–$5p_z$ orbital hybridization, and the spin–orbit interactions.

The spatial charge density distributions, which create the chemical bonds, are responsible for most of the features in the electronic energy spectra and density of states. The bilayer stanine–, germanene– and silicene–graphene composite systems present similar phenomena (the sp^3 and sp^2 bonding on the inorganic and organic honeycomb lattices, respectively, and the interlayer van der Waals interactions). However, for the stanine–graphene system, the σ bonding between two neighboring Sn atoms has a relatively weak charge density, and there exist more obvious sp^3 hybridization effects. Such behaviors could illustrate the existence of low-lying

σ-electronic energy bands (two valence bands and one conduction band at the Γ point in figure 12.11). In addition, the charge densities cannot provide any information about the significant spin–orbit coupling in tinene [36] as they are related closely to each other.

The atom- and orbital-decomposed van Hove singularities, which appear in the low-energy range of $|E| < 1.0$ eV, present significant differences among the tinene–, germanene– and tinene–graphene composite materials. The former two present greatly modified V-shaped structures across the Fermi level. Their densities of states at E_F approach zero, illustrating clear semiconducting behavior (the semiconductors in figures 12.7 and 12.11). Moreover, the special structures of the valence and conduction states are highly asymmetric about E_F, in particular for those due to the σ-electronic states (the (4s, 4p_x, 4p_y) and (5s, 5p_x, 5p_y) orbitals). There are more van Hove singularities arising from the valence energy spectra. Clearly, the characteristic energies between the tinene– and germanene–graphene systems are different to each other, e.g. the merged V-shape and shoulder structure at $E \to 0$ in the former. The above-mentioned features reflect directly the strong coupling effects of the van der Waals interactions, the sp^3 and sp^2 bonding, and the spin–orbit interactions. The varied low-energy densities of states could be verified using high-resolution STS measurements. In addition, the calculated data are not very accurate for the Sn- and Ge-related composites since the large unit cell and the spin-dependent interactions require a large amount of computer time.

There are certain important differences between the bilayer silicene–, germanene– and tinene–graphene composites in terms of the low-energy fundamental properties. The varied phenomena include the two pairs of modified Dirac-cone structures at the K valley or the K and Γ valleys, the zero or observable energy spacings between separated Dirac points at the same valley, the semimetallic behavior of the first system with the valence and conduction band overlap due to the two different Dirac cones, the small indirect gaps in the second and third materials, the distinct chemical bonding in creating the semiconducting properties, the initial σ-electronic valence and conduction energies, the plateau or vanishing density of states across the Fermi level, and the neighboring van Hove singularities of the σ bands. The low-energy electronic spectra appear to be very sensitive to the moiré superlattice, since the zone-folding effects are combined strongly with the significant interlayer atomic interactions, the full-orbital sp^3 hybridization in the silicene, germanene and tinene systems, and their spin–orbit coupling. Only very high-resolution ARPES [37, 38] and STS [39, 40] measurements are able to identify these critical differences.

References

[1] Leven I, Krepel D, Shemesh O and Hod O 2012 Robust superlubricity in graphene/h-BN heterojunctions *J. Phys. Chem. Lett.* **4** 115–20

[2] Mucha-Kruczyński M, Wallbank J R and Fal'Ko V I 2013 Heterostructures of bilayer graphene and h-BN: interplay between misalignment, interlayer asymmetry, and trigonal warping *Phys. Rev.* B **88** 205418

[3] Sakai Y and Saito S 2012 Electronic properties of graphene/h-BN bilayer superlattices *J. Phys. Soc. Jpn* **81** 103701
[4] Behera H and Mukhopadhyay G 2012 Strain-tunable band gap in graphene/h-BN hetero-bilayer *J. Phys. Chem. Solids* **73** 818–21
[5] An X, Liu F, Jung Y J and Kar S 2013 Tunable graphene–silicon heterojunctions for ultrasensitive photodetection *Nano Lett.* **13** 909–16
[6] Nemec N and Cuniberti G 2007 Hofstadter butterflies of bilayer graphene *Phys. Rev.* B **75** 201404
[7] Wurm J, Wimmer M, Baranger H U and Richter K 2010 Graphene rings in magnetic fields: Aharonov–Bohm effect and valley splitting *Semicond. Sci. Technol.* **25** 034003
[8] Dean C R *et al* 2010 Boron nitride substrates for high-quality graphene electronics *Nat. Nanotechnol.* **5** 722
[9] De Juan F, Cortijo A, Vozmediano M A and Cano A 2011 Aharonov–Bohm interferences from local deformations in graphene *Nat. Phys.* **7** 810
[10] Mao J *et al* 2012 Silicon layer intercalation of centimeter-scale, epitaxially grown monolayer graphene on Ru (0001) *Appl. Phys. Lett.* **100** 093101
[11] Pan Y *et al* 2014 Construction of 2D atomic crystals on transition metal surfaces: graphene, silicene, and hafnene *Small* **10** 2215–25
[12] Li G *et al* 2018 Stable silicene in graphene/silicene van der Waals heterostructures *Adv. Mater.* **30** 1804650
[13] Loan P T K, Zhang W, Lin C T, Wei K H, Li L J and Chen C H 2014 Graphene/MoS_2 heterostructures for ultrasensitive detection of DNA hybridisation *Adv. Mater.* **26** 4838–44
[14] Larentis S, Tolsma J R, Fallahazad B, Dillen D C, Kim K, MacDonald A H and Tutuc E 2014 Band offset and negative compressibility in graphene–MoS_2 heterostructures *Nano Lett.* **14** 2039–45
[15] Chang H, Sun Z, Ho K Y F, Tao X, Yan F, Kwok W M and Zheng Z 2011 A highly sensitive ultraviolet sensor based on a facile *in situ* solution-grown ZnO nanorod/graphene heterostructure *Nanoscale* **3** 258–64
[16] Shin D M *et al* 2015 Freestanding ZnO nanorod/graphene/ZnO nanorod epitaxial double heterostructure for improved piezoelectric nanogenerators *Nano Energy* **12** 268–77
[17] Cai Y, Zhang G and Zhang Y W 2015 Electronic properties of phosphorene/graphene and phosphorene/hexagonal boron nitride heterostructures *J. Phys. Chem.* C **119** 13929–36
[18] Padilha J E, Fazzio A and da Silva A J 2015 van der Waals heterostructure of phosphorene and graphene: tuning the Schottky barrier and doping by electrostatic gating *Phys. Rev. Lett.* **114** 066803
[19] Yu D, Yang Y, Durstock M, Baek J B and Dai L 2010 Soluble P3HT-grafted graphene for efficient bilayer−heterojunction photovoltaic devices *ACS Nano* **4** 5633–40
[20] Roy K, Padmanabhan M, Goswami S, Sai T P, Ramalingam G, Raghavan S and Ghosh A 2013 Graphene–MoS_2 hybrid structures for multifunctional photoresponsive memory devices *Nat. Nanotechnol.* **8** 826
[21] Fu H, Zhang J, Ding Z, Li H and Meng S 2014 Stacking-dependent electronic structure of bilayer silicene *Appl. Phys. Lett.* **104** 131904
[22] Padilha J E and Pontes R B 2015 Free-standing bilayer silicene: the effect of stacking order on the structural, electronic, and transport properties *J. Phys. Chem.* C **119** 3818–25
[23] Zhong X, Pandey R and Karna S P 2012 Stacking dependent electronic structure and transport in bilayer graphene nanoribbons *Carbon* **50** 784–90

[24] Brown L, Hovden R, Huang P, Wojcik M, Muller D A and Park J 2012 Twinning and twisting of tri-and bilayer graphene *Nano Lett.* **12** 1609–15
[25] Lai Y H, Ho J H, Chang C P and Lin M F 2008 Magnetoelectronic properties of bilayer Bernal graphene *Phys. Rev.* B **77** 085426
[26] Cahangirov S, Audiffred M, Tang P, Iacomino A, Duan W, Merino G and Rubio A 2013 Electronic structure of silicene on Ag (111): strong hybridization effects *Phys. Rev.* B **88** 035432
[27] Avila J *et al* 2013 Presence of gapped silicene-derived band in the prototypical (3 × 3) silicene phase on silver (111) surfaces *J. Phys. Condens. Matter* **25** 262001
[28] Morell E S, Correa J D, Vargas P, Pacheco M and Barticevic Z 2010 Flat bands in slightly twisted bilayer graphene: tight-binding calculations *Phys. Rev.* B **82** 121407
[29] Uryu S and Ando T 2007 Electronic intertube transfer in double-wall carbon nanotubes with impurities: tight-binding calculations *Phys. Rev.* B **76** 155434
[30] Lin C Y, Wu J Y, Ou Y J, Chiu Y H and Lin M F 2015 Magneto-electronic properties of multilayer graphenes *Phys. Chem. Chem. Phys.* **17** 26008–35
[31] Zhou S Y, Gweon G H, Fedorov A V, First P D, De Heer W A, Lee D H, Guinea F, Castro Neto A H and Lanzara A 2007 Substrate-induced bandgap opening in epitaxial graphene *Nat. Mater.* **6** 770
[32] Brihuega I *et al* 2012 Unraveling the intrinsic and robust nature of van Hove singularities in twisted bilayer graphene by scanning tunneling microscopy and theoretical analysis *Phys. Rev. Lett.* **109** 196802
[33] Martin J, Feldman B E, Weitz R T, Allen M T and Yacoby A 2010 Local compressibility measurements of correlated states in suspended bilayer graphene *Phys. Rev. Lett.* **105** 256806
[34] Chen S C, Wu J Y, Lin C Y and Lin M F 2017 *Theory of Magnetoelectric Properties of 2D Systems* (Bristol: IOP Publishing)
[35] Do T N, Shih P H, Gumbs G, Huang D, Chiu C W and Lin M F 2018 Diverse magnetic quantization in bilayer silicene *Phys. Rev.* B **97** 125416
[36] Matthes L, Pulci O and Bechstedt F 2013 Massive Dirac quasiparticles in the optical absorbance of graphene, silicene, germanene, and tinene *J. Phys. Condens. Matter* **25** 395305
[37] Sprinkle M *et al* 2009 First direct observation of a nearly ideal graphene band structure *Phys. Rev. Lett.* **103** 226803
[38] Johannsen J C *et al* 2013 Direct view of hot carrier dynamics in graphene *Phys. Rev. Lett.* **111** 027403
[39] Li G, Luican A and Andrei E Y 2009 Scanning tunneling spectroscopy of graphene on graphite *Phys. Rev. Lett.* **102** 176804
[40] Brar V W, Zhang Y, Yayon Y, Ohta T, McChesney J L, Bostwick A, Horn K, Rotenberg E and Crommie M F 2007 Scanning tunneling spectroscopy of inhomogeneous electronic structure in monolayer and bilayer graphene on SiC *Appl. Phys. Lett.* **91** 122102

Chapter 13

Concluding remarks

The theoretical framework developed using first-principles calculations appears to be very useful in understanding fully the diverse phenomena of the fundamental properties in silicene-based layered materials. The various 2D emergent condensed-matter systems include the single-layer group-IV monoelements, the sliding bilayer silicenes, the hydrogen-, halogen-, Na-, Mg- and Al-adsorbed silicenes, the B-, C- and N- substituted silicene systems, Si-adlayer and bilayer AB-bt silicene on a Ag (111) substrate, the Si-adsorbed and Si-substituted graphenes, and the silicene–C, –Ge, –Sn and –Pb bilayer composites, which have different critical factors, such as different kinds of adatoms, buckled or planar geometries, stacking symmetries, layer numbers, reconstructions of unit cells, adatom chemisorption, guest-atom substitution and heterojunctions. Our detailed studies and thorough analyses are focused mainly on the optimal lattice symmetries, the atom- and spin-dominated electronic energy spectra, the semiconducting, semimetallic and metallic behaviors (the 2D carrier densities), the well-defined or undefined π- and σ-electronic energy bands, the spatial charge densities and their variation after crystal formation, the interlayer coupling and chemical adsorption and substitution, the atom-, orbital- and spin-decomposed van Hove singularities, the vanishing or finite magnetic moments, and the host- and adatom-induced spin distributions. The key pictures are derived mainly from the single- and multi-orbital hybridization and spin configurations. Some of the calculated results are consistent with up-to-date experimental measurements, while most of them require further detailed examination. The current perspectives could be generalized to other emergent 2D materials, which also need to be studied thoroughly in the future. For example, the few-layer Ge-, Sn-, Pb-, Sb- and Bi-related [1, 2] systems are worthy of systematic study in the near future, since they possess more occupied electron orbitals and significant spin–orbit coupling. Such materials are under current investigation.

doi:10.1088/978-0-7503-3299-6ch13

This book demonstrates that the first-principles method appears to be very successful in studying the geometric, electronic and magnetic properties of few-layer emergent silicene-based materials in the presence or absence of chemical adsorption and substitution, substrates and interlayer coupling. Furthermore, for single-layer group-IV monoelements, the low-lying energy bands near the K/K′ valleys could provide important hopping integrals for the generalized tight-binding model, as shown in chapter 4. The consistency between the two theoretical models illustrates clearly that the π and sp^3 bonding are, respectively, responsible for the low-energy and high-energy physics. The tight-binding model can be used to explore the rich and unique magnetic quantization phenomena in few-layer group-IV systems fully, in particular in systematic studies of carbon-created sp^2-bonding materials (carbon nanotubes, graphene nanoribbons, layered graphenes and graphites; see [3–7]). However, only a few theoretical predictions, which are associated with the Si-, Ge- and Sn-related 2D materials, are produced for the magneto-electronic properties, the selection rules of the inter-Landau-level transitions, the magnetoplasmon modes and their Landau dampings, and the quantum Hall conductivities [1, 8]. On the other hand, the very complicated band structures, which are due to enhanced buckling, important chemical modifications, multi-orbital interlayer coupling, strong substrate effects, greatly enlarged unit cells and significant ferromagnetic configurations, but not weak van der Waals interactions, cannot be simulated using the tight-binding model (or the Hubbard-like model). For example, it is expected to be very difficult to establish phenomenological models for silicene on Ag(111) because of the extremely non-uniform multi-orbital hybridization arising from the intralayer and interlayer Si–Si, Si–Ag and Ag–Ag bonds. That is, first-principles calculations would become the only effective method for understanding their fundamental properties.

Different methods have been utilized to examine the group-IV 2D materials: STM, TEM and LEED for the optimal geometries, STS for the van Hove singularities and ARPES for the occupied valence bands. The top and side views of monolayer and few-layer graphene (the planar honeycomb lattices and the stacking configurations) have been verified using, respectively, STM [9] and TEM [10]. However, the significant substrate effects in non-graphene systems might reveal commensurate or incommensurate moiré superlattices through the combined measurements of STM and LEED [9, 10], e.g. the surface reconstruction structures of (4×4)–$(\sqrt{3} \times \sqrt{3})$–$(2 \times 2)$, (3×3)–(2×2) and $(\sqrt{37} \times \sqrt{37})$ and ($R = 25.3°$), respectively, for silicene–Ag(111)/Ir(111)/ZrB_2 (0001) [11], germanene–Pt(111)/Al (111)/Au(111) [12] and Pb-adlayer–Al_2O_3/Au(111) [13]. STS studies of van Hove singularities have been performed only for graphene [14], for which they are capable of verifying the V-shaped structure with an almost zero density of states at the Fermi level, and the middle-energy π and π^* peaks. Such measurements of other group-IV systems are expected to be available to identify the critical roles of sp^3 and sp^2 bonding and spin–orbit coupling. Low-energy ARPES experiments, which have been conducted on monolayer graphene [15], silicene–Ag(111) [16], germanene–Au (111) [17], Si-adlayer–ZrB_2 (0001) [18] and tinene–Bi_2Te_3 (Pb-adlayer–Au(111) [19]), show a gapless and linear Dirac cone, a separated or modified Dirac cone, multi-

Dirac-cone structures and their complete absence, and p-type semimetals near the Fermi level. The chemical bonding between the adlayer and substrate appears to dominate and varies the geometric and electronic properties greatly.

Group-IV monolayer materials exhibit diverse fundamental properties, owing mainly to the buckled honeycomb structures, multi- and single-orbital hybridization, and significant spin–orbit coupling. The height difference of the A and B sublattices, the lattice constant and the ground state energy become large with an increase of the atomic number. Generally, there are three kinds of band structures corresponding to, respectively, the C-related systems, the Si- and Ge-related systems, and the Sn- and Pb-related systems. The spatial charge density distributions and their variation can be used to thoroughly examine the well-behaved π, σ and sp^3 chemical bonding. Whether the former two are orthogonal to each other depends on the planar or buckling geometries. However, the spin–orbit coupling cannot be identified directly from the vanishing spin configurations and magnetic moments. Graphene has a gapless Dirac-cone structure with a zero density of states at the Fermi level, while silicene and germanene possess slightly separated Dirac points (E_g = 1.55 meV and 24.0 meV) due purely to the spin–orbit interactions. The low-lying energy bands create the stable K and K′ valleys, in particular for further magnetic quantization. Graphene, silicene and germanene exhibit π and π^* bands in the ranges of, respectively, $|E^{c,v}| \approx 3.0$, 1.0 and 0.5 eV. The two σ valence bands, which are associated with the $2p_x$ and $2p_y$, $3p_x$ and $3p_y$, and $4p_x$ and $4p_y$ orbitals, respectively, are initiated from the stable Γ valley at deeper energies. It is relatively easy to observe the π and σ band mixing along MΓ and KΓ only for silicene and germanene. This is due purely the non-orthogonality of the π and σ chemical bonding. For monolayer Sn and Pb materials, the π and σ bands become co-dominant in the low-lying energy bands, revealing frequent mixing, and present $5p_x$- and $6p_x$-splitting, and $5p_y$- and $6p_y$-splitting near the Γ point. These results illustrate clearly the strong coupling of the significant sp^3 chemical bonding and spin–orbit interactions. Also note that all the low-lying band structures of the group-IV monolayer systems could be well characterized by the tight-binding model through fitting with first-principles calculation results [1, 20–22].

The above-mentioned features are consistent with the spatial charge distributions and the van Hove singularities. Specifically, the band-edge states (the critical points), the Dirac-cone form, the extreme form of parabolic dispersion, the saddle point and the constant-energy loop, respectively, create the V-shaped structure, the shoulder, the logarithmically symmetric peak and the prominent peak in the square-root dependence. Electronic and magneto-electronic properties and optical and magneto-optical absorption spectra under external electric and magnetic fields are investigated efficiently using the generalized tight-binding model [1, 22], accompanied by the dynamic Kubo formula [21]. The spin-up and spin-down combined electronic states are split by an electric field because of the destruction of mirror symmetry about the ($z = 0$)-plane. Furthermore, a finite- and zero-gap transition appears at a critical gate voltage. The varied van Hove singularities, absorption structures and Landau levels are very sensitive to different orbital hybridization cases, spin–orbit coupling, gate voltages and magnetic fields, e.g. the sublattice-, orbital-, spin- and valley-dependent Landau level subenvelope functions (diverse magnetic quantization phenomena [23]). In addition, the Coulomb excitations and decay rates have been investigated for graphene [24], silicene [25] and germanene [8], in which there exist certain important similarities and differences, such as the strong doping dependence of the momentum and frequency phase diagrams with different electron–hole regions, plasmon modes and wave-vector-dependent inelastic electron–electron scattering.

Interestingly, the sliding bilayer silicene and graphene systems exhibit diverse geometries, and electronic and magnetic properties, as a result of the significant chemical bonding of the outer four orbitals and spin–orbit coupling. The former and the latter, respectively, possess buckled and planar honeycomb lattices, interlayer distances of ~2.43–2.72 Å and ~3.53–3.35 Å, ground state energy differences between the different configurations of about 0.40 and 0.02 eV, and high and low charge densities between the interlayer spacing. Furthermore, the enhanced buckling angle and height difference between the A and B intralayer sublattices can reach 22° and ~0.88 Å, respectively. These results illustrate clearly that Si-3s, -3p$_x$, -3p$_y$ and 3p$_z$, and C-2p$_z$ play important roles in the interlayer atomic interactions leading to, respectively, completely modified electronic properties and drastic changes to the π and π^* band properties. The electronic energy spectra of bilayer silicenes present multi-constant loops, a complete absence of Dirac-cone structures, semimetals and direct- or indirect-gap semiconductors, parabolic Γ, K and M valleys, M- and M′-related split energy bands (except for the high-symmetry AA and AB stacking cases), negligible or significant spin-configuration effects under non-AB or AB stacking, frequent band mixing, many band-edge states, and undefined π and σ band widths. On the other hand, bilayer graphenes show highly k-dependent valence and conduction bands, dramatic transformations between two pairs of vertical or non-vertical Dirac cones and parabolic bands initiated from the stable K/K′ valleys, sliding-induced low-energy saddle points, an eye-shaped stateless region, split saddle M and M′ points in non-AA and non-AB stacking, degeneracy-enhanced σ bands, direct crossing of the π and σ bands, and well-defined band widths. High-resolution ARPES studies can confirm the cooperative and/or competitive relationships between the π and σ bonding, and the presence or absence of spin–orbit interactions. For the rich and unique van Hove singularities, monolayer silicenes present a V-shaped structure that is zero at E_F, well-behaved π and σ bands by the initial cusp or shoulder, a logarithmic peak and final shoulder, as well as two square-root asymmetric peaks during band mixing, and sliding bilayer silicenes present a finite or vanishing density of states at E_F in a band-gap region, many stacking-induced shoulders, symmetric and asymmetric peaks, and undefined π- and σ-electronic structures. STS measurements are available for distinguishing the dominance of the π and sp^3 bonding on the low-energy physical properties. The above-mentioned main features of the electronic properties can be described well using the four- and single-orbital hybridization for bilayer silicenes and graphenes, respectively. As a result, it is almost impossible to simulate silicenes using phenomenological models [28], while the opposite is true for the graphenes. For example, the magneto-electronic Landau levels [27], magneto-optical absorption spectra [28, 29], magnetoplasmons

[30, 31] and Hall conductivities [32], which have been demonstrated in sliding, twisted and modulated bilayer graphene systems, have been investigated fully using the generalized tight-binding model [33], as well as the single- and many-particle theories [26, 29, 32].

The geometric and electronic properties of hydrogenated silicene systems are varied greatly by the strong cooperation and competition between the sp^3 bonding in the H–C bonds and the sp^2 and sp^3 bonding in the planar or buckled C–C bonds. The chemical and physical properties appear to be very sensitive to the various hydrogen configurations and concentrations. In particular, the H–Si bond is the critical mechanism for creating the enhanced buckled geometries, unusual charge distributions, rich electronic energy spectra and novel van Hove singularities. The high charge transfer between the Si-$3p_z$ and H-1s orbitals leads directly to obvious variation in the carrier densities and indirectly to variation in the bond lengths. There are dramatic transformations in the electronic properties during changes of H adatom chemisorption. The three major discoveries for band structures are the absence or recovery of low-lying π bands, the H-related partially flat bands and the nearly dispersionless band at the Fermi level. The first is determined by the competition between strong H–Si bonds and weak sp^3 hybridization. The band gap, which is determined mainly by the Si-σ orbital, appears at a sufficiently high H adatom concentration, and it is largest in graphene. With a decreasing H concentration, the influence of the Si-π bands recovers gradually, causing the band gap to shrink. H-related partially flat bands originate purely from the very strong chemical bonding in the H and Si atoms, and is a common characteristic of both the top and chair systems at various H concentrations. However, the the nearly dispersionless band at the Fermi level is a distinguishing characteristic of top configurations owing to the destruction of mirror symmetry. The main features of the density of states, such as the recovery of π-band-induced low-lying peaks, the diminishing of H-related strong peaks and the weakened delta-function-like peak at E_F, reflect reduced H concentrations directly. The feature-rich band gaps and energy dispersions, and the energy-dependent strong peaks as van Hove singularities could be confirmed using, respectively, high-resolution ARPES and STS experiments. Such studies provide very useful information about the hydrogen distribution and concentration. In particular, the linear dependence of the special density-of-state peak intensities would become an effective route to detect the H concentration accurately. Hydrogenated silicene systems are highly suitable for applications in energy storage [34], and the tunable electronic properties might have great potential in nano-electronic [35] and nano-photonic devices [36].

The chemical halogenation of silicene can change the geometric, electronic and magnetic properties dramatically, as shown through first-principles calculations. It is very difficult to distinguish the fluorination and non-fluorination effects, owing mainly to the greatly enhanced Si–X chemical bonding in buckled honeycomb lattices. The non-uniform chemical environment is generated in an enlarged unit cell except for the 100% double-sided adsorptions, such as the Si–Si, Si–X and X–X bonds. Generally, the electronic energy spectra consist of silicon-, halogen-, and Si- and X-dominated energy bands. The band properties near the Fermi level are very sensitive to single- or double-sided absorption. The former exhibit p-type metallic behavior due to the extended $3p_z$-π bonding on one side with halogen passivation. However, the latter might be semiconductors or metals, depending on adatom concentrations. Low-lying well-defined valence bands are found only under the highest concentrations. For the single-sided 50% case, the spin-degenerate π-electronic valence band (two spin-split bands) across the Fermi level appears along the KMΓ path, and double σ-orbital valence bands arise in the ΓMK direction. Spin-configuration ferromagnetism is demonstrated clearly by the finite magnetic moments, spin-split energy bands, spatial spin density distributions and spin-decomposed van Hove singularities. It should be noted that the ferromagnetic behaviors survive only in single-sided chemisorption cases within a certain concentration range. There exist certain important differences between halogenated silicene and graphene systems in terms of the p-type metals and semiconductors, the existence or absence of a modified Dirac-cone structure (π bonding), non-magnetic and ferromagnetic behaviors, and their dependence on double- and single-sided adsorption. Specifically, fluorination and non-fluorination can be classified into two different categories according to the main features of the electronic and magnetic properties. For example, the non-fluorinated graphene systems belong to the p-type metals under any chemisorption case [37]. STM, SP-STM, TEM, LEED, ARPES, SP-ARPES, STS and SP-STS (details in chapter 3) could be utilized to examine the VASP calculations and verify the halogenation-varied phenomena in group-IV layered systems.

Sodium, magnetism and aluminum adatoms, which, respectively, possess one, two and three valence electrons in terms of the 3s, $3s^2$ and (3s, $3p_x + 3p_y$) orbitals, are very useful for understanding the diverse adsorption-induced phenomena. Their interactions with the Si-$3p_z$ and Si-sp^3 orbitals account for the fundamental physical quantities. Their geometric, electronic and magnetic properties are in sharp contrast to each other and to those of graphene-related systems, owing mainly to the different adatom–Si and adatom–C interlayer chemical bonding. The important differences include the optimal hollow- or valley-site positions, different adsorption energies, different adatom–silicon bond lengths, degrees of enhanced buckling, the highest saturation adsorption, free carrier densities of 2D conduction electrons in the absence of sensitive estimations, the adatom-induced blue shift of the Fermi levels and irregular doping behaviors, slightly, strongly or completely modified Dirac-cone structures (the Na, Mg and Al adsorption cases, respectively), well-behaved or undefined π and σ valence states (the former in Na- and the latter in Mg- and Al-adsorbed silicenes), adatom-dominated valence and/or conduction bands, spin-degenerate or spin-split electronic energy spectra across E_F, adatom-enriched charge density distributions, ferromagnetic or non-magnetic spin distributions under certain Mg and Al adsorption configurations (finite or vanishing magnetic moments) and the adsorption-varied van Hove singularities. The critical mechanisms are deduced to arise from the moiré superlattices and the different single- and multi-orbital hybridization cases in adatom–Si, Si–Si and adatom–adatom bonds (e.g. the 3s–$3p_z$, 3s–$3p_z$ and (3s, $3p_x + 3p_y$)–sp^3 orbital hybridization in the Na–, Mg– and Al–Si chemical bonds, respectively). The low-lying electronic states are initiated from the stable K or Γ valleys, and are determined mainly by the zone-folding effect. The hollow and valley adsorption, which corresponds to, respectively, the Na and Mg cases and the Al case, creates extremely non-uniform environments within the moiré superlattices. Their direct combination with the atomic configurations is responsible for the diverse adsorption-induced phenomena. Interestingly, all the alkali–adatom adsorptions can create 2D conduction electrons, while the free carrier densities are lower than those in graphene-related systems [38]. Also, Mg and Al adsorption generate n-type doping behavior only under low-concentration cases. However, the drastic changes in the low-lying energy subbands make the density estimation of conduction carriers almost impossible for the three kinds of metal–adatom adsorption. Among these three, the Al adsorption configurations present the strongest chemical modifications, such as more oscillatory and partially flat energy dispersions across the Fermi level, the complete disappearance of the modified Dirac-cone structures at sufficiently high adatom concentrations, and the largest charge density difference in the Al–Si and Al–Al bonds. The main reason arises from the three orbitals of (3s, $3p_x + 3p_y$) in each Al adatom. The ferromagnetic configurations are shown to survive only in specific Mg and Al adsorptions, but not in the Na cases. This indicates that the spin-dependent many-particle interactions might play a critical role in reducing the total ground state energy. The guest- or host-atom-dominated spin distributions are consistent with the spin-split density of states across the Fermi level. Such highly non-homogeneous spin distributions will lead to high barriers in simulating the low-lying band structures through the tight-binding model [27, 39]. It should be noted that there exist certain important differences between the Na-, Mg- and Al-adsorbed silicenes and graphenes. Interestingly, the optimal position of Al adatom adsorptions on the graphene surface is situated at the hollow site [40], as observed in the Na and Mg cases [41, 42]. For example, it is very easy to characterize the slightly modified Dirac-cone structures and π- and σ-electronic states in a planar carbon honeycomb lattice without any spin-related magnetic properties [43–46]. Furthermore, the σ bands exhibit only red-shift phenomena for three kinds of metal–adatom adsorptions on monolayer graphene. Similar studies on the metal–adatom adsorption effects could be generalized for the Ge-, Sn- and Pb-monolayer systems.

Boron, carbon and nitrogen substitution can vary the geometric, electronic and magnetic properties of silicene-related binary compounds greatly, since the different guest atoms are able to provide three, four and five valence electrons, respectively. For the B and C cases, the magnetic spin configurations disappear completely through the reasonably well-defined π and σ chemical bonding. On the other hand, N-related systems present complicated four-orbital hybridization and semimetallic or semiconducting ferromagnetic configurations under certain nitrogen substitution cases. The moiré superlattices can create planar or buckled structures, non-uniform bond lengths in an enlarged unit cell (coexistent X–X, X–Si and Si–Si bonds), and the typical meta, ortho (the highest ground state energy), para (the lowest ground state energy) and single configurations. Specifically, the N–N bonds are unstable except for in the 2:4 ortho case. Among the three kinds of guest atoms, the carbon-related bond lengths are the shortest and thus correspond to the lowest ground state energies, while the boron-related bond lengths are longest and correspond to the highest ground state energies, making these the most stable systems. Interestingly, C, B and N guest-atom substitution generates diverse band properties across the Fermi level associated with, respectively, the semiconducting, p-type, and semiconducting and semimetallic behaviors. n-type metals cannot be created by such guest atoms. The largest band gap arises in the full carbon atom substitution, e.g. $E_g \sim 2.53$ eV in the pure 100% SiC compound. When the π and π^* energy subbands dominate the low-lying electronic structures, they are initiated from the K/K′ or Γ valleys, become a saddle structure centered at the M point and then end at the Γ valley. The initial symmetry points are determined mainly by the zone-folding effect. Their energy widths are identified clearly for the C- and B-substituted systems under any case, illustrating well-behaved π chemical bonding. Similar results are obtained from the first σ-electronic valence subbands. However, it is very difficult to characterize the π- and σ-electronic states well in the nitrogen substitution cases. The energy spectra of the electrons and holes are highly asymmetric for boron atom substitution, owing mainly to the red shift of the Fermi level and the great reduction of the π-electronic chemical bonding, and for nitrogen atom substitution owing to the four–four impure or pure orbital hybridization. For C-substituted silicenes, the modified Dirac-cone structure is recovered gradually for a sufficiently low concentration, and the V-shaped van Hove singularity across the Fermi level for concentrations lower than or equal to 12.5%. Degenerate threshold σ bands, which are deeper than −1.0 eV, appear at the Γ point. The π and σ energy subbands show frequent crossing and little anticrossing along the KΓ and MΓ directions, except that the full substitution presents only the former behavior and thus pure sp^2 bonding (non-sp^3 and non-hybridized sp^2 bonding). Generally, the electronic states of boron-related silicenes could be regarded as zone-folding states with p-type doping, e.g. the Fermi level below the modified Dirac points, and the initial π and σ energy subbands at, respectively, the K/Γ and Γ points. On the other hand, the nitrogen-substituted silicenes display quite complicated band structures, such as the strongly oscillatory wave-vector dependence at low energy, undefined π and σ subbands, and the spin-split energy spectra (semimetallic and semiconducting ferromagnetic configurations under specific cases). Obtaining multi-orbital hopping integrals and spin-dependent many-particle interactions in the tight-binding model would become almost impossible for the nitrogen-substituted silicene systems [47]. Five kinds of van Hove singularities, the V-shaped structure, shoulders, logarithmically symmetric peaks, delta-function-like peaks and square-root asymmetric peaks correspond to, respectively, the linear Dirac cone, extreme band-edge states of parabolic dispersions, saddle points, partially flat bands and constant-energy loops. Whether the vanishing or finite density of states could survive within an energy range covering the Fermi level determines the three types of band properties mentioned above. High-resolution ARPES, SP-ARPES, STS and SP-STS measurements are capable of examining the wave-vector-dependent π- and initial σ-electronic energy spectra, and the band crossing and anti-crossing phenomena. That is, they could provide much useful information on the complex relationships between the π, σ, sp^3 and multi-orbital bonding. There are certain important differences between the B-,

C- and N-substituted silicene systems, as a result of their distinct electron affinities, atomic configurations, ionization energies and orbital hybridization. They are suitable for exploring varied physical, chemical and material phenomena fully, e.g. the unusual magnetic quantization close to the Fermi level [46], and the E_F-enriched optical absorption spectra and Coulomb excitations [25]. Most importantly, boron substitution is predicted to create very high 2D free hole densities the same as those of the guest atoms, leading to potential applications in nanoscale electronic devices. Whether suitable tight-binding models related to the multi-orbital hopping integrals of 2s, $2p_x$, $2p_y$ and $2p_z$, and 3s, $3p_x$, $3p_y$ and $3p_z$ can simulate the first-principles results is worthy of further careful investigation.

The differences and similarities between the chemisorption and substitution effects are demonstrated clearly in silicon-doped graphene systems, providing full information about the single- and multi-orbital hybridization. Significant chemical bonding is proposed to account for the Si-induced chemical modifications in graphene systems. The calculated results show clearly that free carriers and band gaps can be created by chemical adsorption and substitution, respectively. Different modification approaches for silicon guest atoms are able to modulate the semiconducting, metallic and semimetallic behaviors in a way that is difficult to observe using other kinds of guest atoms [49, 50]. For example, the alkali adatoms can generate a high free electron density under a rigid band shift of graphene [41]. The Si-adsorbed and Si-substituted graphenes could serve as potential candidates in nanoscale applications [51]. The chemisorption and substitution of Si guest atoms on monolayer graphene present unusual geometric properties, which are reflected directly in the spatial charge distributions. Si-adsorbed graphene is a non-buckled honeycomb lattice under an optimal bridge-site position. The planar structure indicates very little variation in the σ bonding of the C-2s, $-2p_x$ and $-2p_y$ orbitals; therefore, they barely take part in the Si–C bonds. This adsorption configuration is similar to that of graphene oxide [45], while there is slight buckling in the latter. The Si–C bond length is about 2.1–2.5 Å, leading to significant multi-orbital hybridization. Furthermore, the C–C bond lengths remain almost the same. According to detailed analyses of the 3D charge density distributions, their spatial variation, and the atom- and orbital-projected densities of states, the sp^3–p multi-orbital hybridization is deduced to determine the chemical Si–C bonds. The theoretical predictions for the bridge site and the Si–C and C–C bond lengths could be examined using high-resolution STM, TEM and LEED measurements. Also, the Si-substituted graphene systems present planar geometries under multi-orbital chemical bonding in the Si–C, C–C and Si–Si bonds. The Si–C bond length is about 1.62–1.83 Å, shorter than those in Si-adsorbed systems. This means that the four orbitals in the host and guest atoms are hybridized strongly with one another. STM experiments are available for examining the predicted Si–C, C–C and Si–Si bond lengths. Any chemisorption or substitution of Si guest atoms alters the unusual band structure of monolayer graphene completely, in particular for zero-gap semiconducting behavior and the linear Dirac cone due to π bonding of the $2p_z$ orbitals. In the 100% double- and single-sided adsorption cases, the Si-adsorbed graphene systems are semimetals with free conduction electrons and valence holes. For the low-concentration systems, they belong to the p-type metals with only free valence holes. In general, the Dirac-cone structure near the stable valley is distorted seriously after Si chemisorption. There are more valence and conduction energy subbands, which are accompanied by various band-edge states within a whole energy spectrum, e.g. the emergent low-lying energy bands along the KΓ and MΓ directions. However, the σ bands, which arise from the $2p_x$ and $2p_y$ orbitals of the carbon atoms, exhibit a rigid red shift of 1 and 0.5 eV for the 100% and lower-concentration cases, respectively. The above-mentioned important results reveal the critical mechanisms directly, namely, the multi-orbital hybridization of sp^3–p in Si–C bonds, sp^3–sp^3 in Si–Si bonds and sp^2–sp^2 in C–C bonds. Such types of chemical bonding arise mainly from the four Si- and C-orbitals, as a result of the bridge-site adsorption positions, and the binding energies and charge distributions of the separated orbitals. High-resolution ARPES measurements are very useful for verifying the low-lying valence bands near to or crossing the Fermi level along ΓK and ΓM and the rigid σ bands initiated from the Γ valley. On the other hand, all substitution cases result in semiconducting behavior with a finite or zero band gap. The Dirac-cone structure presents deviation from the Γ point, strong distortion, or even complete destruction. The number of valence and conduction energy subbands remains the same after chemical substitution; furthermore, they are co-dominated by the Si guest atoms and C host atoms. The main features of band structures in Si-substituted graphene systems appear to be dominated by the sp^2–sp^2 and p–p orbital hybridization in the Si–C, C–C and Si–Si bonds. Many unusual van Hove singularities in the atom- and orbital-decomposed densities of states are created under Si guest-atom chemisorption and substitution. Pristine monolayer graphene presents only a V-shaped structure with a vanishing value at the Fermi level, logarithmically symmetric π and π^* peaks at −2.41 and 1.81 eV, respectively, and a shoulder structure at −3.01 eV, of which the first two correspond to π bonding and the latter corresponds to the σ bonding of carbon atoms. The strong π-bonding evidence is destroyed completely by any chemisorption cases except for the very dilute Si–adatom case. They are replaced by a finite density of states at $E_F = 0$, many shoulders and peak structures. Furthermore, they are co-dominated by the four Si-3s, $-3p_x$, $-3p_y$ and $3p_z$ orbitals and the single C-$2p_z$ orbital, since their contributions are merged together. Specifically, the σ-band shoulder arises at −4.21, −4.11 or −3.5 eV, and the 0.5–1.0 eV red shift is related closely to the distinct ionization energies of the Si and C atoms. The above-mentioned significant features further support and illustrate the sp^3–p and sp^2–sp^2 multi-orbital hybridization in the Si–C bonds and C–C bonds, respectively. Concerning the substitution cases, the density of states is zero within a finite energy range centered at the Fermi level, except for a few systems with deformed V-shaped structures at E_F. All the Si-substituted graphene systems are finite- or zero-gap semiconductors. The special structures, which originate from the Si-$3p_z$ and C-$2p_z$ orbitals, appear simultaneously. Also, similar behavior is revealed in the Si-$3p_x$ and $-3p_y$ and C-$2p_x$ and $2p_y$ orbitals. These further illustrate the critical single- and multi-orbital hybridization in Si–C bonds. The predicted van Hove singularities could be verified using high-resolution STS experiments [52].

The fundamental properties are changed dramatically through the significant silicene–substrate chemical bonding, zone-folding effects, enhanced buckling and stacking configurations. Sufficient substrate silver layers in the VASP calculations are utilized to simulate stable geometric structures. It appears that extremely non-uniform Si–Si, Si–Ag and Ag–Ag bonds and interlayer distances could survive under a specific moiré superlattice (3×3 and 4×4), illustrating clearly the very complicated intralayer and interlayer hopping integrals related to neighboring atoms [53]. The spin-dependent single- and many-particle interactions behave similarly. The various interlayer distances of the neighboring layers are comparable to one another. Moreover, buckled structures, which are characterized by a height difference of the A and B sublattices, are revealed in monolayer silicene, bilayer silicene and first-layer silver depending on, respectively, the intralayer mixed sp^2–sp^3 bonding, the interlayer sp^3 bonding and the interlayer multi-orbital hybridization in the heterojunction. The theoretical predictions, conducted on the optimal geometries of Si-adlayer and AB-bt bilayer silicene on Ag(111) with 3×3 and 4×4 superlattices, could be examined using STM, TEM and LEED (details in section 3.2). These two compounds might exhibit similar band structures, including many critical points in the energy–wave-vector space, greatly enhanced asymmetry of the electron and hole bands about the Fermi level, the absence of modified Dirac-cone structures, low-lying non-uniform or oscillatory energy dispersions, strongly anisotropic Fermi-momentum states, semimetallic behaviors with low free carrier densities, Si-, Ag-, and Si- and Ag-dominated band structures due to different types of chemical bonding, and undefined π- and σ-band widths. It should be noted that the very complicated band structures could be classified into certain energy ranges according to the Si and Ag atom dominance. From the ARPES and STS perspectives, such measurements cannot identify the modified Dirac-cone structure, V-shaped structure, band gap, direction-dependent Fermi momenta, and π- and σ-band widths. Based on the atom-dominated band structures, atom- and orbital-decomposed van Hove singularities under different energy ranges, as well as the spatial charge density distributions, the (3s, $3p_x + 3p_y$, $3p_z$)–(3s, $3p_x + 3p_y$, $3p_z$), (3s, $3p_x + 3p_y$)–(4d, 5p, 5s) and 4d–4d multi-orbital hybridization is found to play a important role in the Si–Si, Si–Ag and Ag–Ag bonds, respectively. Most importantly, there exist certain important differences between bilayer and monolayer silicene on Ag(111) in terms of the interlayer Si–Si and Si–Ag chemical bonding and the spin-dependent magnetic properties. Interestingly, the bilayer silicene on Ag(111) belongs to the ferromagnetic semimetals and has spin-split electronic states cross the Fermi level, and the monolayer system belongs to the non-magnetic semimetals and has spin-degenerate electronic states across the Fermi level. It is almost impossible to evaluate the 2D conduction electron density in detail from the direction-dependent Fermi momenta in the presence or absence of spin splitting. The spin configuration in bilayer AB-bt-Si–Ag(111), which presents a rather non-uniform spin-up and spin-down arrangement within an enlarged unit cell, originates mainly from the six second-layer silicon atoms. Furthermore, the $3p_x$, $3p_y$ and $3p_z$ orbitals of Si atoms make the most important contributions to the ferromagnetic properties associated with the spin-created many-particle interactions. The predicted net magnetic moment could be verified using high-resolution SQUID measurements [54]. The unusual electronic and magnetic properties would be very difficult to simulate through the Hubbard-like tight-binding model because of the overly complicated multi-orbital hopping integrals and atom-, orbital- and spin-dependent on-site Coulomb potentials. It is also noted that few-layer silicene systems, with or without significant spin–orbit coupling, do not produce any magnetism except under external magnetic fields, chemical modification and substrate effects. In addition, experimental observations have identified certain stable moiré superlattices, which are expected to display similar behaviors in the fundamental properties. However, the different substrate effects would vary the physical, chemical and material phenomena greatly in other emergent layered materials, e.g. the diverse phenomena in few-layer graphene, germanene and stanene on different heterojunctions.

Bilayer composite systems, Si-, Ge- and Sn-adlayers on graphene, present diverse fundamental properties. The critical mechanisms are the weak but significant van der Walls interactions in the non-uniform enlarged unit cells, the sp^3 bonding on silicene, germanene and tinene, the sp^2 bonding on graphene and the important spin–orbit coupling. The first factor is revealed in the interlayer distances, which are comparable to those in bilayer graphene systems [13]. Concise physical and chemical pictures are further confirmed from the atom-dominated band structures, the total charge density distributions and their variation after coupling, and the atom- and orbital-decomposed van Hove singularities. However, the ferromagnetic and anti-ferromagnetic configurations are absent in the three kinds of composite systems. The silicene–, germanene– and tinene–graphene bilayer materials possess, respectively, the smallest moiré superlattices of $(\sqrt{3} \times \sqrt{3})$–$(\sqrt{7} \times \sqrt{7})$ ($R = 10.9°$), (3×3)–(5×5) ($R = 0°$) and $(\sqrt{7} \times \sqrt{7})$–$(5 \times 5)$ ($R = 19.1°$), in which the interlayer orbital bonding originates principally from the $3p_z$–$2p_z$, $4p_z$–$2p_z$, $5p_z$–$2p_z$, and $6p_z$–$2p_z$ single-orbital hybridization. This indicates the existence of very complicated interlayer hopping integrals because of the various effective atomic interactions in the two different honeycomb lattices. Specifically, the heavy tin atoms create the largest buckling (the most obvious sp^3 hybridization) and the strongest spin–orbit coupling. The low-lying valence and conduction bands are dominated by two modified Dirac cones due to the intralayer distorted π bonding and/or the σ-electronic states. Such electronic spectra are initiated from the K/K′ and Γ valleys, depending the zone-folding effect. The initial π and the first and second σ valence bands could be characterized approximately from the complex band structures and orbital-projected density of states. However, it is almost impossible to define all the band widths well for any chemical bonding as a result of there being too many energy subbands. Subband anti-crossing and crossing behaviors occur frequently below $E_v^{\sim} - 1.0$ eV. Interestingly, there exist important differences between the bilayer silicene–, germanene– and tinene–graphene composites, such as the two pairs of modified Dirac cones at the K valley or the K and Γ valleys, the zero or finite energy spacings between separated Dirac points at the same valley, the semimetallic property of the silicene-related system with valence and conduction band overlap due to two different Dirac cones (the coexistence of free electrons and holes), the very small indirect gaps in the germane- and tinene-related materials, different types of chemical bonding creating different semiconducting behaviors, the initial σ-electronic valence and conduction energies, a plateau or vanishing density of states across the Fermi level and the low-lying van Hove singularities of the σ bands. Such varied phenomena are determined mainly by the strengths of sp^3 bonding and spin–orbit interactions.

The above-mentioned features of the electronic properties could be verified using high-resolution ARPES [55] and STS [56] measurements. In addition to the first-principles method, the single- and many-particle theories are worthy of systematic development, even in the presence of many A and B sublattices (complicated tight-binding functions). For example, the generalized tight-binding model [57] might be suitable and its reliable hopping integrals could be achieved in the presence of single-orbital interlayer hybridization [58]. Obviously, there are a lot of A and B atoms in any moiré superlattices for the group-IV-based bilayer composite systems. The Bloch wave functions, which are calculated from the generalized tight-binding model, consist of the amplitudes (the subenvelope functions) associated with many sublattice-dependent tight-binding functions. Under magnetic quantization [59], the magneto-electronic states are expected to exhibit complicated localization behaviors with different or the same oscillation modes in various A and B sublattices, and thus rich and unique optical absorption spectra and quantum Hall transport.

References

[1] Kaloni T P, Modarresi M, Tahir M, Roknabadi M R, Schreckenbach G and Freund M S 2015 Electrically engineered band gap in two-dimensional Ge, Sn, and Pb: a first-principles and tight-binding approach *J. Phys. Chem.* C **119** 11896–902

[2] Sun L, Lin Z, Peng J, Weng J, Huang Y and Luo Z 2014 Preparation of few-layer bismuth selenide by liquid-phase-exfoliation and its optical absorption properties *Sci. Rep.* **4** 4794

[3] Lin C Y, Chen R B, Ho Y H and Lin M F 2018 *Electronic and Optical Properties of Graphite-Related Systems* (Boca Raton, FL: CRC Press)

[4] Lin C Y, Do T N, Huang Y K and Lin M F 2017 *Optical Properties of Graphene in Magnetic and Electric Fields* (Bristol: IOP Publishing)

[5] Chen S C, Wu J Y, Lin C Y and Lin M F 2017 *Theory of Magnetoelectric Properties of 2D Systems* (Bristol: IOP Publishing)

[6] Lin C Y, Wu J Y, Chiu C W and Lin M F 2019 *Coulomb Excitations and Decays in Graphene-Related Systems* (Boca Raton, FL: CRC Press)

[7] Lin C Y, Do T N, Wu J Y, Shih P H, Lin S Y, Ho C H and Lin M F 2019 *Diverse Quantization Phenomena in Emergent Layered Materials* (Boca Raton, FL: CRC Press)

[8] Shih P H, Chiu Y H, Wu J Y, Shyu F L and Lin M F 2017 Coulomb excitations of monolayer germanene *Sci. Rep.* **7** 40600

[9] Biedermann L B, Bolen M L, Capano M A, Zemlyanov D and Reifenberger R G 2009 Insights into few-layer epitaxial graphene growth on 4*H*-SiC ($000\bar{1}$) substrates from STM studies *Phys. Rev.* B **79** 125411

[10] Stobinski L, Lesiak B, Malolepszy A, Mazurkiewicz M, Mierzwa B, Zemek J, Jiricek P and Bieloshapka I 2014 Graphene oxide and reduced graphene oxide studied by the XRD, TEM and electron spectroscopy methods *J. Electron. Spectrosc. Relat. Phenom.* **195** 145–54

[11] Jamgotchian H, Colignon Y, Hamzaoui N, Ealet B, Hoarau J Y, Aufray B and Bibérian J P 2012 Growth of silicene layers on Ag(111): unexpected effect of the substrate temperature *J. Phys. Condens. Matter* **24** 172001

[12] Lin C H *et al* 2018 Single-layer dual germanene phases on Ag(111) *Phys. Rev. Mater.* **2** 024003

[13] Yakes M and Tringides M C 2011 Probing the buried Pb/Si (111) interface with SPA LEED and STM on Si(111)–Pb$\alpha\sqrt{3}\times\sqrt{3}$ *J. Phys. Chem.* A **115** 7096–104

[14] Brihuega I, Mallet P, González-Herrero H, De Laissardière G T, Ugeda M M, Magaud L, Gómez-Rodríguez J M, Ynduráin F and Veuillen J Y 2012 Unraveling the intrinsic and robust nature of van Hove singularities in twisted bilayer graphene by scanning tunneling microscopy and theoretical analysis *Phys. Rev. Lett.* **109** 196802

[15] Sprinkle M *et al* 2009 First direct observation of a nearly ideal graphene band structure *Phys. Rev. Lett.* **103** 226803

[16] Cahangirov S, Audiffred M, Tang P, Iacomino A, Duan W, Merino G and Rubio A 2013 Electronic structure of silicene on Ag(111): strong hybridization effects *Phys. Rev.* B **88** 035432

[17] Dávila M E and Le Lay G 2016 Few layer epitaxial germanene: a novel two-dimensional Dirac material *Sci. Rep.* **6** 20714

[18] Lee C C, Fleurence A, Yamada-Takamura Y, Ozaki T and Friedlein R 2014 Band structure of silicene on zirconium diboride (0001) thin-film surface: convergence of experiment and calculations in the one-Si-atom Brillouin zone *Phys. Rev.* B **90** 075422

[19] Crepaldi A, Pons S, Frantzeskakis E, Calleja F, Etzkorn M, Seitsonen A P, Kern K, Brune H and Grioni M 2013 Combined ARPES and STM study of Pb/Au(111) moiré structure: one overlayer, two symmetries *Phys. Rev.* B **87** 115138

[20] Bravo S, Correa J, Chico L and Pacheco M 2018 Tight-binding model for opto-electronic properties of penta-graphene nanostructures *Sci. Rep.* **8** 11070

[21] Do T N *et al* 2019 Peculiar optical properties of bilayer silicene under the influence of external electric and magnetic fields *Sci. Rep.* **9** 36547

[22] Chiu Y H, Lai Y H, Ho J H, Chuu S and Lin M F 2008 Electronic structure of a two-dimensional graphene monolayer in a spatially modulated magnetic field: Peierls tight-binding model *Phys. Rev.* B **77** 045407

[23] Lin C Y, Wu J Y, Chiu Y H and Lin M F 2015 Magneto-electronic properties of multilayer graphenes *Phys. Chem. Chem. Phys.* **17** 26008

[24] Ho J H, Lu C L, Hwang C C, Chang C P and Lin M F 2006 Coulomb excitations in AA- and AB-stacked bilayer graphites *Phys. Rev.* B **74** 085406

[25] Wu J Y, Chen S C and Lin M F 2014 Temperature-dependent Coulomb excitations in silicene *New J. Phys.* **16** 125002

[26] Do T N, Shih P H, Gumbs G, Huang D, Chiu C W and Lin M F 2018 Diverse magnetic quantization in bilayer silicene *Phys. Rev.* B **97** 125416

[27] Huang Y K, Chen S C, Ho Y H, Lin C Y and Lin M F 2014 Feature-rich magnetic quantization in sliding bilayer graphenes *Sci. Rep.* **4** 7509

[28] Ho Y H, Chiu C W, Su W P and Lin M F 2014 Magneto-optical spectra of transition metal dichalcogenides: a comparative study *Appl. Phys. Lett.* **105** 222411

[29] Ho Y H, Chiu Y H, Lin D H, Chang C P and Lin M F 2010 Magneto-optical selection rules in bilayer Bernal graphene *ACS Nano* **4** 1465

[30] Chung H C, Su W P and Lin M F 2013 Electric-field-induced destruction of quasi-Landau levels in bilayer graphene nanoribbons *Phys. Chem. Chem. Phys.* **15** 868

[31] Wu J Y, Chen S C, Do T N, Su W P, Gumbs G and Lin M F 2018 The diverse magneto-optical selection rules in bilayer black phosphorus *Sci. Rep.* **8** 13303

[32] Do T N, Chang C P, Shih P H, Wu J Y and Lin M F 2017 Stacking-enriched magneto-transport properties of few-layer graphenes *Phys. Chem. Chem. Phys.* **19** 29525

[33] Sharpe A L, Fox E J, Barnard A W, Finney J, Watanabe K and Kastner M A 2019 Emergent ferromagnetism near three-quarters filling in twisted bilayer graphene *Science* **10** 1126

[34] Hussain T, Kaewmaraya T, Chakraborty S and Ahuja R 2013 Functionalization of hydrogenated silicene with alkali and alkaline earth metals for efficient hydrogen storage *Phys. Chem. Chem. Phys.* **15** 18900–5

[35] Tao L, Cinquanta E, Chiappe D, Grazianetti C, Fanciulli M, Dubey M, Alessandro M and Akinwande D 2015 Silicene field-effect transistors operating at room temperature *Nat. Nanotechnol.* **10** 227

[36] Khanikaev A B and Shvets G 2017 Two-dimensional topological photonics *Nat. Photonics* **11** 763–73

[37] Tran N T T, Nguyen D K, Glukhova O E and Lin M F 2017 Coverage-dependent essential properties of halogenated graphene: a DFT study *Sci. Rep.* **7** 17858

[38] Tran N T T, Lin S Y, Lin C Y and Lin M F 2018 *Geometric and Electronic Properties of Graphene-related Systems: Chemical Bonding Schemes* (Boca Raton, FL: CRC Press)

[39] Tsai S J, Chiu Y H, Ho Y H and Lin M F 2012 Gate-voltage-dependent Landau levels in AA-stacked bilayer graphene *Chem. Phys. Lett.* **550** 104–10

[40] Lin S Y, Lin Y T, Tran N T T, Su W P and Lin M F 2017 Feature-rich electronic properties of aluminum-absorbed graphenes *Carbon* **120** 209–18

[41] Wang L, Wu Y, Chen F and Yang X 2014 Photocatalytic enhancement of Mg-doped ZnO nanocrystals hybridized with reduced graphene oxide sheets *Progress Nat. Sci.: Mater. Int.* **24** 6–12

[42] Choi S M and Jhi S H 2009 Electronic property of Na-doped epitaxial graphenes on SiC *Appl. Phys. Lett.* **94** 153108

[43] Tran N T T, Lin S Y, Lin Y T and Lin M F 2016 Chemical bondings induced rich electronic properties of oxygen absorbed few-layer graphenes *Phys. Chem. Chem. Phys.* **18** 4000–7

[44] Tran N T T, Lin S Y, Glukhova O E and Lin M F 2016 π-bondings-dominated energy gaps in graphene oxides *RSC Adv.* **6** 24458–63

[45] Tran N T T, Lin S Y, Glukhova O E and Lin M F 2015 Configuration-induced rich electronic properties of bilayer graphene *J. Phys. Chem.* C **119** 10623–30

[46] Tran N T T, Dahal D, Gumbs G and Lin M F 2017 Adatom doping-enriched geometric and electronic properties of pristine graphene: a method to modify the band gap *Struct. Chem.* **28** 1311–8

[47] Sivek J, Sahin H, Partoens B and Peeters F M 2013 Adsorption and absorption of boron, nitrogen, aluminum, and phosphorus on silicene: stability and electronic and phonon properties *Phys. Rev.* B **87** 085444

[48] Wu J-Y, Chen S-C, Gumbs G and Lin M-F 2017 Field-induced diverse quantizations in monolayer and bilayer black phosphorus *Phys. Rev.* B **95** 115411

[49] Chan K T, Neaton J B and Cohen M L 2008 First-principles study of metal adatom adsorption on graphene *Phys. Rev.* B **77** 235430

[50] Sun M, Ren Q, Zhao Y, Chou J P, Yu J and Tang W 2017 Electronic and magnetic properties of 4d series transition metal substituted graphene: a first-principles study *Carbon* **120** 265–73

[51] Chen Y, Gao B, Zhao J X, Cai Q H and Fu H G 2012 Si-doped graphene: an ideal sensor for NO or NO_2-detection and metal-free catalyst for N_2O-reduction *J. Mol. Model.* **18** 2043–54

[52] Brar V W *et al* 2007 Scanning tunneling spectroscopy of inhomogeneous electronic structure in monolayer and bilayer graphene on SiC *Appl. Phys. Lett.* **91** 122102

[53] Feng B, Ding Z, Meng S, Yao Y, He X, Cheng P, Chen L and Wu K 2012 Evidence of silicene in honeycomb structures of silicon on Ag(111) *Nano Lett.* **12** 3507–11

[54] Zhang L D, Yang F and Yao Y 2015 Possible electric-field-induced superconducting states in doped silicene *Sci. Rep.* **5** 8203

[55] Sakamoto K, Setvin M, Mawatari K, Eriksson P E J, Miki K and Uhrberg R I G 2009 Electronic structure of the Si(110)–(16 × 2) surface: high-resolution ARPES and STM investigation *Phys. Rev.* B **79** 045304

[56] Lauffer P, Emtsev K V, Graupner R, Seyller T, Ley L, Reshanov S A and Weber H B 2008 Atomic and electronic structure of few-layer graphene on SiC(0001) studied with scanning tunneling microscopy and spectroscopy *Phys. Rev.* B **77** 155426

[57] Korshunov M M, Gavrichkov V A, Ovchinnikov S G, Nekrasov I A, Pchelkina Z V and Anisimov V I 2005 Hybrid LDA and generalized tight-binding method for electronic structure calculations of strongly correlated electron systems *Phys. Rev.* B **72** 165104

[58] Markiewicz R S, Sahrakorpi S, Lindroos M, Lin H and Bansil A 2005 One-band tight-binding model parametrization of the high-T_c cuprates including the effect of k_z dispersion *Phys. Rev.* B **72** 054519

[59] Chen S C, Wu J Y and Lin M F 2018 Feature-rich magneto-electronic properties of bismuthene *New J. Phys.* **20** 062001

Chapter 14

Open issues and potential applications

The theoretical framework developed using first-principles calculations could be generalized further to other mainstream materials. It is able to explore the varied fundamental properties associated with charges and spin configurations, clarify the critical issues, and provide concise physical, chemical and material pictures. Only certain typical silicene-based materials have been chosen herein to illustrate the basic behaviors. More rich and unique phenomena are expected to be revealed in the other layered systems. They are worthy of next-step systematic investigations using various numerical methods and phenomenological models (single- and many-particle models), and even the direct or indirect combination of these methods. For example, up-to-date VASP studies have been combined successfully with the generalized tight-binding model, leading to a full understanding of the diverse magnetic quantization phenomena in many graphene-related sp^2-bonding systems [1, 2]. The main reason is that the extremely non-uniform interlayer hopping integrals which appear in a greatly enlarged unit cell arise only from the single C-$2p_z$ orbitals.

Non-uniform physical and chemical environments are created easily by modulating the pristine honeycomb lattices of group-IV 2D materials. Many different experimental synthesis methods have produced specific vacancies (figure 14.1) [3], deformed lattices (e.g. hexagons replaced by pentagons and heptagons in figure 14.2; Stone–Wales defects in [4]) and adatom impurities [5]. However, only a few theoretical predictions have been produced for the defect-enriched fundamental properties of the layered group-IV systems [6, 7]. The defect-related C, Si, Ge, Sn and Pb 2D systems are highly suitable for model studies. Obviously, the various defects are capable of inducing the destruction of lattice symmetry, different site energies (a change of ionization potential) and non-uniform atomic interactions (position-dependent hopping integrals). The strong dependence of the electronic properties on the type, distribution and concentration of defects requires thorough

doi:10.1088/978-0-7503-3299-6ch14

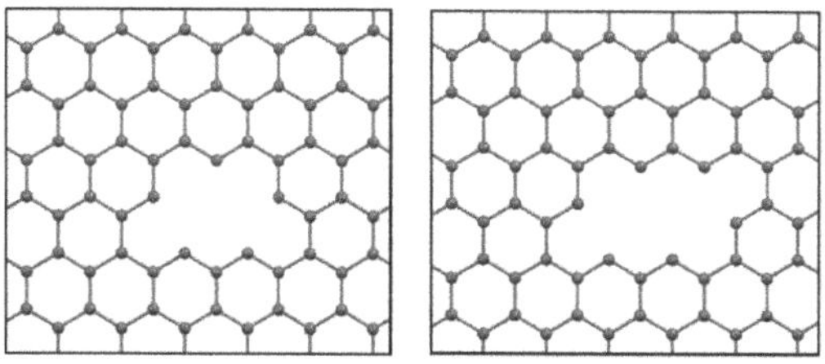

Figure 14.1. The distinct vacancy structures in graphene-related systems: (a) TV1 and (b) TV2.

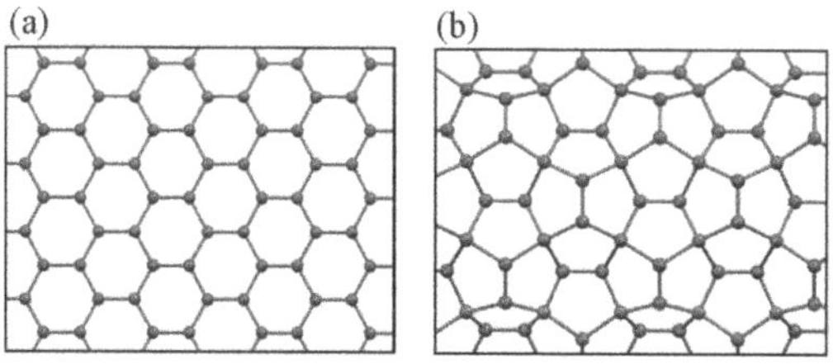

Figure 14.2. (a) A perfect honeycomb lattice with hexagons and (b) a deformed honeycomb lattice with pentagons and heptagons (Stone–Wales defects).

study using numerical and phenomenological methods. For example, whether the semiconductor–semimetal transitions, the localizations of wave functions, and the valley- and spin-dependent states arise under various defects could be explored in detail. The cooperative or competitive relationships between the dominant and significant chemical bonding, as well as the defect-induced spin configurations, are proposed for understanding the diverse electronic properties. Such effects are enhanced greatly by the external electric and magnetic fields, e.g. drastic changes in the magnetic quantization in terms of distortions and mixing of the Landau-level wave functions, changes of the localization n ranges, a reduced state degeneracy, and irregular field-dependent energy spectra with crossing and anti-crossing behaviors [1, 2].

The sliding bilayer systems appear to be quite different between silicene (chapter 5) and graphene [8] in terms of their electronic properties, which depend strongly on the buckled or planar honeycomb lattices, multi- or single-orbital hybridization, dominant or observable interlayer hopping integrals. The electronic properties are predicted to exhibit three kinds of Landau levels (the well-behaved, perturbed and undefined modes), owing mainly to the dramatic transformation between two vertical and non-vertical Dirac cones and two pairs of parabolic valence and conduction bands. Furthermore, there exist extra magneto-optical selection rules [9] and unusual quantum Hall effects (e.g. integer or non-integer conductivities [10, 11]). For the magneto-optical selection rules, recent theoretical predictions show more complicated magnetic quantization phenomena in the AA- and AB-bt silicene materials, even for only the first pair of irregular energy bands [12]. It is relatively easy to obtain reliable parameters for the tight-binding model for the low-lying energy bands in few-layer graphene systems [13, 14], since such electronic structures are determined mainly by the dominant intralayer hopping integral and the significant interlayer integrals due to the C-2p_z orbitals. However, suitable atomic interactions of bilayer silicenes are very difficult to obtain from their unusual electronic energy spectra using first-principles calculations (figures 5.2(a)–(l)) because there are too many critical mechanisms. The quantization phenomena in few-layer Si, Ge, Sn and Pb systems are the next focus of study, in particular for the magneto-electronic properties, optical absorption spectra, Hall conductivities and magnetoplasmons.

The composite bilayer materials in which each layer has a buckled honeycomb lattice are highly suitable for studying complicated physical, chemical and material phenomena [15, 16], such as Sn-, Pb-, Sb- and Bi-adlayers on a silicene or germanene surface. The fundamental properties are expected to be varied greatly by buckling, stacking configurations, moiré superlattices, atomic configurations and spin arrangements [17]. Generally, there are more interlayer atomic interactions and even their values are larger than the intralayer values, as observed in bilayer silicene systems (details in section 5.1). That is, the distinct geometric symmetries of two coupled honeycomb lattices, which are due to the buckling and stacking effects, make the interlayer hopping integrals become more complicated and thus dominate the low-lying electronic states. The different bond lengths in pristine subsystems will create an enlarged unit cell with a non-uniform chemical environment; therefore, certain interlayer atomic interactions are changed dramatically. The multi-orbital hybridization and the site energies depend strongly on the kind of atom, e.g. the significant differences between silicene, germanene and tinene [18]. Monolayer group-IV and group-V materials, for example monolayer Sn, Pb, Sb and Bi, possess quite large spin–orbit coupling. The above-mentioned intrinsic interactions need to be taken into account simultaneously for the first-principles method. The geometric, electronic and magnetic properties can be investigated thoroughly using VASP calculations (section 2.1). Whether the interlayer attractive interactions belong to the van der Waals interactions or to sufficiently strong multi-orbital chemical bonding, and whether spin–orbit coupling survives in a composite system will be the focus of future studies. The phenomenological model—the tight-binding model—appears to be too cumbersome to account for the fundamental properties.

Dimension-varied phenomena are a mainstream topic. The 0D quantum dots (figure 14.3(a)), 1D nanoribbons (figure 14.3(b)) and cylindrical nanotubes (not shown) present finite-sized confinement. In general, an open boundary includes the achiral (zigzag or armchair) or chiral edge structures. Such dangling bonds exist in the presence or absence of adatom decorations, e.g. the frequent hydrogen passivations in the experimental growth of graphene nanoribbons [19]. The very active surface and boundary provide outstanding environments for studying the rich and unique fundamental properties. To date, systematic studies have been performed on graphene nanoribbons using the phenomenological models [20] and first-principles calculations [21]. The theoretical frameworks could be further generalized to 1D Si-, Ge-, Sn- and Pb-related nanoribbons and nanotubes. The quantum confinements with open or periodical boundary conditions, edge structures, multi-orbital hybridization, spin–orbit coupling, and external electric and magnetic fields

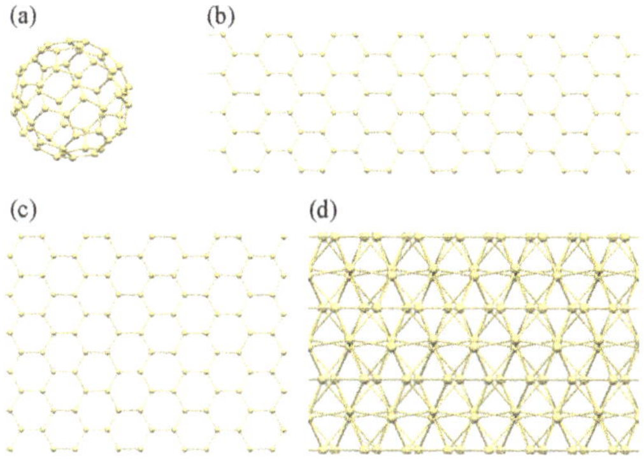

Figure 14.3. Silicon-based materials: (a) 0D quantum dot, (b) 1D nanoribbon, (c) 2D layer and (d) 3D bulk.

will need to be taken into account simultaneously. The magnetic quantization is expected to be enriched greatly and varied by the rather strong cooperation and competition between these critical factors. Using the generalized tight-binding model [20], the 1D Landau subbands, with k_x energy dispersions along the periodical direction, will present multi-sublattice-, valley-, multi-orbital- and spin-dependent subenvelope functions, in particular the two latter which are absent in graphene nanoribbons. Moreover, the first-principles calculations are useful in exploring edge decoration [22], surface chemisorption [23] and guest-atom substitution [24] in detail. Specifically, for zigzag Si, Ge, Sn and Pb nanoribbons, two kinds of spin-dependent interactions, spin–orbit coupling and spin-configuration-induced on-site Coulomb interactions (Hubbard-like interactions), need to be measured to clarify their critical roles in the geometric, electronic and magnetic properties. The direct combination of the first-principles method and the tight-binding model would be useful in solving these open issues.

The dimension crossover phenomena between layered and bulk Si, Ge, Sn and Pb materials (figures 14.3(c) and (d)) are the focus of next-step research; that is, a dramatic transformation from sp^2 into sp^3 chemical bonding is expected to appear during the increase of layer number. How to characterize the time-dependent growth process will become an interesting research topic. It is well known that 3D silicon is used frequently in the semiconductor industry [25], as a result of its flexible characteristics under specific ion implantation (e.g. oxygen, boron and phosphorous). With the shrinking of device size, the lower dimensions will induce certain fundamental properties, such as surface and boundary reconstructions, edge structures and chemical bonding. These interesting and unconventional properties cannot be solved from the phenomenological models. However, electronic and molecular dynamics calculations are capable of exploring these rich and unique phenomena fully. The former provides reliable and suitable predictions in terms of the interface-dependent chemical bonds, carrier transfer and density of states. The latter is responsible for the unusual geometric structures and diffusion behaviors based on full information of the thermal budget, pressure and time. Under rapid progress of chemical and physical modification, the low-dimension silicon-related materials are expected to have quite good potential in the next-generation semiconductor industry [26]. Both methods would be very powerful in predicting varied nanoelectronic phenomena.

In nature, infinite materials do not exist, indicating that any 3D bulk materials possess boundaries or surfaces. From the theoretical viewpoint, such systems are regarded as periodic systems regardless of the edge effects, since the number of total atoms is much higher than that near specific surfaces. On the other hand, quantum confinement plays a critical role in the presence of finite widths, or it will dominate or vary the fundamental physical, chemical and material properties [27]. Clearly, the low-dimensional materials (0D, 1D and 2D [28, 29]) have attracted a lot of theoretical and experimental attention in the basic and applied sciences [28–33]. How the dimension crossover between such materials arises remains an outstanding issue. For example, how thin a 3D system behaves like a 2D one, and how many layers of a 2D material resembles the bulk. While a thinner bulk material can approach a multi-layer system, the former has a layer-dependent geometric structure; furthermore, the interlayer binding energy is much smaller than the intralayer one, such as in graphite [34, 35], black phosphorus [36], MoS_2 [37], etc. The weaker interlayer atomic interactions illustrate clearly that the layer–layer distance is large, and a 2D system could be obtained easily by stripping from the bulk. The best example is the first 2D few-layer graphenes produced by mechanical exfoliation of 3D graphite in 2004 [38]. In contrast, for the pure silicon-related materials, the thinner bulk systems and multi-layer systems are not similar to one another. That is to say, layered silicenes are completely different to bulk silicon in all the fundamental properties. The Si–Si chemical bonds in silicene exhibit well-defined π and σ bonding, which are, respectively, parallel and perpendicular to the surface. This contrasts sharply with the sp^3 bonding in diamond-like bulk silicon [39]. It is impossible to produce 2D silicene materials from stripping bulk silicon. Current experimental syntheses show that few-layer silicenes can be grown successfully on different substrates, as discussed in chapter 11. The significant substrate-induced effects and the important interlayer interactions of buckled honeycomb lattices need to be taken into consideration simultaneously. How to identify the dramatic changes in the layer- and stacking-dependent fundamental properties is being studied, e.g. the sensitive dependence of the dominant chemical bonding (sp^2 or sp^3) and band gap on the layer number.

The B-, C- and N-substituted silicene systems are outstanding candidates for studying fundamental properties under external fields, in particular for different magnetic quantization cases. The generalized tight-binding model, as developed in previous studies [1, 2], is able to clarify the varied magnetic phenomena fully. However, the nearest-neighbor hopping integrals, which are associated with the Si–Si, Si–X and X–X bonds, need to be fitted from the low-lying energy bands near the K and/or Γ valleys (e.g. figures 9.2, 9.6 and 9.10). Then, the moiré superlattice and the Peierls-phase period due to the vector potential are included in the calculations simultaneously. The physical and chemical environment in a superlattice unit cell is non-uniform; therefore, it would be very interesting to understand the unusual relationships between the sublattice-dependent magnetic subenvelope functions.

Their main features, the valley and spin dependence, the main and side oscillation modes, and the non-crossing, crossing and anti-crossing behaviors, dominate the magnetic quantization phenomena, such as the magneto-optical selection rules [9], quantum Hall conductivities [10], and inter-Landau-level transitions and magneto-plasmons [40].

The buckled or planar honeycomb lattices in layered group-IV materials, which can provide very active environments, are highly suitable for exploring unusual surface physics. During the experimental growth of a sample or chemical adatom chemisorption, the unique geometric structures might produce the formation of nanoclusters or nanoislands with specific adatom configurations according to the neighboring chemical environment. For example, the bismuth adatoms are absorbed successfully at a planar surface of slightly deformed monolayer graphene, supported by a corrugated buffer layer on a 4H-SiO_2 (0001) substrate, as identified from STM measurements [41]. A large-scale hexagonal array of Bi adatoms is revealed clearly at room temperature; furthermore, they are aggregated in triangular and rectangular nanoclusters of uniform size through a further annealing process. STS measurements of the dI/dV spectrum confirm the Dirac point, free conduction electrons and Bi-related structures [41, 42]. These results have shed light on controlling the nucleation of adatoms and subsequent growth of nanostructures on graphene surfaces. How the energetically favorable distribution configurations of Bi adatoms depend on the substrate effect and the buffer layer has been explored fully using first-principles calculations [42]. Through detailed analyses, the important factors, the various atomic sites and hexagonal positions, include the hollow, top and bridge sites and the non-uniform van der Waals interactions. Two different kinds of bismuth distributions could be achieved, namely uniform or aggregated adatom arrangements with critical energy barriers induced by the buffer layer. The aggregated adatom distributions are deduced to be associated with the large-scale hexagonal symmetry, Bi coverage and relatively few vacancies. The chemisorption of large adatoms on buckled few-layer materials is a relatively easy method to create more distribution configurations, e.g. diverse nanostructures with the various member numbers in B-adsorbed silicene systems. However, the main barriers to VASP calculations lie in the supercell effect of absorption adatoms and the layer number of the substrate, such as the moiré superlattice in Bi–silicene–Ag(111). The stable and metastable geometric symmetries, the guest- and host-atom-dominated energy bands, and the atom- and orbital-decomposed densities of states are required for further theoretical predictions for the surface nanostructures.

The composite two-monolayer systems, associated with the same or different elements, are capable of creating rich low-energy physics phenomena under the various moiré superlattices, e.g. the twisted bilayer graphene systems [43], the BN-graphene materials [44], and the group-IV and group-V coupled monolayers (details in chapter 12). The first-principles method is successful in determining the enlarged unit cells, the interlayer chemical bonding and the atom- and orbital-dominated electronic properties. The low-lying electronic states will play a critical role in the varied behaviors, e.g. the magnetic quantization in many subgroups of Landau levels [45], the magneto-optical selection rules of absorption spectra [46] and the single-particle and collective Coulomb excitations [47]. In general, there are a lot of non-uniform chemical bonds in the enlarged unit cells. Empirical formulas which are reliable in characterizing the intralayer and interlayer hopping integrals are required for the tight-binding model. They have been achieved in the theoretical predictions for the sp^2-based pure carbon systems, e.g. good parameters of the tight-binding model for the multi-walled carbon nanotubes, sliding bilayer graphenes and twisted graphenes. The well-defined π and σ bonding, and the dominant interlayer $2p_z$-orbital interactions appear to be responsible for the excellent fitting. On the other hand, the non-graphene subsystems might induce hybridized orbital coupling in all the atomic interactions and the layer-dependent spin configurations. Such mechanisms are accompanied by the zone-folding effect; therefore they need to be considered simultaneously for the good fitting of the hopping integrals in the phenomenological models.

First-principles calculations are able to explore the varied quantum confinements fully through a uniform electric field. Gate voltages can be applied to many 2D, 1D and 0D condensed-matter systems along the finite-width directions. For example, 1D graphene nanoribbons [48], carbon nanotubes [49] and few-layer graphenes [50] in the presence of gate voltages have attracted a lot of theoretical and experimental attention. It is well known that this field creates Coulomb potential differences on the sublattice planes with different heights for a perpendicular electric field and on the lattice sites with different positions for a parallel electric field. Unusual semiconductor–metal and semiconductor–semimetal transitions are revealed in, respectively, the 1D [51] and few-layer graphene-related systems [52]. The non-graphene group-IV materials have buckled structures even in monolayer systems; therefore, they are suitable for studying the composite effects due to the hopping integrals, spin–orbit coupling and electric field. The destruction of mirror symmetry appears to lead to spin-split electronic states, a finite- and zero-gap transition, and enhanced asymmetry of the electron and hole bands [53]. In general, the tight-binding model might be too cumbersome to study the fundamental properties of few-layer systems, owing mainly to the complicated intrinsic interactions. For bilayer AA-bt, AA-bb, AB-bt and AB-bb silicene systems [54, 55], the current parameters in the Hamiltonians are not reliable for the two pairs of valence and conduction bands nearest to the Fermi level. The first-principles method is the only way to study the electronic-field-varied phenomena.

The first-principles method is very reliable in exploring in detail the varied fundamental properties produced by guest-atom intercalation into 2D few-layer materials with available interlayer spacings. The various graphite intercalation compounds have attracted a lot of experimental [56, 57] and theoretical [58, 59] attention for about forty years. Electrons are either transferred to the carbon layers from the intercalants, corresponding to the donor-type compounds such as the graphite alkali-intercalation compounds, or from the carbon layers to the intercalants, corresponding to the acceptor-type compounds, such as the $FeCl_3$ compounds [60, 61]. Such systems can achieve an electrical conductivity as good as copper under proper intercalation, e.g. transport properties of stage-1 LiC_6 [62] (MC_8; M = K, Rb and Cs [63]). Specifically, they exhibit the low-temperature superconductivity of the ~1 K transition temperature and thus belong to the traditional BCS superconductors [64]. The diverse metallic behaviors appear to arise from the extended π bonding on the planar honeycomb lattices after proper chemical intercalation.

For the VASP calculations, the van der Waals forces in layered graphite compounds need to be taken into account, in which the semi-empirical DFT-D2 correction of Grimme [65] is able to characterize correctly the significant atomic interactions between two graphene layers or the graphene and intercalant layers. However, the other group-IV layered materials are expected to present more complicated intercalation-induced phenomena, owing mainly to the buckling lattice symmetries, the different type of chemical bonding and the significant spin-dependent interactions. Some focuses of study include: the presence of buckled or unbuckled host- and guest-atom layers, metallic or insulating properties, conducting π bonding and intercalant-created bonding, and well-behaved π and σ bands, and whether spin-dependent band structures, spatial distribution densities and van Hove singularities can survive under different guest-atom intercalation cases.

The rich and unique energy spectra and wave functions in layered materials will be revealed as the special absorption structures of optical transitions. The vertical excitations at the long wavelength limit ($\mathbf{q} \rightarrow 0$), in which the occupied valence states are excited to the unoccupied conduction states under the same wave vectors, are characterized by van Hove singularities with a high density of states and specific wave functions. The first-principles method [66] and the tight-binding model [67], accompanied by the dynamic Kubo formula, have been utilized successfully for studying the diverse absorption phenomena in few-layer graphene systems with different stacking configurations. The calculated results include the real and imaginary parts of the frequency-dependent transverse dielectric function, the reflectance and transmission spectra, and the energy loss function or the screened response function at $\mathbf{q} \rightarrow 0$. In general, the phenomenological model can account for the low- and middle-energy optical excitation spectra due to the π and π^* bands. For example, monolayer graphene exhibits a featureless and linear spectral function at $\omega < 2$ and a ~5 eV prominent π peak arising from, respectively, the Dirac-cone structure and the π-electronic saddle-point structures. The higher-frequency excitation behaviors, which are associated with the ω-electronic states and energy spectra, are more reliable through the first-principles calculations. This method is very suitable for few-layer non-graphene group-IV materials, since their low-lying band structures are co-dominated by the π- and σ-electronic orbitals, e.g. the significant hybridization of the π and σ bands in sliding bilayer silicene systems (figures 5.2(a)–(l)). Compared to graphene systems, the Si-, Ge-, Sn- and Pb-related materials are expected to present more absorption structures during optical excitation because of the significant sp^3 bonding, greater oscillatory energy dispersions, frequent band anti-crossing and non-negligible spin–orbit coupling. Whether the specific optical selection rules can survive depends on the sublattice-, layer- and spin-dependent subenvelope functions, and this will be a focus of future study. In addition, only the optical and magneto-optical absorption spectra of the first pair of valence and conduction bands in AA-bt and AB-bt bilayer silicene systems have been investigated using the tight-binding model [12].

The charge screening ability is a very interesting topic in 2D materials [68], since it dominates the quasiparticle energy spectra and decay rates [69]. The dynamic and static responses are defined well by the longitudinal dielectric function [70] and the dimensionless screened response function (the energy loss function) with strong momentum and frequency dependence. The 2D random-phase approximation (RPA) has been modified successfully to explore the single-particle and collective Coulomb excitations in layered systems fully [71], in which the Bloch wave functions of electronic states are decomposed into a linear superposition of layer- and sublattice-dependent tight-binding functions. The RPA is combined directly with the generalized tight-binding model under any external fields, and is highly suitable for studying magneto-electronic excitations (e.g. the inter-Landau-level transitions and magnetoplasmon modes under strong competition between the longitudinal Coulomb forces and the transverse magnetic forces). Systematic investigations, which cover all the π-electronic Coulomb excitations in the range of $0 \leqslant \omega \leqslant 10$, have been conducted on AA-, AB- and ABC-stacked graphites [72], 2D few-layer graphene systems with different layer numbers and stacking configurations [73], and 1D multi-walled carbon nanotubes [74]. The dimension-, layer-number- and stacking-enriched properties are clearly revealed to have diverse momentum- and frequency-dependent phase diagrams with electron–hole excitation regions and dispersion relations of plasmon modes. The calculated results are responsible for the up-to-date EELS measurements [75, 76]. The above-mentioned theoretical framework is suitable for monolayer non-graphene group-IV systems, but not for few-layer systems. Basic issues also remain within the cumbersome tight-binding model with very complicated interlayer hopping integrals, multi-orbital hybridization, and layer- and sublattice-dominated spin–orbit interactions; that is, it is almost impossible to obtain reliable Hamiltonians in layered materials. On the other hand, the RPA is linked directly with the first-principles method, in which the former needs to be modified to match the 2D geometries. When the 2D theoretical models can be established well in numerical programs, diverse Coulomb excitation phenomena and decay rates due to different types of chemical bonding (multi-orbital hybridization) will be revealed in various 2D materials, e.g. complex Landau-damping regions, accompanied by acoustic and optical magnetoplasmon modes in few-layer Si-, Ge-, Sn- and Pb-related materials, and their propagation on real honeycomb lattices.

In addition to layered graphenes and silicenes, the Ge-, Sn- and Pb-based 2D materials also belong to the mainstream condensed-matter systems. All of them are able to provide sufficient physical and chemical pictures to understand the varied fundamental properties fully, since the critical mechanisms, planar and buckled structures, stacking configurations, single- and multi-orbital hybridization, and spin–orbit coupling exhibit obvious variation with the atomic number. Many barriers to research, which arise mainly from the experimental growth and measurement methods and the theoretical model calculations, have been found in silicene-based materials, as examined in this book. For example, the band structures are predicted to be very complicated for bilayer silicene systems, such that it is very difficult to simulate the stacking-configuration-enriched Hamiltonians. The current tight-binding model with reliable parameters, is only for the first pair of valence and conduction bands in the AA-bt and AB-bt bilayer silicenes [12]. Such electronic states create rich magnetic quantization phenomena, unusual magneto-electronic Landau levels and unique magneto-optical selection rules. From the theoretical viewpoint, the lattice symmetries, the different site energies due to various orbitals, the complex intralayer and interlayer hopping integrals associated with distinct orbitals, and the layer-dependent spin configurations will require more important parameters in the phenomenological models, which are currently responsible for only few predicted results.

How to simulate the significant intrinsic interactions efficiently becomes one focus of study. With an increasing atomic number, calculation difficulties grow quickly under enhanced non-uniform environments and intrinsic interactions. Germanene is a counter-part to silicene, in which the former has stronger buckling and spin–orbit interactions. Generally, the single $4p_z$- or $3p_z$-orbital Hamiltonian might be good for the low-energy physical properties of monolayer systems. Both systems are predicted to show unique magneto-electronic properties [77], inter-Landau-level absorption spectra [78], magnetoplasmon modes with significant electron–hole damping [79] and inelastic e–e scattering rates [80]. The sp^3 bonding needs to be included for the bilayer and trilayer germanene systems; therefore, there are more than ten parameters in the theoretical models. Most importantly, obtaining reasonable magnitudes for all the non-negligible interactions in few-layer Si, Ge, Sn and Pb materials is worthy of systematic effort.

Silicene-based 2D magnets will become important emergent layered materials, since they can provide unique magnetic configurations and spintronics [81]. In general, these properties arise from the strong chemisorption or intercalation of rare-earth metal adatoms on or into a few-layer silicon structure, e.g. several monolayers of $GdSi_2$ and $EuSi_2$. Such materials have been synthesized successfully from a Si (111) substrate using molecular beam epitaxy. The experimental observations show that the $GdSi_2$ and $EuSi_2$ possess, respectively, planar and buckled honeycomb lattices (figures 14.4(a) and (b)). Interestingly, the bulk and 2D systems exhibit antiferromagnetic and ferromagnetic configurations, respectively. Furthermore, their mutual transitions are very sensitive to the layer number, temperature and

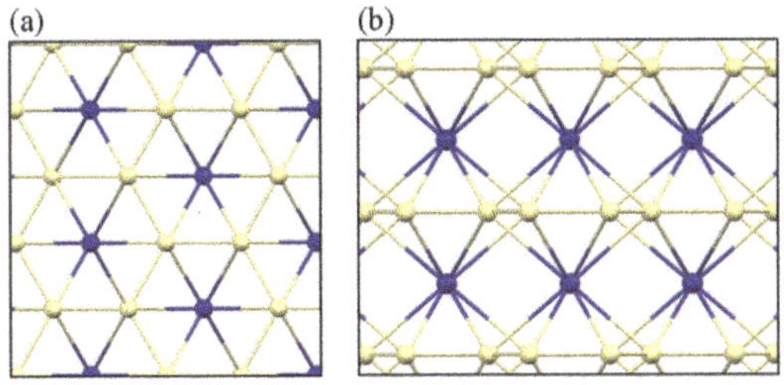

Figure 14.4. (a) Top and (b) side views of $GdSi_2$–$EuSi_2$ layered materials (rare-earth-intercalated silicene systems).

external magnetic-field strength. The unusual magnetic properties can be investigated using only the first-principles method, but not the Hubbard-like tight-binding model. The spin-induced on-site Coulomb interactions and the atom-dependent spin–orbit coupling might be the key mechanisms. Also, the band properties, the metallic or semiconducting behaviors, are determined mainly by the multi-orbital hybridization in the Si–Si, RE–Si and RE–RE bonds (3s, $3p_x$, $3p_y$, $3p_z$) in Si and seven-*f* orbitals in RE [82]. These chemical pictures are expected to be strongly dependent on the concentration and distribution of the adsorbed adatoms. Clarifying the complicated cooperative and competitive relationships between the charge- and spin-induced intrinsic interactions will be the focus of near-future study.

Interestingly, the group-V monoelements, except for N and As, can create layered structures through different growth methods. To date, the first-principles predictions for monolayer arsenene have not been verified by experimental synthesis. Few-layer phosphorene (figures 14.5(a) and (b)) [83], antimonene (figure 14.5(c)) [84] and bismuthene (figure 14.5(d)) [85] materials have attracted a lot of basic research attention both experimentally and theoretically, in particular for the P- and Sb-related systems. For example, 2D phosphorene systems have been synthesized successfully using mechanical cleavage, liquid exfoliation and mineralizer-assisted transport reactions [86]. To date, such few-layer materials have presented two kinds of allotropes, of which the black [87] and blue [88] ones are identified to be the stable structures under normal experimental conditions. They possess, respectively, hexagonal and puckered lattices (which could be regarded as deformed hexagons). Transport experiments show clearly that a phosphorene-based field-effect transistor presents an on/off ratio of 10^5 and a super-high carrier mobility of $\sim 10^3$ $(V\ s)^{-1}$ at $T = 300$ K, indicating the important role such materials will play in next-generation electronic devices [89]. The low-lying energy bands without the spin–orbit coupling are highly anisotropic, as reflected directly in the recent measurements of optical spectra [86] and transport properties [90]. Monolayer black phosphorene has a middle-energy gap of ~1.5–2.0 eV at the Γ point, which is quite different from the

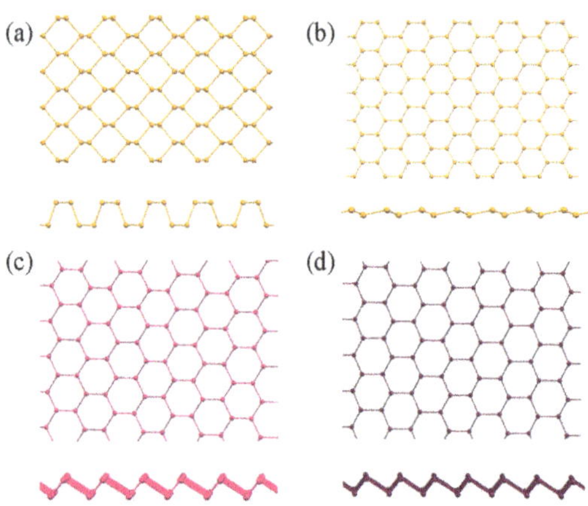

Figure 14.5. Monolayer group-V materials: (a) black phosphorene, (b) blue phosphorene, (c) antimonene and (d) bismuthine.

narrow or zero gaps related to the K/K′ points in group-IV systems. The low-lying energy dispersions dominated by $3p_z$ orbitals can be characterized by a four-band model with complicated hopping integrals. The low-energy electronic structures are manipulated easily using gate voltages and magnetic fields, e.g. the monotonic increase of the E_g gap with V_z in monolayer phosphorene, the unusual transition from a finite-gap to a gapless material in the bilayer system [91], the highly asymmetric Landau-level energy spectrum about the Fermi level, and the spin-degenerate configuration. Similarly, rich and unique fundamental properties have been observed and are predicted for antimonene- [84] and bismuthene-related [85] emergent materials in the presence of strong spin–orbit coupling. The significant differences between group-V and group-IV layered materials originate principally from the lattice symmetries, multi-orbital hybridization (sublattice- and layer-dependent hopping integrals) and spin–orbit interactions. The future research focuses might include the geometric-structure-, chemical-modification- and external-field-dependent fundamental properties and potential applications.

The theoretical framework of the first-principles calculations could be further developed for the functional materials and/or the complex compounds with large unit cells (moiré superlattices). For example, Li-ion-based batteries (figure 14.6(a)) and solar cells (figure 14.6(b)) are very popular in the basic sciences [92] and in well-established commercial applications [93]. Generally, these highly efficient batteries consist of a cathode, an anode and an electrolyte, in which the solid-state materials correspond to, respectively, $Li_3Ti_3O_6$ (figure 14.7(a)) [58, 94], $Li_4Ti_5O_{12}$, (figure 14.7(b)), and Li_2SnO_3 (figure 14.7(c)). (figure 14.7(b)) [95]. To date, knowledge of the multi-orbital hybridization and spin configurations has been absent in understanding the main features of the band structures and the density of states fully, since many atoms (electronic orbitals) can make significant contributions to the fundamental physical and chemical properties. That is to say, for the complex atom- and orbital-decomposed physical properties, it is very difficult to obtain highly accurate results and thus complete detailed analyses. According to the first-principles calculations on the varied chemical bonds in optimal geometries, the unusual

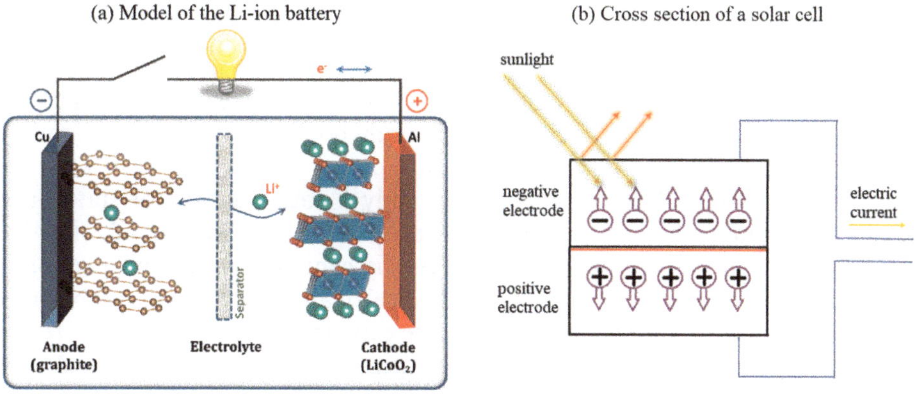

Figure 14.6. Schematics of (a) Li^+-based batteries and (b) solar cells.

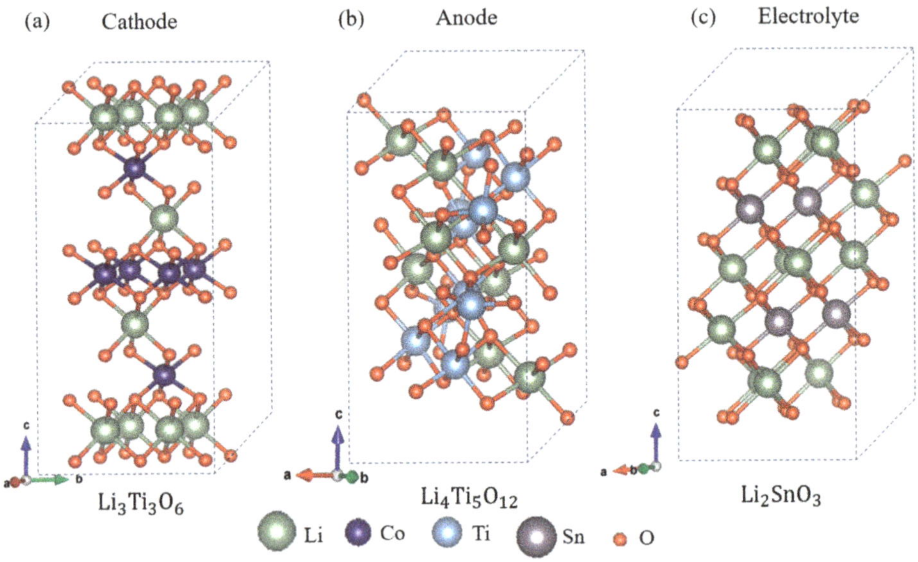

Figure 14.7. A typical solid-state (a) cathode, (b) anode and (c) electrolyte.

atom-dominated energy bands, the complex charge or spin density distributions, and the rich atom-, orbital- and spin-projected van Hove singularities, these properties might be sufficient for proposing concise chemical pictures for the significant multi-orbital hybridization under various types of bonding, and will be the subject of future study.

The layered rock-salt type structure materials, which consist of $LiTO_2$ (T = transition metal), have been investigated extensively, both experimentally [96] and theoretically [94], owing mainly to their high potential for application as a cathode material in lithium-ion-based batteries [93]. For example, ternary LiFe–Co–NiO_2 compounds are utilized frequently as commercial solid-state cathodes. During the charging and discharging processes, the lithium ions escape the original cathode, propagate along the electrolyte, intercalate into the graphite-based anode, and then return through the opposite path. There appear to be dramatic and flexible transformations in the observable structural variation in cathode lattice symmetry; that is, its deformed lattice structure plays a critical role in battery performance [97]. Previous first-principles calculations show clearly that a perfect lattice structure of $LiTO_2$ might be affected strongly by the grain boundaries (e.g. a certain twin boundary in $LiCoO_2$), and extra substitution or vacancy defects (e.g. $LiNi_xMn_{1-x}O_2$). Most theoretical predictions present only a rough density of states in the absence of band structures, where the density of states reveals only the local contributions and ignores the Li parts, and it cannot identify the real orbital hybridization in various chemical bonds. The calculated band structures below the Fermi level should be very important from the experimental perspective, and can be examined directly using ARPES measurements (details in section 3.3). The $Li_{12}Co_{12}O_{24}$ compound has been chosen as a typical material in current VASP calculations. This system has the smallest unit cell, with twelve atoms in a triangular lattice, and it consists of three cobalt oxide layers intercalated by Li atoms along the *z*-axis (figure 14.7(a)).

The current results for the cathode materials of Li^+-based batteries, as shown in figures 14.8(a) and (b), present rich and unique electronic energy spectra. All the intrinsic interactions and the critical mechanisms, which are related closely to the quasiparticle charge distributions and spin configurations, will be clarified fully through the theoretical framework developed in this book. To simulate the charging and discharging processes in a Li^+-based battery, the lithium concentrations are also reduced to 75%, 50% and 25%.

The $Li_{12}Fe_{12}$–Co_{12}–$Ni_{12}O_{24}$ compounds might be semiconductors with energy gaps of $E_g \sim 1.01$–1.10 eV or gapless metals in the presence of free conduction electrons. However, the lower-Li-concentration cases belong only to the metals. The semiconducting or metallic behaviors should be related closely to the occupied and unoccupied 3d orbitals. The rich phenomena, which originate principally from the band properties with free carrier densities, the atom- and spin-dominated valence and conduction bands, the complex charge density distributions in various chemical bonds, the spin magnetic configurations, and the atom-, orbital- and spin-decomposed densities of states, are worthy of systematic investigation. The direct association of complete data and careful analyses can provide the first step toward a full understanding of this functional material and promote its application. To specify concise multi-orbital hybridization cases in various chemical bonds, self-developed numerical programs need to be completed carefully for the projections of Bloch wave functions into the atom- and orbital-dependent contributions, e.g. one, three and five categories of s, p and d orbitals in the van Hove singularities, respectively, in particular for the d orbitals due to transition metal atoms. Moreover, the calculated results could be generalized to more complicated compounds with guest alkali and transition atoms, e.g. the fundamental physical, chemical and material properties of $Li_xK_{1-x}CoO_2$ and $LiCo_xMn_{1-x}O_2$.

The ternary intermetallic compounds, $RE_2T_3X_5$ (RE = rare-earth element, X = transition metals, Ga, Si, Ge or Sn), exhibit rich and unique physical, chemical and material phenomena, such as diverse stable geometric phases [98], unusual superconductivity behaviors in the transition temperature range of $T_c \sim 1$–10 K [98], anomalous magnetoresistance [99], meta-magnetic transitions [100] and

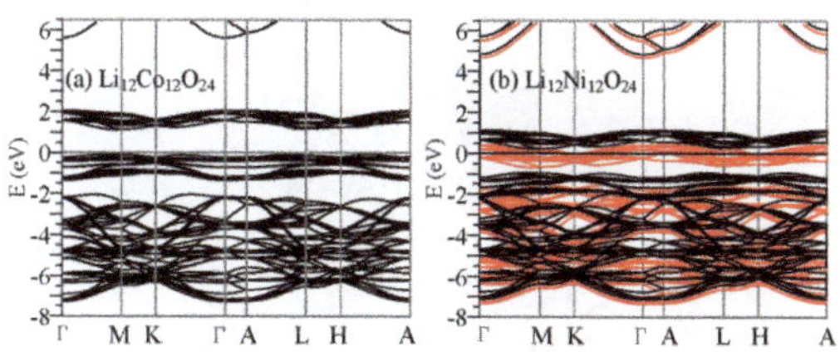

Figure 14.8. Valence and conduction bands for 3D (a) $Li_{12}Co_{12}O_{24}$ and (b) $Li_{12}Ni_{12}O_{24}$ compounds along the high-symmetry points in the first Brillouin zone.

antiferromagnetic or ferromagnetic spin configurations [100], many oscillatory valence and conduction energy spectra along the high-symmetry points with significant band overlaps, and charge density waves. Only the detailed first-principles calculations are capable of illustrating these very complicated fundamental properties clearly, since such materials possess many atoms in a primitive unit cell [98], and important multi-orbital hybridization of different chemical bonds and atom-induced spin configurations. Up-to-date x-ray patterns show the most stable structures for the orthorhombic $U_2Co_3Si_5$-type compounds (figure 14.9) [101], metastable structures for the tetragonal $Sc_2Fe_3Si_5$-type [101] and monoclinic $Lu_2Co_3Si_5$-type [102] compounds, and very few cases for the orthorhombic $Y_2Rh_3Sn_5$-type [103] and $Yb_2Pt_3Sn_5$-type [104] compounds. Most importantly, such geometric structures have a large unit cell with many atoms, in which the different types of chemical bonding can generate non-uniform active environments. This leads to high numerical barriers to identifying concise multi- and single-orbital hybridization in the various chemical bonds and spin-dependent ferromagnetism or anti-ferromagnetism through first-principles methods. The theoretical framework which is developed in the current book could be utilized to explore the varied geometric, electronic and magnetic properties of the intermetallic $RE_2T_3X_5$ compounds in detail. The critical mechanisms, which can be examined thoroughly using the optimal lattice symmetry, atom- and spin-dominated energy bands, spatial charge distributions, spin densities and magnetic moments, and the atom-, orbital- and spin-decomposed van Hove singularities, are under current investigation. For example, specific chemical bonds and orbital hybridization cases will be proposed to determine the valence and conduction band overlaps across the Fermi level (figure 14.10). It should be noted that the seven categories of 4f orbitals in the rare-earth atoms, five categories of 3d orbitals in the transition metal atoms, and three categories of 3p orbitals and one category of 3s orbitals in the Si atoms might become high barriers to numerical calculations under the projections of the Bloch wave functions. In addition, it might be almost impossible to characterize the low-energy electronic structures using the tight-binding model with a high-dimensional Hamiltonian matrix.

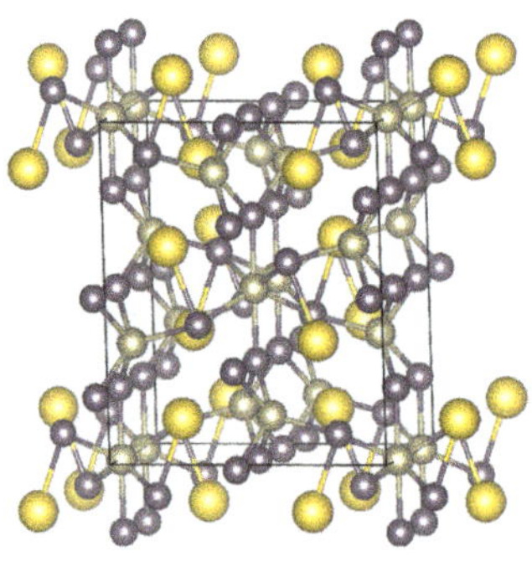

Figure 14.9. The ternary intermetallic compounds for the orthorhombic $U_2Co_3Si_5$-type compound.

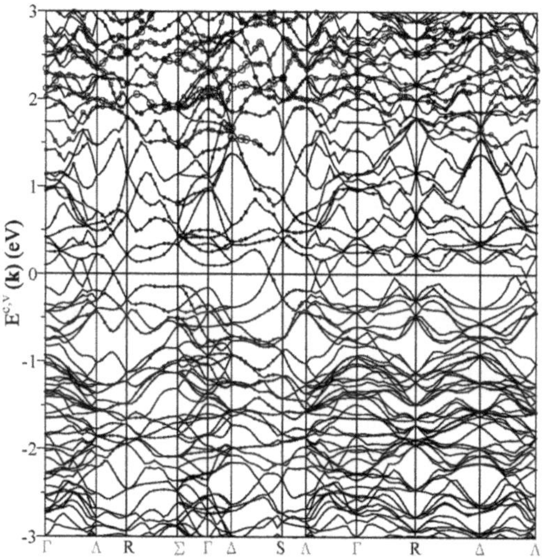

Figure 14.10. The band structure of the $RE_2T_3X_5$ compound along the high-symmetry points in the first Brillouin zone.

Among the group-IV monolayer systems, silicene and germanene possess similar fundamental properties, as verified by previous experimental [105] and theoretical [106] studies on the sp^3-bonding in bulk silicon and germanium. For example, the stable 3D compounds include the ternary intermetallic $RE_2T_3Si_5$–Ge_5 [98], the anode materials of graphite–Si and –Ge composites [107], the Li_2SiO_3 and Li_2GeO_3 electrolytes [108], and RESi or REGe adlayers on a Si[111] or Ge[111] substrate. Furthermore, the 1D silicene and germanene nanoribbons and nanotubes, with strong quantum confinement effects, are expected to present rich and diverse phenomena. To date, the silicene- and germanene-based layered materials have shown unusual electronic properties and magnetic spin configurations; therefore, they are predicted to possess high potential in various electronic device applications, such as nanoelectronic, optical, biological, medical and energy-storage devices. Many efforts have been focused on the intrinsic and spin field-effect transistors (FETs), e.g. the successful fabrication of a silicene-based FET. The alkali ions could be retained on the weakly buckled surface of silicene, germanene and stanine. The pure silicene materials and the silicene–graphite composites could serve as promising anodes in lithium-ion-based batteries, as evaluated from their electrochemical performance in terms of carrier mobility, chemical stability and metal–Si interaction. In short, the silicon- and germanium-related condensed-matter systems are worthy of complete study, and further results will be presented in the next book through our current investigations.

How to understand the relationships between the first-principles method and the tight-binding model could be regarded an open issue forever. It is well known that both have many advantages and drawbacks. In general, the functional condensed-matter systems, e.g. chemically or physically modified materials, possess complicated geometries and intrinsic interactions; therefore, the first-principles method is highly suitable for studying their geometric, electronic and magnetic properties, as identified in this book. However, no researcher can use this method to solve the magnetic-field-induced quantum phenomena; that is, with the first-principles method it is almost impossible to understand the interesting magneto-electronic states [1, 2], magneto-optical absorption spectra [9] and quantum Hall transport [10]. As a result, we need to obtain reliable parameters for the generalized tight-bind model. Many difficulties arise in the simultaneous consideration of the site energies, multi-orbital hopping integrals, spin–orbit interactions and magnetic-field effects. If we are capable of overcoming these problems, new diverse phenomena will be revealed in future studies.

In addition to the VASP calculations, other *ab initio* simulation packages are also available for exploring the fundamental properties in detail. Quantum ESPRESSO (opEn-Source Package for Research in Electronic Structure, Simulation and Optimization) [109] is based on density functional theory, plane waves and pseudopotentials. It is capable of performing many *ab initio* calculations, such as structural optimization, molecular dynamics, and spectroscopic property and material modeling under nanoscale environments. Another open-source package is the EXCITING code [110]. This is an all-electron full-potential computer package for first-principles calculations, which is constructed from the augmented plane waves and the local orbitals. Its advantages are focused on the excited-state properties; thus, this method can evaluate the optical absorption spectra of various materials through both the time-dependent density functional theory (TDDFT) and the many-body perturbation theory. The latter involves solving the Bethe–Salpeter equation (BSE) under the G_0W_0 approximation and thus includes the excitonic effects (many-body scattering). This code is a reliable method for understanding the momentum-dependent electronic excitations of 2D emergent materials, e.g. the real and imaginary parts of dielectric functions, and the energy loss spectra in few-layer graphene and phosphorene systems. VASP, ESPRESSO and EXCITING will be utilized to investigate the fundamental properties of germanene-based materials in ongoing investigations.

Multi-layer silicene, germanene, tinene and plumbene, which have been synthesized successfully in experimental laboratories [111–114], are expected to present diverse physical, chemical and material phenomena in future studies. Similar systems are observed easily in graphene-related systems, such as layered graphenes, graphene nanoribbons with specific layer numbers and stacking configurations, multi-walled coaxial carbon nanotubes, and a single- and multi-walled carbon nanotube bundles. Interestingly, the honeycomb lattices of the carbon atoms remain the same under the weak, but important, van del Waals interactions. That is, they possess perfect the hexagonal subsystems after interlayer coupling, except for the chemical adsorption and substitution and the various defects. Both phenomenological models and numerical simulations are very successful in illustrating the optimal geometries, band structures, densities of states, absorption spectra, magneto-optical selection rules, Coulomb excitations and quantum Hall transport. For the other group-IV layered materials, only the quantum Hall transport is available for investigating certain physical properties, including the geometric, electronic and optical properties, but not the magnetic quantization-induced phenomena. According to the detailed first-principles calculations and analyses in chapters 4 and 5, a trilayer system would have many metastable geometric structures created by the strong coupling effects due to enhanced and varied buckling and various defects. The critical mechanisms might lie in the obvious non-orthogonality of the π and σ bonding, the intralayer and multi-orbital hybridization, and the layer- or sublattice-dependent spin–orbit interactions. Many suitable intrinsic parameters need to be acquired for the tight-binding model, such that it is still almost impossible to establish a well-defined Hamiltonian matrix for a layer number of three or more.

References

[1] Chen S C, Wu J Y, Lin C Y and Lin M F 2017 *Theory of Magnetoelectric Properties of 2D Systems* (Bristol: IOP Publishing)

[2] Lin C Y, Do T N, Huang Y K and Lin M F 2017 *Optical Properties of Graphene in Magnetic and Electric Fields* (Bristol: IOP Publishing)

[3] Hashimoto A, Suenaga K, Gloter A, Urita K and Iijima S 2004 Direct evidence for atomic defects in graphene layers *Nature* **430** 870

[4] Shunhong Z, Jian Z, Qian W, Xiaoshuang C, Yoshiyuki K and Puru J 2015 Penta-graphene: a new carbon allotrope *Proc. Natl Acad. Sci.* **112** 2372–7

[5] Wehling T O, Katsnelson M I and Lichtenstein A I 2009 Impurities on graphene: midgap states and migration barriers *Phys. Rev.* B **80** 085428

[6] Gao J, Zhang J, Liu H, Zhang Q and Zhao J 2013 Structures, mobilities, electronic and magnetic properties of point defects in silicene *Nanoscale* **5** 9785–92

[7] Padilha J E and Pontes R B 2016 Electronic and transport properties of structural defects in monolayer germanene: an *ab initio* investigation *Solid State Commun.* **225** 38–43

[8] Son Y W, Choi S M, Hong Y P, Woo S and Jhi S H 2011 Electronic topological transition in sliding bilayer graphene *Phys. Rev.* B **84** 155410

[9] Ho Y H, Chiu Y H, Lin D H, Chang C P and Lin M F 2010 Magneto-optical selection rules in bilayer Bernal graphene *ACS Nano* **4** 1465–72

[10] Zhang Y, Tan Y W, Stormer H L and Kim P 2005 Experimental observation of the quantum Hall effect and Berry's phase in graphene *Nature* **438** 201

[11] Gusynin V P and Sharapov S G 2005 Unconventional integer quantum Hall effect in graphene *Phys. Rev. Lett.* **95** 146801

[12] Do T N, Shih P H, Gumbs G, Huang D, Chiu C W and Lin M F 2018 Diverse magnetic quantization in bilayer silicene *Phys. Rev.* B **97** 125416

[13] Reich S, Maultzsch J, Thomsen C and Ordejon P 2002 Tight-binding description of graphene *Phys. Rev.* B **66** 035412

[14] Mak K F, Sfeir M Y, Misewich J A and Heinz T F 2010 The evolution of electronic structure in few-layer graphene revealed by optical spectroscopy *Proc. Natl Acad. Sci.* **107** 14999–5004

[15] Leven I, Krepel D, Shemesh O and Hod O 2012 Robust superlubricity in graphene/h-BN heterojunctions *J. Phys. Chem. Lett.* **4** 115–20

[16] An X, Liu F, Jung Y J and Kar S 2013 Tunable graphene–silicon heterojunctions for ultrasensitive photodetection *Nano Lett.* **13** 909–16

[17] Tozzini V and Pellegrini V 2011 Reversible hydrogen storage by controlled buckling of graphene layers *J. Phys. Chem.* C **115** 25523–8

[18] Kaloni T P, Schreckenbach G, Freund M S and Schwingenschlögl U 2016 Current developments in silicene and germanene *Phys. Status Solidi–Rapid Res. Lett.* **10** 133–42
[19] Lu Y H *et al* 2009 Effects of edge passivation by hydrogen on electronic structure of armchair graphene nanoribbon and band gap engineering *Appl. Phys. Lett.* **94** 122111
[20] Chung H C, Chang C P, Lin C Y and Lin M F 2016 Electronic and optical properties of graphene nanoribbons in external fields *Phys. Chem. Chem. Phys.* **18** 7573–616
[21] Lin S Y, Tran N T T, Chang S L, Su W P and Lin M F 2018 *Structure- and Adatom-enriched Essential Properties of Graphene Nanoribbons* (Boca Raton, FL: CRC Press)
[22] Chang S L, Lin S Y, Lin S K, Lee C H and Lin M F 2014 Geometric and electronic properties of edge-decorated graphene nanoribbons *Sci. Rep.* **4** 6038
[23] Silva-Tapia A B, García-Carmona X and Radovic L R 2012 Similarities and differences in O_2 chemisorption on graphene nanoribbon vs carbon nanotube *Carbon* **50** 1152–62
[24] Kawai S *et al* 2018 Multiple heteroatom substitution to graphene nanoribbon *Sci. Adv.* **4** eaar7181
[25] Schulz M 1999 The end of the road for silicon? *Nature* **399** 729
[26] O'Hare A, Kusmartsev F V and Kugel K I 2012 A stable 'flat' form of two-dimensional crystals: could graphene, silicene, germanene Be minigap semiconductors? *Nano Lett.* **12** 1045–52
[27] De Padova P *et al* 2010 Evidence of graphene-like electronic signature in silicene nanoribbons *Appl. Phys. Lett.* **96** 261905
[28] Tang Q and Zhou Z 2013 Graphene-analogous low-dimensional materials *Prog. Mater Sci.* **58** 1244–315
[29] Balandin A A and Nika D L 2012 Phononics in low-dimensional materials *Mater. Today* **15** 266–75
[30] Cresti A, Nemec N, Biel B, Niebler G, Triozon F, Cuniberti G and Roche S 2008 Charge transport in disordered graphene-based low dimensional materials *Nano Res.* **1** 361–94
[31] Dissado L A and Hill R M 1984 Anomalous low-frequency dispersion. Near direct current conductivity in disordered low-dimensional materials *J. Chem. Soc. Faraday Trans. 2* **80** 291–319
[32] Dresselhaus M S *et al* 2007 New directions for low-dimensional thermoelectric materials *Adv. Mater.* **19** 1043–53
[33] Gao E Q, Cheng A L, Xu Y X, He M Y and Yan C H 2005 From low-dimensional manganese (II) azido motifs to higher-dimensional materials: structure and magnetic properties *Inorg. Chem.* **44** 8822–35
[34] Benedict L X, Chopra N G, Cohen M L, Zettl A, Louie S G and Crespi V H 1998 Microscopic determination of the interlayer binding energy in graphite *Chem. Phys. Lett.* **286** 490–6
[35] Spanu L, Sorella S and Galli G 2009 Nature and strength of interlayer binding in graphite *Phys. Rev. Lett.* **103** 196401
[36] Shulenburger L, Baczewski A D, Zhu Z, Guan J and Tomanek D 2015 The nature of the interlayer interaction in bulk and few-layer phosphorus *Nano Lett.* **15** 8170–5
[37] Ebnonnasir A, Narayanan B, Kodambaka S and Ciobanu C V 2014 Tunable MoS_2 bandgap in MoS_2–graphene heterostructures *Appl. Phys. Lett.* **105** 031603
[38] Geim A K and Novoselov K S 2010 The rise of graphene *Nanoscience and Technology: A Collection of Reviews from Nature Journals* (Singapore: World Scientific) pp 11–9

[39] Moon J H, Yang S, Dong W, Perry J W, Adibi A and Yang S M 2006 Core–shell diamond-like silicon photonic crystals from 3D polymer templates created by holographic lithography *Opt. Express* **14** 6297–302
[40] Poumirol J M *et al* 2013 Magnetoplasmons in quasineutral epitaxial graphene nanoribbons *Phys. Rev. Lett.* **110** 246803
[41] Chen H H, Su S H, Chang S L, Cheng B Y, Chong C W, Huang J C A and Lin M F 2015 Long-range interactions of bismuth growth on monolayer epitaxial graphene at room temperature *Carbon* **93** 180–6
[42] Lin S Y, Chang S L, Chen H H, Su S H, Huang J C and Lin M F 2016 Substrate-induced structures of bismuth adsorption on graphene: a first principles study *Phys. Chem. Chem. Phys.* **18** 18978–84
[43] Havener R W, Zhuang H, Brown L, Hennig R G and Park J 2012 Angle-resolved Raman imaging of interlayer rotations and interactions in twisted bilayer graphene *Nano Lett.* **12** 3162–7
[44] Liu Z *et al* 2013 In-plane heterostructures of graphene and hexagonal boron nitride with controlled domain sizes *Nat. Nanotechnol.* **8** 119
[45] Chen Z G *et al* 2014 Observation of an intrinsic bandgap and Landau level renormalization in graphene/boron–nitride heterostructures *Nat. Commun.* **5** 4461
[46] Du L *et al* 2018 Robust spin–valley polarization in commensurate MoS_2/graphene heterostructures *Phys. Rev.* B **97** 115445
[47] Gumbs G, Iurov A, Wu J Y, Lin M F and Fekete P 2016 Plasmon excitations of multi-layer graphene on a conducting substrate *Sci. Rep.* **6** 21063
[48] Ilyasov V V, Nguyen C V, Ershov I V and Hieu N N 2015 Effect of electric field on the electronic and magnetic properties of a graphene nanoribbon/aluminium nitride bilayer system *RSC Adv.* **5** 49308–16
[49] O'keeffe J, Wei C and Cho K 2002 Bandstructure modulation for carbon nanotubes in a uniform electric field *Appl. Phys. Lett.* **80** 676–8
[50] Liang X Y, Ding N, Ng S P and Wu C M L 2017 Adsorption of gas molecules on Ga-doped graphene and effect of applied electric field: a DFT study *Appl. Surf. Sci.* **411** 11–7
[51] Li Y, Zhou Z, Shen P and Chen Z 2009 Spin gapless semiconductor−metal−half-metal properties in nitrogen-doped zigzag graphene nanoribbons *ACS Nano* **3** 1952–8
[52] Aghaei S M, Monshi M M, Torres I, Zeidi S M J and Calizo I 2018 DFT study of adsorption behavior of NO, CO, NO_2, and NH_3 molecules on graphene-like BC_3: a search for highly sensitive molecular sensor *Appl. Surf. Sci.* **427** 326–33
[53] Peng X, Wei Q and Copple A 2014 Strain-engineered direct-indirect band gap transition and its mechanism in two-dimensional phosphorene *Phys. Rev.* B **90** 085402
[54] Mohan B, Kumar A and Ahluwalia P K 2013 A first principle calculation of electronic and dielectric properties of electrically gated low-buckled mono and bilayer silicene *Physica* E **53** 233–9
[55] Liu J and Zhang W 2013 Bilayer silicene with an electrically-tunable wide band gap *RSC Adv.* **3** 21943–8
[56] Vogel F L 1977 The electrical conductivity of graphite intercalated with superacid fluorides: experiments with antimony pentafluoride *J. Mater. Sci.* **12** 982–6
[57] Noel M and Santhanam R 1998 Electrochemistry of graphite intercalation compounds *J. Power Sources* **72** 53–65

[58] Okamoto Y 2013 Density functional theory calculations of alkali metal (Li, Na, and K) graphite intercalation compounds *J. Phys. Chem.* C **118** 16–9
[59] Thinius S, Islam M M, Heitjans P and Bredow T 2014 Theoretical study of Li migration in lithium–graphite intercalation compounds with dispersion-corrected DFT methods *J. Phys. Chem.* C **118** 2273–80
[60] Inagaki M, Wang Z D, Okamoto Y and Ohira M 1987 The synthesis of $NiCl_2$–$FeCl_3$–graphite intercalation compounds *Synth. Met.* **20** 9–13
[61] Yan Z, Zhuxia Z, Tianbao L, Xuguang L and Bingshe X 2008 XPS and XRD study of $FeCl_3$–graphite intercalation compounds prepared by arc discharge in aqueous solution *Spectrochim. Acta* A **70** 1060–4
[62] Hadigol M, Maute K and Doostan A 2015 On uncertainty quantification of lithium-ion batteries: application to an LiC_6/$LiCoO_2$ cell *J. Power Sources* **300** 507–24
[63] Potter M E, Johnson W D and Fischer J E 1981 Transport properties of alkali metal–graphite intercalation compounds *Solid State Commun.* **37** 713–8
[64] Smith R P, Weller T E, Howard C A, Dean M P, Rahnejat K C, Saxena S S and Ellerby M 2015 Superconductivity in graphite intercalation compounds *Physica* C **514** 50–8
[65] Grimme S 2006 Semiempirical GGA-type density functional constructed with a long-range dispersion correction *J. Comput. Chem.* **27** 1787–99
[66] Kresse G and Hafner J 1993 *Ab initio* molecular dynamics for liquid metals *Phys. Rev.* B **47** 558
[67] Charlier J C, Michenaud J P, Gonze X and Vigneron J P 1991 Tight-binding model for the electronic properties of simple hexagonal graphite *Phys. Rev.* B **44** 13237
[68] Popov V V, Polischuk O V and Shur M S 2005 Resonant excitation of plasma oscillations in a partially gated two-dimensional electron layer *J. Appl. Phys.* **98** 033510
[69] Polini M, Asgari R, Borghi G, Barlas Y, Pereg-Barnea T and MacDonald A H 2008 Plasmons and the spectral function of graphene *Phys. Rev.* B **77** 081411
[70] Hill A, Mikhailov S A and Ziegler K 2009 Dielectric function and plasmons in graphene *Europhys. Lett.* **87** 27005
[71] Ho J H, Lu C L, Hwang C C, Chang C P and Lin M F 2006 Coulomb excitations in AA- and AB-stacked bilayer graphites *Phys. Rev.* B **74** 085406
[72] Chen R B, Chiu C W and Lin M F 2015 Magnetoplasmons in simple hexagonal graphite *RSC Adv.* **5** 53736–40
[73] Zhu J J, Badalyan S M and Peeters F M 2013 Plasmonic excitations in Coulomb-coupled N-layer graphene structures *Phys. Rev.* B **87** 085401
[74] Schoenenberger C, Bachtold A, Strunk A C, Salvetat J P and Forro L 1999 Interference and interaction in multi-wall carbon nanotubes *Appl. Phys.* A **69** 283–95
[75] Siokou A, Ravani F, Karakalos S, Frank O, Kalbac M and Galiotis C 2011 Surface refinement and electronic properties of graphene layers grown on copper substrate: an XPS, UPS and EELS study *Appl. Surf. Sci.* **257** 9785–90
[76] Generalov A V and Dedkov Y S 2012 EELS study of the epitaxial graphene/Ni (111) and graphene/Au/Ni (111) systems *Carbon* **50** 183–91
[77] Chung H C, Chiu C W and Lin M F 2019 Spin-polarized magneto-electronic properties in buckled monolayer gas *Sci. Rep.* **9** 2332
[78] Tabert C J and Nicol E J 2013 Valley–spin polarization in the magneto-optical response of silicene and other similar 2D crystals *Phys. Rev. Lett.* **110** 197402

[79] Wei W, Dai Y, Huang B and Jacob T 2013 Many-body effects in silicene, silicane, germanene and germanane *Phys. Chem. Chem. Phys.* **15** 8789–94

[80] Shih P H, Chiu C W, Wu J Y, Do T N and Lin M F 2018 Coulomb scattering rates of excited states in monolayer electron-doped germanene *Phys. Rev.* B **97** 195302

[81] Zhao J *et al* 2016 Rise of silicene: a competitive 2D material *Prog. Mater Sci.* **83** 24–151

[82] Larson P, Lambrecht W R, Chantis A and van Schilfgaarde M 2007 Electronic structure of rare-earth nitrides using the LSDA+ U approach: importance of allowing 4f orbitals to break the cubic crystal symmetry *Phys. Rev.* B **75** 045114

[83] Cai Y, Zhang G and Zhang Y W 2014 Layer-dependent band alignment and work function of few-layer phosphorene *Sci. Rep.* **4** 6677

[84] Gibaja C *et al* 2016 Few-layer antimonene by liquid-phase exfoliation *Angew. Chem. Int. Ed.* **55** 14345–9

[85] Lu L *et al* 2018 Few-layer bismuthene: sonochemical exfoliation, nonlinear optics and applications for ultrafast photonics with enhanced stability *Laser Photonics Rev.* **12** 1700221

[86] Woomer A H, Farnsworth T W, Hu J, Wells R A, Donley C L and Warren S C 2015 Phosphorene: synthesis, scale-up, and quantitative optical spectroscopy *ACS Nano* **9** 8869–84

[87] Jain A and McGaughey A J 2015 Strongly anisotropic in-plane thermal transport in single-layer black phosphorene *Sci. Rep.* **5** 8501

[88] Sun M, Chou J P, Yu J and Tang W 2017 Electronic properties of blue phosphorene/graphene and blue phosphorene/graphene-like gallium nitride heterostructures *Phys. Chem. Chem. Phys.* **19** 17324–30

[89] Liu H, Neal A T, Zhu Z, Luo Z, Xu X, Tománek D and Ye P D 2014 Phosphorene: an unexplored 2D semiconductor with a high hole mobility *ACS Nano* **8** 4033–41

[90] Zhou X Y *et al* 2015 Landau levels and magneto-transport property of monolayer phosphorene *Sci. Rep.* **5** 12295

[91] Dai J and Zeng X C 2014 Bilayer phosphorene: effect of stacking order on bandgap and its potential applications in thin-film solar cells *J. Phys. Chem. Lett.* **5** 1289–93

[92] Li H, Wang Z, Chen L and Huang X 2009 Research on advanced materials for Li-ion batteries *Adv. Mater.* **21** 4593–607

[93] Van Schalkwijk W and Scrosati B 2002 Advances in lithium ion batteries introduction *Advances in Lithium-Ion Batteries* (Boston, MA: Springer) pp 1–5

[94] Koyama Y, Tanaka I and Nishitani S R 1999 First principles study on factors determining battery voltages of $LiMO_2$ (M = Ti–Ni) *Jpn J. Appl. Phys.* **38** 4804

[95] Ziebarth B, Klinsmann M, Eckl T and Elsässer C 2014 Lithium diffusion in the spinel phase $Li_4Ti_5O_{12}$ and in the rocksalt phase $Li_7Ti_5O_{12}$ of lithium titanate from first principles *Phys. Rev.* B **89** 174301

[96] Thackeray M M, Kang S H, Johnson C S, Vaughey J T, Benedek R and Hackney S A 2007 Li_2MnO_3-stabilized $LiMO_2$ (M = Mn, Ni, Co) electrodes for lithium-ion batteries *J. Mater. Chem.* **17** 3112–25

[97] Wang C, Du G, Ståhl K, Huang H, Zhong Y and Jiang J Z 2012 Ultrathin SnO_2 nanosheets: oriented attachment mechanism, nonstoichiometric defects, and enhanced lithium-ion battery performances *J. Phys. Chem.* C **116** 4000–11

[98] Nakashima M *et al* 2005 A change of electronic state tuned by pressure: pressure-induced superconductivity of the antiferromagnet $Ce_2Ni_3Ge_5$ *J. Phys. Condens. Matter* **17** 4539

[99] Malliakas C D and Kanatzidis M G 2009 A double charge density wave in the single tellurium square net in $Cu_{0.63}EuTe_2$? *J. Am. Chem. Soc.* **131** 6896–7

[100] Salvador J R, Gour J R, Bilc D, Mahanti S D and Kanatzidis M G 2004 Stabilization of new forms of the intermetallic phases β-$RENiGe_2$ (RE = Dy, Ho, Er, Tm, Yb, Lu) in liquid indium *Inorg. Chem.* **43** 1403–10

[101] Chabot B and Parthe E 1984 $Ce_2Co_3Si_5$ and $R_2Ni_3Si_5$ (R ≡ Ce, Dy, Y) with the orthorhombic $U_2Co_3Si_5$-type structure and the structural relationship with the tetragonal $Sc_2Fe_3Si_5$-type structure *J. Less-Common Met.* **97** 285–90

[102] Venturini G, Meot-Meyer M, Mareche J F, Malaman B and Roques B 1986 De nouveaux isotypes de $U_2Co_3Si_5$ ou $Lu_2Co_3Si_5$ dans les systems R-T-Ge (R= elements des terres rares; T= Ru, Co, Rh, Ir). Supraconductivite de $Y_2Ir_3Ge_5$ *Mater. Res. Bull.* **21** 33–9

[103] Gamża M B, Gumeniuk R, Burkhardt U, Schnelle W, Rosner H, Leithe-Jasper A and Ślebarski A 2017 Coexistence of magnetic order and valence fluctuations in the Kondo lattice system Ce_2 Rh_3Sn_5 *Phys. Rev.* B **95** 165142

[104] Kußmann D, Pöttgen R and Kotzyba G 2000 New $Yb_2Pt_3Sn_5$ type stannides *J. Solid State Chem.* **150** 112–20

[105] Borensztein Y, Prévot G and Masson L 2014 Large differences in the optical properties of a single layer of Si on Ag (110) compared to silicene *Phys. Rev.* B **89** 245410

[106] Wang S 2011 A comparative first-principles study of orbital hybridization in two-dimensional C, Si, and Ge *Phys. Chem. Chem. Phys.* **13** 11929–38

[107] Li W, Sun X and Yu Y 2017 Si-, Ge-, Sn-based anode materials for lithium-ion batteries: from structure design to electrochemical performance *Small Methods* **1** 1600037

[108] Xiao R, Li H and Chen L 2015 Candidate structures for inorganic lithium solid-state electrolytes identified by high-throughput bond-valence calculations *J. Materiomics* **1** 325–32

[109] Giannozzi P *et al* 2009 QUANTUM ESPRESSO: a modular and open-source software project for quantum simulations of materials *J. Phys. Condens. Matter* **21** 395502

[110] Gulans A *et al* 2014 Exciting: a full-potential all-electron package implementing density-functional theory and many-body perturbation theory *J. Phys. Condens. Matter* **26** 363202

[111] Liu C C, Feng W and Yao Y 2011 Quantum spin Hall effect in silicene and two-dimensional germanium *Phys. Rev. Lett.* **107** 076802

[112] Dávila M E, Xian L, Cahangirov S, Rubio A and Le Lay G 2014 Germanene: a novel two-dimensional germanium allotrope akin to graphene and silicene *New J. Phys.* **16** 095002

[113] Cai B, Zhang S, Hu Z, Hu Y, Zou Y and Zeng H 2015 Tinene: a two-dimensional Dirac material with a 72 meV band gap *Phys. Chem. Chem. Phys.* **17** 12634

[114] Zhao H *et al* 2016 Unexpected giant-gap quantum spin Hall insulator in chemically decorated plumbene monolayer *Sci. Rep.* **6** 20152

IOP Publishing

Silicene-Based Layered Materials
Essential properties
Shih-Yang Lin, Hsin-Yi Liu, Duy Khanh Nguyen, Ngoc Thanh Thuy Tran, Hai Duong Pham, Shen-Lin Chang, Chiun-Yan Lin and Ming-Fa Lin

Chapter 15

Problems

The featured problems regarding the physical, chemical and material properties (described below) can be solved using the generalized tight-binding model and the first-principles method (VASP calculations).

1. Determine the low-energy electronic properties using a phenomenological model (the tight-binding model [1]) for buckled monolayer silicene or germanene: (i) plot a honeycomb lattice, (ii) evaluate the first Brillouin zone, (iii) build a single-3p_z-orbital Hamiltonian in the absence or presence of spin–orbit coupling [2], (iv) diagonalize this Hermitian matrix to obtain the energy spectra and wave functions, (v) discuss their main features and (vi) investigate the spin-dependent effects.
2. Single-layer silicene exists in an external gate voltage of V [3]. Investigate the coupled effects on the electronic properties due to the combination of the spin–orbit interaction and electric field.
3. Illustrate clearly the important differences between monolayer graphene [4] and silicene (problems 1 and 2) in terms of the main features of the band structures.
4. Rich quantization phenomena are clearly revealed in monolayer silicene when a uniform perpendicular magnetic field is applied to this system. A generalized tight-binding model, which was applied successfully to graphene systems [5], can be utilized to calculate (i) the discrete Landau-level energies and (ii) localized oscillation modes, and (iii) explore the strong splitting effects induced by the perpendicular electric fields.
5. Use the first-principles method to study the quantum-confinement effects in 1D silicene and germanene nanoribbons with hydrogen passivation [6]. (i) Define armchair, zigzag and chiral systems from the open edge structures, (ii) investigate the modulated bond lengths, (iii) evaluate the featured energy spectra, (iv) discuss the unusual wave functions, (v) explore

doi:10.1088/978-0-7503-3299-6ch15

the strong dependence of band gaps on the boundaries and ribbon widths, (vi) determine the important types of chemical bonding from the spatial charge density distributions, (vii) identify the multi-orbital hybridizations through the atom- and orbital-decomposed van Hove singularities of the density of states, and (viii) clarify the dimensionality-enriched phenomena.

6. Carry out the VASP calculations for studying the diverse phenomena of 1D silicene and germanene nanotubes [7]. (i) Plot armchair, zigzag and chiral hollow cylinders with buckled A and B sublattices, (ii) investigate the band structures and wave functions under periodic boundary conditions, in particular for the standing waves along the azimuthal direction, (iii) explore the roles of orbital hybridization on a cylindrical surface, (iv) evaluate the radius- and chiral-angle-dependent band gaps, and (v) compare the significant similarities and differences between silicene nanotubes and nanoribbons.
7. Alkali-atom and alkali-ion chemical modifications (chemisorption and substitution [8]) of 2D silicene, 1D nanoribbons and 1D nanotubes are useful in understanding the anode material of lithium-ion-based batteries. Explore (i) the optimal positions of lithium-ion, (ii) the dimensionality- and edge-varied adatom distribution configurations, (iii) the structure- and atom-dependent free carrier densities, (iv) the orbital mixing of silicon and alkali atoms, and (v) the spin configurations on the open edges of silicene nanoribbons.
8. Ferromagnetic configurations are associated with the chemical adsorption of transition and rare-earth metal adatoms on silicene-related surfaces [9]. Investigate (i) the adatom- and lattice-enriched arrangements, (ii) the atom- and orbital-induced magnetic moments, (iii) the spin-degenerate and spin-split energy bands, (iv) the metallic free carriers and the semiconducting band gaps, (iv) the silicon and adatom-created spin density distributions, (v) the complex chemical bonding in the spatial charge density distributions and (vi) the many van Hove singularities to determine the critical multi-orbital hybridizations of active chemical bonds.
9. The rich and unique properties of binary 3D silicon carbide (SiC) [10] are created by the variation of silicon and carbon concentrations and various dimensionalities, being focused mainly on (i) the modulated chemical bonds, (ii) band gaps, (iii) feature energy spectra and wave functions, (iv) atom- and orbital-generated charge densities, (v) concise multi-orbital hybridizations, (vi) distinct special structures of van Hove singularities, and (vii) significant differences among 3D, 2D and 1D compounds.
10. The 3D ternary LiSi, Ge and SnO compounds act as electrolytes in lithium-ion-based batteries [11]. The VASP numerical simulations are able to provide (i) extremely non-uniform physical and chemical environments (e.g. the great modulation of bond lengths), (ii) strong three-atom dependence for each energy band and (iii) the different contributions of oxygen orbitals from the spatial charge densities and van Hove singularities.

11. To determine the diverse phenomena of 3D silicon, germanium and diamond [12], as mono-element crystals, detailed VASP calculations are conducted on (i) the symmetric geometries with sp^3 chemical bonding, (ii) the atom- and orbital-varied energy bands, (iii) the rich charge density distributions, (iv) the various van Hove singularities, and (v) the similarities and differences between the three kinds of materials, in particular for the significant effects due to spin–orbit coupling. Also, detailed comparisons are made for the first-principles method and the tight-binding model in terms of the fundamental properties, e.g. the hole bands nearest to the Fermi level.
12. The clear identification of multi-orbital hybridization in 3D SiO_2 has been shown in [13]. This material serves frequently as a very stable substrate; therefore, its intrinsic properties play a critical role in crystal growth. The first-principles simulations can solve (i) the optimal crystal structure, (ii) the distinct Si-, O-, and Si- and O-dominated energy ranges, (iii) the highly anisotropic charge density due to the strong chemical bonding, and (iv) the concise orbital hybridization related to the atom- and orbital-projected densities of states. SiO_2 is completely different to $LiSiO_2$ [14], for example, in terms of geometric and electronic properties.

References

[1] Lu C L, Chang C P, Huang Y C, Lu J M, Hwang C C and Lin M F 2006 Low-energy electronic properties of the AB-stacked few-layer graphites *J. Phys. Condens. Matter* **18** 5849

[2] Türeci H E and Alhassid Y 2006 Spin–orbit interaction in quantum dots in the presence of exchange correlations: an approach based on a good-spin basis of the universal Hamiltonian *Phys. Rev.* B **74** 165333

[3] Dyrdał A and Barnaś J 2012 Intrinsic spin Hall effect in silicene: transition from spin Hall to normal insulator *Phys. Status Solidi (RRL)–Rapid Res. Lett.* **6** 340–2

[4] Novoselov K S, Geim A K, Morozov S, Jiang D, Katsnelson M I, Grigorieva I, Dubonos S V and Firsov A A 2005 Two-dimensional gas of massless Dirac fermions in graphene *Nature* **438** 197

[5] Lin C Y, Do T N, Huang Y K and Lin M F 2017 *Optical Properties of Graphene in Magnetic and Electric Fields* (Bristol: IOP Publishing)

[6] Fang D Q, Zhang S L and Xu H 2013 Tuning the electronic and magnetic properties of zigzag silicene nanoribbons by edge hydrogenation and doping *RSC Adv.* **3** 24075–80

[7] Salimian F and Dideban D 2018 A silicene nanotube field effect transistor (SiNTFET) with an electrically induced gap and high value of I_{on}/I_{off} *ECS J. Solid State Sci. Technol.* **7** M1–5

[8] Hussain T, Chakraborty S and Ahuja R 2013 Metal-functionalized silicene for efficient hydrogen storage *ChemPhysChem* **14** 3463–6

[9] Sun X, Wang L, Lin H, Hou T and Li Y 2015 Induce magnetism into silicene by embedding transition-metal atoms *Appl. Phys. Lett.* **106** 222401

[10] Kim S, Ihm J, Choi H J and Son Y W 2008 Origin of anomalous electronic structures of epitaxial graphene on silicon carbide *Phys. Rev. Lett.* **100** 176802

[11] Khorassani A, Izquierdo G and West A R 1981 The solid electrolyte system, Li_3PO_4, Li_4SiO_4 *Mater. Res. Bull.* **16** 1561–7

[12] Hoh H Y, Ouyang T, Sullivan M B, Wu P, Nesladek M and Loh K P 2010 A HREELS and DFT study of the adsorption of aromatic hydrocarbons on diamond (111) *Langmuir* **26** 3286–91

[13] Fan X F, Zheng W T, Chihaia V, Shen Z X and Kuo J L 2012 Interaction between graphene and the surface of SiO_2 *J. Phys. Condens. Matter* **24** 305004

[14] Dunham S T 1989 Interstitial kinetics near oxidizing silicon interfaces *J. Electrochem. Soc.* **136** 250–4

Lightning Source UK Ltd.
Milton Keynes UK
UKHW051254231220
375661UK00002B/47

9 780750 332972